Recent Studies of Hypothalamic Function

Proceedings of the International Symposium on
Recent Studies of Hypothalamic Function. Calgary, May 28–31, 1973

Recent Studies of Hypothalamic Function

Editors: K. LEDERIS and K. E. COOPER, Calgary

137 figures and 17 tables, 1974

S. Karger · Basel · München · Paris · London · New York · Sydney

Previous publications in this field

Brain-Endocrine Interaction. Median Eminence: Structure and Function
International Symposium on Brain-Endocrine Interaction, Munich, 1971
Editors: K. M. KNIGGE, D. E. SCOTT and A. WEINDL, Rochester, N.Y.
XII + 368 p., 177 fig., 16 tab., 2 cpl., 1972
ISBN 3-8055-1257-0

Brain-Pituitary-Adrenal Interrelationships
International Symposium on Brain-Pituitary-Adrenal Interrelationships, Cincinnati, Ohio, 1972
Editors: A. BRODISH, Cincinnati, Ohio, and E. S. REDGATE, Pittsburgh, Pa.
XII + 340 p., 182 fig., 29 tab., 1973
ISBN 3-8055-1457-3

S. Karger · Basel · München · Paris · London · New York · Sydney
Arnold-Böcklin-Strasse 25, CH-4011 Basel (Switzerland)

Printed in Switzerland by Graphische Anstalt Schüler AG, Biel
ISBN 3-8055-1694-0

Contents

Metabolic and Behavioral Aspects of Hypothalamic Function

Hypothalamus in Thermoregulation

List of Contributors and Participants

Contributors

ADAIR, E., John B. Pierce Foundation, 290 Congress Avenue, New Haven, CT 06519 (USA)

BALA, R. M., Division of Medicine, University of Calgary, Calgary, Alberta T2N 1N4 (Canada)

BECK, J. C. (Chairman), Royal Victoria Hospital, Room 15, 4 Main, Montreal 112, Quebec (Canada)

BLIGH, J., Agricultural Research Council, Institute of Animal Physiology, Babraham, Cambridge CB2 4AT (England)

COOPER, K., Division of Medical Physiology, University of Calgary, Calgary, Alberta T2N 1N4 (Canada)

CROSS, B. A., Department of Anatomy, Medical School, Bristol BS8 1TD (England)

EISENMAN, J. S., Mount Sinai Medical School, 5th Avenue and 100th Street, New York, NY 10029 (USA)

FORTIER, C., Department of Physiology, Laval University, Quebec City, P.Q. 10e (Canada)

FRIESEN, H., Royal Victoria Hospital, Room 19, 4 Main, Montreal 112, Quebec (Canada)

GRANT, G., The Salk Institute, P.O. Box 1809, San Diego, CA 92112 (USA)

HARDY, J. D., John B. Pierce Foundation, 290 Congress Avenue, New Haven, CT 06519 (USA)

HAYWARD, J., Department of Neurology-RNRC, UCLA School of Medicine, Los Angeles, CA 90024 (USA)

HILTON, S. M., Department of Physiology, The Medical School, Vincent Drive, Birmingham B15 2TJ (England)

KASTIN, A. J., Endocrinology Section, Veterans Administration Hospital, 1601 Perdido Street, New Orleans, LA 70140 (USA)

KRAICER, J., Department of Physiology, Queen's University, Kingston, Ontario (Canada)

LEDERIS, K., Division of Pharmacology and Therapeutics, University of Calgary, Calgary, Alberta T2N 1N4 (Canada)

LEGAN, S. J., Reproductive Endocrinology Program, Department of Pathology, The University of Michigan, Ann Arbor, MI 48104 (USA)

MACLEAN, P. D., Laboratory of Brain Evolution and Behaviour, National Institute of Mental Health, Building 110, NIHAC, Bethesda, MD 20014 (USA)

MCCANN, S. M., Department of Physiology, Southwestern Medical School, Dallas, TX 75235 (USA)

MCKENZIE, J. M., Department of Medicine, Royal Victoria Hospital, 687 Pine Avenue West, Montreal 112, P.Q. (Canada)

MOGENSON, G., Departments of Physiology and Psychology, University of Western Ontario, London 72, Ontario (Canada)

MROSOVSKY, N., Departments of Zoology and Psychology, University of Toronto, Toronto, Ontario (Canada)

MYERS, R. D., Laboratory of Neuropsychology, Purdue University, Lafayette, IN 47097 (USA)

RAISMAN, G., Department of Human Anatomy, University of Oxford, South Parks Road, Oxford OX1 3QX (England)

ROSENDORFF, C., Department of Physiology, University of Witwatersrand, Medical School, Hospital Street, Johannesburg (South Africa)

RUOFF, H.-J., Pharmakologisches Institut der Universität Tübingen, D-74 Tübingen, Wilhelmstrasse 56 (FRG)

SACHS, H., Roche Institute of Molecular Biology, Nutley, NJ 07110 (USA)

SCHARRER, B., Department of Anatomy, Albert Einstein College of Medicine, 1300 Morris Park Avenue, Bronx, NY 10461 (USA)

SMITH, O. A., Regional Primate Research Center SJ-50, University of Washington, Seattle, WA 98195 (USA)

VAN LOON, G. R., Clinical Investigation Unit, Toronto General Hospital, Toronto, Ontario M5G 1L7 (Canada)

VEALE, W., Division of Medical Physiology, University of Calgary, Calgary, Alberta T2N 1N4 (Canada)

WAYNER, M. J., Brain Research Laboratory, Syracuse University, 601 University Avenue, Syracuse, NY 13210 (USA)

WEITZMAN, E. D., Montefiore Hospital and Medical Center, 111 East 210th Street, Bronx, NY 10467 (USA)

Participants

ABRAHAMS, V. C., Department of Physiology, Queen's University, Kingston, Ontario (Canada)

AYLESWORTH, S., 71 Glouchester Cresc. S.W., Calgary, Alberta T3E 4V3 (Canada)

BASHFORTH, G., University of Saskatchewan, Saskatoon, Saskatchewan (Canada)

BEHRMAN, H. R., Research Division, Merck & Co. Inc., Rahway, New Jersey (USA)

BERZINS, R., University of Saskatchewan, Saskatoon, Saskatchewan (Canada)

BRYSON, W., William S. Merrell Company, Toronto, Ontario (Canada)

CHRISTOPHERSON, R., Department of Animal Science, University of Alberta, Edmonton, Alberta (Canada)

CIOE, J., Department of Psychology, University of Western Ontario, London, Ontario N6A 3K7 (Canada)

COTTLE, W., Department of Physiology, University of Alberta, Edmonton, Alberta (Canada)

CHURCH, R., Division of Medical Biochemistry, University of Calgary, Faculty of Medicine, Calgary, Alberta T2N 1N4 (Canada)

CRIM, L., Marine Sciences Research Laboratory, Memorial University of Newfoundland, St. John's, Newfoundland (Canada)

DAVIS, S. L., Department of Animal Industries, College of Agriculture, University of Idaho, Moscow, ID 83843 (USA)

DOWNMAN, C., Royal Free Hospital, School of Medicine, 8 Hunter Street, London WC1N 1PB (England)

DUNSMORE, C., 19 Baker Crescent N.W., Calgary, Alberta T2L 1R3 (Canada)

FAIERS, A., Department of Physiology, University of Western Ontario, London, Ontario N6A 3K7 (Canada)

FRYER, J., Department of Zoology, University of Alberta, Edmonton, Alberta (Canada)

GOBA, H., 413 Tegler Building, 10189–101 Street, Edmonton, Alberta T5J 0T8 (Canada)

GUDAUSKAS, J., 1030 Gretna Green Way, Los Angeles, CA 90049 (USA)

HENDERSON, N. E., Department of Biology, University of Calgary, Alberta T2N 1N4 (Canada)

HENINGER, R., Department of Zoology, Brigham Young University, 575 Widtsoe Building, Provo, UT 84602 (USA)

HEPBURN, A. L., 359 Palliser Square, 115 – 9 Avenue S.E., Calgary, Alberta T2G 0P5 (Canada)

ISABIRYE, J., University of Saskatchewan, Saskatoon, Saskatchewan (Canada)

KARSCH, F., Reproductive Endocrinology Program, Department of Pathology, University of Michigan, Medical School, Ann Arbor, MI 48104 (USA)

KREGZDE, J., 12341 Chianti, Los Alamitos, Calif. (USA)

LEMON, P., Department of Pharmacology and Therapeutics, University of Manitoba, Winnipeg, Manitoba R3P 0B6 (Canada)

MANNS, J., Department of Veterinary Physiology, University of Saskatchewan, Saskatoon, Saskatchewan (Canada)

MARTIN, J., Division of Neurology, McGill University, Montreal General Hospital, Montreal, Quebec (Canada)

MCMAHON, J. P., 190 A Cain Avenue, De Ridder, LA 70634 (USA)

MILLER, J., Department of Physiology, University of British Columbia, Vancouver, B.C. (Canada)

MULLIN, W., Department of Physiology, University of Manitoba, Winnipeg, Manitoba R3E 0B3 (Canada)

MURPHY, B., Department of Biology, University of Idaho, Moscow, ID 83843 (USA)

MUTCH, C., Department of Biology, University of Calgary, Alberta T2N 1N4 (Canada)

PAGE, R., M.S. Hershey Medical Centre, Hershey, PA 17033 (USA)
PATRICK, S. J., Department of Biochemistry, Dalhousie University, Halifax, Nova Scotia (Canada)
PEITCHINIS, J., School of Nursing, University of Calgary, Calgary, Alberta T2N 1N4 (Canada)
PETER, R. E., Department of Zoology, University of Alberta, Edmonton, Alberta (Canada)
PETRALI, E. A., Department of Psychiatry, University Hospital, Saskatoon, Saskatchewan (Canada)
PHILLIPS, T., Department of Psychology, University of British Columbia, Vancouver 8, B.C. (Canada)
REEL, J. R., Parke-Davis Research Laboratories, 2800 Plymouth Road, Ann Arbor, MI 48106 (USA)
REICHERT, H., Psychiatric Research Unit, University Hospital, Saskatoon, Saskatchewan (Canada)
RENAUD, L., Montreal General Hospital, Montreal, P.Q. (Canada)
RIPPEL, R. H., Abbott Laboratory, North Chicago, Ill. (USA)
ROBERTSON, F., 3720 Underhill Drive, Calgary, Alberta T2N 4G1 (Canada)
RUF, K., Department of Neurosciences, McMaster University Medical Centre, 1200 Main Street West, Hamilton 16, Ontario (Canada)
SHERWOOD, N., Biology Department, University of Victoria, Victoria, B.C. (Canada)
SMITH, C., Department of Biology, University of Calgary, Calgary, Alberta T2N 1N4 (Canada)
STEINBERG, H. H., 122 South Michigan Avenue, Suite 1811, People's Gas Building, Chicago, IL 60603 (USA)
THOMPSON, J. R., Department of Animal Science, University of Alberta, Edmonton, Alberta (Canada)
USHER, D., Department of Pharmacology, Faculty of Medicine, University of Ottawa, Ottawa, Ontario (Canada)
VAN PETTEN, G. R., Department of Pharmacology and Therapeutics, University of Calgary, Calgary, Alberta T2N 1N4 (Canada)
WANG, L., Department of Zoology, University of Alberta, Edmonton, Alberta (Canada)
WARD, R., Medical Clinic of Castlegar, Box 669, Castlegar, B.C. (Canada)
WHITHEAD, S., Neurological Sciences, McMaster University, Hamilton, Ontario (Canada)
WILSON, N., Faculty of Medicine, Department of Physiology, University of British Columbia, Vancouver 8, B.C. (Canada)
WISHART, T., Department of Psychology, University of Saskatchewan, Saskatoon, Saskatchewan (Canada)
WU, P. H., Psychiatric Research Unit, University Hospital, Saskatoon, Saskatchewan (Canada)
WYSE, G., Department of Pharmacology and Therapeutics, University of Calgary, Calgary, Alberta T2N 1N4 (Canada)

Preface

The diverse functions of the hypothalamus have, in recent years, been shown to be clearly inter-related. The hypothalamic control of any one physiological function cannot any longer be considered in isolation. For example, during the reproductive cycle in man, which is under hypothalamic control, water and salt balance and body temperature are also regulated and emotional changes are often apparent. Thus, the research endocrinologist needs to consult with the neurochemist-neurophysiologist, the behavioral scientist and the clinical investigator.

The symposium was part of the opening celebrations of the new Health Sciences Centre which houses the Medical School at Calgary in which research and teaching have both been organized on an interdisciplinary basis. At Calgary a nucleus of hypothalamic researchers had already come together. This setting provided a reason for an attempt to gather leading investigators concerned with endocrine, metabolic and behavioral research, who usually work in parallel but are not always cognizant of likely relationships between divergent functional aspects. It is felt that this integration has been of mutual benefit to the workers from different fields.

Another aspect of this symposium, namely the publication of the proceedings within a few months after the presentation of the reports, should also be beneficial to those interested in the diverse hypothalamic functions: the progress in investigations at present is such that new knowledge presented today may become obsolete within a year or two. Realization of this and appropriate co-operation by the Messrs. S. Karger AG will, no doubt, be appreciated by interested readers when the proceedings arrive on the bookshelves.

The proceedings are presented in the form and in the sequence in which the symposium was held. The section dealing with 'Hypothalamus and hormones' includes all presentations in 3 sessions dealing with the endocrine aspects. Equally, the reports given in sessions on 'Metabolic and behavioral' functions and in 'Thermoregulation' are grouped under separate headings. The Summary represents an abbreviated edited form of the final sessions of the symposium, in which general aspects of endocrine and nonendocrine functions of the hypothalamus were considered in several formal presentations by members of the panel and the resulting general discussion.

The planning of the symposium was the joint effort of the members of the Symposium Organizing Committee and the Central Opening Ceremonies Committee, to whom our thanks are due. Especially, the input by Dr. WARREN VEALE, before, during and after the symposium is greatly appreciated.

The symposium could not have been held without the generous financial support from the Medical Research Council of Canada and contributions from many organizations and firms (Astra Chemicals Ltd., Ayerst Laboratories, Beckman Instruments Inc., Boehringer Ingelheim (Canada) Ltd., Burroughs Wellcome Co., Garworth, Walter A. Carveth Ltd., Fisher Scientific Co. Ltd., Mead Johnson (Canada), The William S. Merrell Co., The Mogul Corporation, Sandoz Pharmaceuticals, Schering Corporation Ltd., Warner-Chilcott Laboratories Co. Ltd., E. R. Squibb & Sons Ltd., Wild of Canada Ltd., Winthrop Laboratories, Wyeth Limited, Abbott Laboratories Ltd., Smith, Kline & French of Canada Ltd., Merck-Frosst Laboratories and Pfizer). Special thanks are due to Messrs. Merck-Frosst (Canada) who supported the session on 'Metabolic-behavioral aspects of hypothalamic function' in its entirety.

We are grateful for the smooth day-to-day operation of the symposium to Miss BETTY BUCHANAN from the Department of Continuing Education, University of Calgary, and her assistants, especially to Mrs. SANDRA DEMCHUK; to Mr. V. JACKSON and Mr. STEWART EATON for excellent audio-visual services and other physical arrangements.

Miss LOUISE WORKMAN, assisted by Miss LINDA SMYTH, Mrs. JEAN GOLDIE, Miss JUDY STEARNS and Miss MARIANNE JOCHUMSEN carried the main secretarial burden in the organization of this symposium and, especially, in the patient and rapid preparation of manuscripts for the press. Last, but not least, the patience and indulgence of our wives during the period before and during the symposium, and the editorial work, are greatly appreciated.

K. LEDERIS

Calgary, June 1973 K. E. COOPER

Introductory Remarks

Recent Studies of Hypothalamic Function
Int. Symp. Calgary 1973, pp. 1–7 (Karger, Basel 1974)

The Concept of Neurosecretion Past and Present

Berta Scharrer

Department of Anatomy, Albert Einstein College of Medicine, New York, N.Y.

It seems appropriate that this Symposium on Hypothalamic Function (commemorating the official opening of the University of Calgary Health Sciences Center) be initiated by a discussion of the phenomenon of neurosecretion. Its central role in neuroendocrinology is now firmly established, and it may be worthwhile to consider the gradual evolvement of our current concept of neurosecretion.

The discovery that certain hypothalamic neurons specialize in secretory activity to a degree comparable to that of endocrine gland cells, and the suggestion that this activity is related to hypophysial function, was made by Ernst Scharrer [14] in a teleost fish, *Phoxinus laevis*. Like an earlier description of 'gland-like' nerve cells in the spinal cord of skates by Speidel [20], this discovery was based on cytological criteria and was, therefore, not only received with skepticism but, for the most part, with outright rejection. The proposition that members of a class of cells as readily defined as neurons are capable of functioning as glands of internal secretion, and that they deserve a special name ('neurosecretory cells'), required more than morphological documentation.

For years the search for the functional role of neurosecretory centers by classical endocrinological methods remained unrewarding, and much time was devoted to indirect approaches such as the determination of cytophysiological correlates. One of the early criticisms, according to which the cytological characteristics of neurosecretory cells were judged to be nothing more than signs of *post mortem* changes or degenerative processes, could be countered by demonstrating the virtually universal occurrence of neurosecretory phenomena in all classes of animals from coelenterates to mammals, including man. The remarkable degree of analogy [17] revealed by this broadly based comparative approach presented us not merely with a wide choice of experimental animals, but with insights that could not have been obtained from mammalian material alone.

An important step forward occurred after the introduction by BARGMANN [1] and his collaborators of histological procedures originally designed by GOMORI for the demonstration of pancreatic β-cells. By selectively staining the secretory material throughout the entire neuron, these methods clearly revealed the neurosecretory fiber tracts linking the cells of origin with distinctive storage and release sites for their specific products, i.e. the neurohypophysis and analogous structures. The functional implications of these spatial relationships became increasingly apparent and gave rise to the concept of 'neurosecretory systems'. The best-known among these structural and functional units, and the one with which we shall be concerned in this symposium, is the hypothalamic-hypophysial system of higher vertebrates.

I. Hypothalamic Origin of Posterior Lobe Hormones

The first phase of investigative effort centered on the neurosecretory cells of the supraoptic and paraventricular nuclei whose axons form the hypothalamo-neurohypophysial tract and terminate in the posterior lobe. The key to the interpretation of this and corresponding organ systems was that the neuron terminals are not in synaptic contact with other neurons or non-neural effector cells. Instead, such terminals, laden with secretory material, showed a close affiliation with vascular channels. It was, therefore, reasonable to propose that neurochemical messengers manufactured in the perikarya of neurosecretory neurons become bloodborne to control effector cells at some distance from the storage and release sites, termed 'neurohemal organs' by KNOWLES. Expressed in different and more specific terms, the stainable secretory material with its hormone content in the neurohypophysis was now judged to be of hypothalamic origin. Vasopressin and oxytocin thus became prototypes of a new class of neurochemical mediators, justifiably called 'neurohormones', and the posterior pituitary was demoted from the rank of an endocrine gland in its own right to that of a storage depot [4].

Evidence for the intraneuronal transport of the active principles, together with their carrier substances (neurophysins), was obtained by severance of the neurosecretory pathway [10]. In this and other experimental tests (for example, severe dehydration followed by rehydration of rats) the degree of pharmacologically determined hormone activity of tissue extracts proved to parallel the amounts of neurosecretory material they contained. At this point everything seemed to fall into place, and several formerly paradoxical aspects, such as the excessive 'innervation' of the posterior pituitary or the remission of diabetes insipidus in hypophysectomized animals, had found an explanation. The neurosecretory neuron was accepted as a new and distinctive cell type with dual, i. e. neural and glandular, properties. Its separation from conventional neurons seemed clear-cut. The crucial issue in the definition of what is now often referred to as 'classical neurosecretory neuron' was that it fails to engage in synaptic chemical transmission, but instead manufactures (and discharges) substantial amounts of peptidergic neurohormones in the manner of classical protein-secreting cells.

But why do neurons deviate so profoundly from the norm in order to provide for the endocrine requirements of terminal target cells such as those of the kidney or mammary gland? The raison d'être for the neurosecretory neuron seems to lie elsewhere.

II. Hypothalamic-Adenohypophysial Axis

Owing to its highly specialized dual properties, the neurosecretory neuron was recognized to be singularly endowed for bridging the gap between neural and endocrine regulatory centers, each of which operates in its own way [15]. With this important conceptual step, neuroendocrine research had entered another phase, the main emphasis now being directed to the mechanism of control over adenohypophysial function. Again, a combination of diverse experimental approaches brought clarification of the problem which turned out to be more complex than originally thought. Since a discussion of the spectrum of neuroendocrine mediation is part of the article which follows, only a few highlights of the accomplishments of this research period are briefly referred to here. Among these are the elucidation of the role of the hypophysial portal system [8] and of the identities of the neurochemical mediators regulating the various adenohypophysial functions [5, 11]. Most of these hypophysiotropic ('regulating' or 'releasing') factors were found to represent a distinctive class of neurohormones of hypothalamic origin, released in the median eminence for transport to their destinations via the short 'directed' route of the portal system.

Subsequent electron-microscopic and cytochemical evidence revealed the existence of adaptive specializations, enabling some of these peptidergic (and possibly aminergic) factors to reach adenohypophysial effector sites by a nonvascular route. These forms of operation, being neither neurohormonal nor neurohumoral, established the existence of an intermediate category of neurochemical communication and gave rise to modulations in the definition of the neurosecretory neuron [12, 13].

III. Present State of Neurosecretory Research

The fact that interest in the phenomenon of neurosecretion remains as strong as ever is documented by the enormously increasing literature in this field. It would be impossible, in this brief retrospective survey, even to touch upon every facet that is currently under investigation. Accounts of some of the pertinent details will be found in the following articles.

In brief, major advances in methodological as well as conceptual terms are concerned with the mechanism of release of neurosecretory mediators and the neurophysiological parameters governing this process. Sophisticated and painstaking efforts at the biochemical and immunoenzyme histochemical levels are adding to our knowledge of the intraneuronal localization of neurosecretory products; the energetics of intraaxonal transport; the differential rates of secretion characteristic of individual neurohypophysial hormones and their respective neurophysin carriers; the special attributes of aminergic neurosecretory elements and the isolation and synthetic preparation of hypophysiotropic factors. The molecular events at the sites of action of neurosecretory mediators are just beginning to be understood. Finally, a new peptide (coherin), with the property of inducing rhythmic contraction of the gastrointestinal tract, has been found in the bovine neurohypophysis [7]. This substance, which is not bound to neurophysin, invites further study.

Conclusions

The concept of neurosecretion, once considered heretical, has long since acquired not only respectability but considerable biological significance. Recognition of the diverse modes of operation of neurosecretory neurons has clarified their relationship with more conventional types, and has thus extended and deepened the scope of neurochemical mediation in general.

Since all neurons share the capacity to synthetize and release distinctive chemical mediators, the existing dichotomy should be viewed as a matter of degree. It derives from the fact that, in the course of phylogeny, the classical 'neurosecretory neuron' has developed its secretory activity to the point where it takes precedence over all of the other functions of the cell. This specialization enables the nervous system to communicate by means of neurohormonal as well as neurohumoral signals, and the class of neurosecretory neurons takes its place on one side of a spectrum the other of which is occupied by conventional neurons. Furthermore, the different levels of specialization within the class seem to hold unequal rank in terms of functional significance as well as frequency.

The position at the nonconventional end is taken up by the first-order systems in which neurohormonal commands reach terminal target cells directly via the general circulation. As mentioned earlier, in higher animals the necessity for this relatively primitive mechanism is not readily apparent, but its existence can be interpreted as a carry-over from an early state in the evolution of the endocrine system.

Next in line is the group of neurosecretory neurons that have risen to a key position commensurate with their dual capacities, i. e. that of serving as 'final common path' by which the nervous system accomplishes its liaison with the endocrine apparatus. The only question here is whether the several 'semiprivate' hormonal and nonhormonal pathways used for this step represent necessary adaptations to special situations, or whether they are perhaps interchangeable. At any rate their availability, together with that of some non-neurosecretory nervous input, appears to fulfill the complex requirements of the neuroendocrine axis.

An examination of the 'conventional' side of the range reveals structural and functional digressions from the pattern of orthodox neural transmission which, although less prominent, parallel some of those first recognized within the group of neurosecretory neurons. Thus, the two sides of the neurochemical spectrum are neither rigidly uniform nor separated by as clear-cut a line of demarcation as originally conceived. Instead, there is an intermediate zone where one neuron type gradually blends into another. The striking features distinguishing classical neurosecretory from conventional neurons have now become 'part of a whole' and serve to underscore not merely the existence but the remarkable degree of flexibility inherent in neurochemical communication.

Summary

A brief survey has been presented of the gradual evolvement of the current concept of neurosecretion. It began with the discovery by ERNST SCHARRER [14] of hypothalamic neurons the cytological characteristics of which indicated their engagement in secretory activity to a degree comparable to that of endocrine gland cells. The elucidation of the functional role of these highly specialized 'neurosecretory neurons' took many years and required the involvement of a variety of disciplines.

The first phase resulted in the demonstration of the hypothalamic origin of the 'posterior lobe hormones' and their storage in, and release from, the posterior lobe of the hypophysis.

During the second phase, investigative efforts focusing on the role of neurosecretion in the control of adenohypophysial function established the existence of another class of neurohormones, the hypophysiotropic ('regulating' or 'releasing') factors. These advances underscored the importance of the neuroendocrine axis, and directed attention to special adaptations of the nervous system apparently enhancing its efficiency. On the basis of electron-microscopic and cytochemical evidence, the existence of neuroendocrine signals mediated by other than peptidergic blood-borne factors became apparent.

Issues of special interest in the current phase of inquiry are the neurophysiological events involved in neurosecretory activity, the biochemical characteristics of neurosecretory messengers and their carrier substances (neurophysins), and the molecular interactions at the receptor sites. The borderline between neurosecretory and conventional neurons has become an intermediate zone. Nevertheless, the classical neurosecretory cell retains its special position within the spectrum as a neuron which engages in secretory activity to a degree above and beyond that of other neurons, and whose chemical mediators can, and most frequently do, act as neurohormones.

References

Aside from some of the classical contributions cited for historical reasons, this list by necessity contains only a small selection of references intended to provide the reader with sufficient guidelines for further orientation in this field.

1 BARGMANN, W.: Über die neurosekretorische Verknüpfung von Hypothalamus und Neurohypophyse. Z. Zellforsch. *34:* 610–634 (1949).

2 BARGMANN, W.; HANSTRÖM, B.; SCHARRER, B. und SCHARRER, E. (eds.): Zweites Internationales Symposium über Neurosekretion, Lund, 1957, p. 125 (Springer, Berlin 1958).

3 BARGMANN, W. and SCHARRER, B. (eds.): Aspects of neuroendocrinology. 5th Int. Symp. on Neurosecretion, p. 380 (Springer, Berlin 1970).
4 BARGMANN, W. and SCHARRER, E.: The site of origin of the hormones of the posterior pituitary. Amer. Sci. *39:* 255–259 (1951).
5 BLACKWELL, R. E. and GUILLEMIN, R.: Hypothalamic control of adenohypophysial secretions. Annu. Rev. Physiol. *35:* 357–390 (1973).
6 GABE, M.: Neurosecretion. Int. Ser. Monogr. Biol., vol. *28:* p. 872 (Pergamon Press, Oxford 1966).
7 GOODMAN, I. and HIATT, R. B.: Coherin: A new peptide of the bovine neurohypophysis with activity on gastrointestinal motility. Science *178:* 419–421 (1972).
8 HARRIS, G. W.: Neural control of pituitary gland (Arnold, London 1955).
9 HELLER, H. and CLARK, R. B. (eds.): Neurosecretion. 3rd Int. Symp. on Neurosecretion. Mem. Soc. Endocrin. vol. 12, p. 455 (Academic Press, London 1962).
10 HILD, W.: Experimentell-morphologische Untersuchungen über das Verhalten der 'Neurosekretorischen Bahn' nach Hypophysenstieldurchtrennungen, Eingriffen in den Wasserhaushalt und Belastung der Osmoregulation. Virchows Arch. path. Anat. *319:* 526–546 (1951).
11 SCHALLY, A. V.; ARIMURA, A., and KASTIN, A. J.: Hypothalamic regulatory hormones. Science *179:* 341–350 (1973).
12 SCHARRER, B.: General principles of neuroendocrine communication; in SCHMITT The neurosciences. Second study program, pp. 519–529 (Rockefeller University Press, New York 1970).
13 SCHARRER, B.: Neuroendocrine communication (neurohormonal, neurohumoral, and intermediate); in ARIËNS KAPPERS and SCHADÉ Progress in brain research, vol. 38, pp. 7–18 (Elsevier, Amsterdam 1972).
14 SCHARRER, E.: Die Lichtempfindlichkeit blinder Elritzen (Untersuchungen über das Zwischenhirn der Fische. I.). Z. vergleich. Physiol. *7:* 1–38 (1928).
15 SCHARRER, E.: The general significance of the neurosecretory cell. Scientia *46:* 177–183 (1952).
16 SCHARRER, E. (ed): Neurosecrezione. Proc. Ist Int. Symp. on Neurosecretion, Naples 1953. Pubbl. Staz. Zool. Napoli *24:* suppl., p. 98 (1954).
17 SCHARRER, E.: The concept of analogy. Pubbl. Staz. Zool. Napoli *28:* 204–213 (1956).
18 SCHARRER, E.: The final common path in neuroendocrine integration. Arch. Anat. micr. Morph. exp. *54:* 359–370 (1965).
19 SCHARRER, E. and SCHARRER, B.: Neuroendocrinology, p. 289 (Columbia University Press, New York 1963).
20 SPEIDEL, C. C.: Gland-cells of internal secretion in the spinal cord of the skates. Publ. No. 13, pp. 1–31 (Carnegie Institution, Washington 1919).
21 STUTINSKY, F. (ed.): Neurosecretion. 4th Int. Symp. on Neurosecretion, p. 253 (Springer, Berlin 1967).

Author's address: Dr. BERTA SCHARRER, Department of Anatomy, Albert Einstein College of Medicine, 1300 Morris Park Avenue, *Bronx, NY 10461* (USA)

Recent Studies of Hypothalamic Function
Int. Symp. Calgary 1973, pp. 8–16 (Karger, Basel 1974)

The Spectrum of Neuroendocrine Communication[1]

BERTA SCHARRER

Department of Anatomy, Albert Einstein College of Medicine, New York, N.Y.

The effectiveness of the integrative functions of the body depends on appropriate mechanisms for the exchange of information between the neural and the endocrine control systems. The most central, and therefore crucial, step in this chain of events is that by which the nervous system addresses itself to the first way-station in the endocrine hierarchy. It is the 'final common path' which handles the integrated commands resulting from a multitude of afferent signals intended for endocrine centers, or for delivery directly to nonendocrine 'terminal target cells' [13, 15–17].[2]

Analysis of this neuroendocrine link has revealed a remarkable capacity of the nervous system to adapt itself to the special requirements of hormone-producing effector cells. Therefore, a review of recent progress in the area of hypothalamic function has much to contribute to our knowledge of the existing spectrum of neurochemical communication and, thus, to modern neurobiology in general. The present discussion is intended to go beyond the framework of neuroendocrine interaction, and to spotlight parallelisms with other known instances of nonconventional neural activity [11, 12, 14]. It should, perhaps, be stated here that many a lead in the detection of these special mechanisms has come from electron-microscopic data, and that some of them have not yet been subjected to experimental analysis.

1 This review was aided by USPHS Grants NB–00840 and 5P01–NS–07512.

2 The literature in this field is so voluminous (see WEITZMAN [20]) that a detailed documentation of the statements made in the present text must be dispensed with.

I. Neural Control of Endocrine Cells

A. Conventional Synaptic Signals

In examining the versatility of the nervous apparatus we need not dwell on the fact that the great majority of neuronal signals, including those of the central autonomic system, is of the conventional type. What has become evident only recently, however, is that chemical synaptic transmission seems to participate to some degree in the dispatch of neural commands to endocrine cells. Aside from physiological evidence, this conclusion is supported by the electron-microscopic demonstration of 'innervation', i.e. of junctional complexes in a variety of endocrine tissues, including those of the adenohypophysis [15]. Most of these resemble adrenergic synapses except that some of their ultrastructural characteristics tend to be missing.

However, this evidence does not argue against either the desirability or the existence of neuronal adaptations to the special requirements of effector cells that are geared to respond to hormonal rather that neurohumoral input.

B. Nonconventional Mediation

It is difficult to conceive how the complex functions of the pituitary could be monitored by synaptic intervention alone, i.e. by the kind of strictly localized signals of exceedingly short duration which are so effective in interneuronal communication. How does the organism cope with this situation?

The existence of neurosecretory neurons in hypothalamic nuclei, such as the arcuate, that direct their axons to be pituitary provides the answer (see preceding article). This spatial arrangement permits peptidergic neurochemical mediators, produced in appropriate quantity in the perikarya of these specialized neurons, to reach the adenohypophysis without a major detour. Part of the transport, like that in the hypothalamic-neurohypophysial system, occurs via the intraneuronal proximo-distal pathway. The distances between the axon terminals, i.e. the storage and release sites of neurosecretory materials, and their endocrine 'targets' vary, as do the extracellular pathways. A special vascular channel, the hypophysial portal system, provides a semiprivate vehicle to the anterior pituitary of higher and some lower vertebrates.

The neurochemical messengers availing themselves of this extracellular route are, by definition, neurohormones and have the attributes of endocrine principles. They can reach multiple effector cells simultaneously, and their effect is much more sustained than that of neurotransmitters. By being able to bypass the general circulation, they are neither diluted nor delayed. On all accounts, then, this appears to be the perfect solution for the neurochemical control of the pituitary and its analogs, and yet it is not universally used.

We know of several variants in which such hypophysiotropic (regulating, or releasing) factors do not become blood-borne, but take alternate routes to reach their goals. In these forms of 'directed delivery', the extracellular pathway is considerably shortened or even virtually absent.

In certain teleost species, neurosecretory terminals are separated from adenohypophysial receptor cells by nothing more than a layer of extracellular stroma. In others, peptidergic fibers are even in direct contact with the endocrine cells [2, 19]. The same spatial intimacy occurs in the mammalian pars intermedia [3]. Thus, we are dealing with a series of individual modifications in which some representatives attain the same degree of 'privacy' as that provided by standard types of synapses.

Whether or not the two morphologically comparable junctions hold different positions in the temporal spectrum cannot be decided as long as the mode of operation of peptidergic ('neurosecretomotor') junctions remains unknown. Until we can demonstrate mechanisms of inactivation, of the same efficiency as those typical of neurotransmitters, for proteinaceous neurochemical mediators, we may entertain the thought that the latter handle localized signals that are comparatively more prolonged.

We are equally uninformed about neural signals that are transmitted via an intervening layer of stroma. It would seem, though, that their characteristics are determined not only by the closeness of the effector site, but also by the physico-chemical properties of this extracellular material. By temporarily sequestering the active principles released from terminal and preterminal areas of the axon, this compartment can make them available to small groups of effector cells gradually and relatively slowly [1].

In future efforts to explore the distinctive features of this stromal compartment, the use of insect material should prove useful since these invertebrates, because of their lack of a capillary system, greatly depend on a network of stromal channels for the dissemination of their messenger substances. At any rate, this mode of communication is neither neurohormonal nor neurosecretomotor, but somewhere in-between. It seems to operate with neural signals of relatively long latency and extended duration.

A variant that is still largely unexplored concerns neurosecretory (B-type) fibers in which the neurochemical mediator is aminergic instead of peptidergic. Their existence can be demonstrated with highly specific fluorescence histochemical methods for catecholamines as well as by certain ultrastructural procedures [4, 6]. The blood-borne products of some of these nonclassical neurosecretory neurons seem to be brought into action under circumstances that call for relatively short-term responses.

Finally, mention should at least be made of a potential extracellular pathway for neurochemical messengers, i.e. the cerebrospinal fluid. Information on possible receptor sites for substances transported by this medium is, however, still lacking.

II. Control of Nonendocrine Terminal Effector Cells

If we now briefly turn from endocrine to other non-neuronal effector cells, we find certain parallelisms with the patterns discussed thus far. Again the mechanisms range from neurohumoral to neurohormonal. Well-known examples of the latter are the effects of vasopressin and oxytocin on terminal target cells, such as those of the kidney or the mammary gland. No special vascular pathway is involved here, but otherwise these activities correspond to those of hypophysiotropic hormones.

Conventional neuroeffector junctions are the standard equipment of striated musculature. The signals are as speedy and localized as those in other cholinergic synapses. However, some striated muscle fibers, at least among insects, appear to be 'innervated' by peptidergic fibers which can be compared to those forming neurosecretory contact sites in the mammalian pars intermedia.

An example of the intermediate type of arrangement discussed earlier is found in the nerve supply to certain elements of the smooth and striated musculature. Here, too, several cells can share the neural messenger substance released into an intervening stromal compartment by one axon 'en passant', the difference being that in this case the mediator is not peptidergic as in the pituitary but presumably aminergic [1].

A parallelism of another kind is seen in the fact that in both exocrine and endocrine cells synaptic elements deviate somewhat in their ultrastructure from the more conventional types [see, for example, ref. No. 8].

III. Nonconventional Neuron-to-Neuron Signals

Since synaptic transmission seems to be the method of choice for the transfer of chemical information between neurons, departures from this universal pattern might come as something of a surprise, except when one or even both of the interacting neurons are of a nonconventional type.

A. Control of Neurosecretory Neurons

Because of the dual nature of neurosecretory neurons, their response to neural signals affecting both their neuronal and glandular activities may be expected to differ from those of conventional neurons.

The relatively slow action potential of the neurosecretory neuron appears to be responsible for initiating the release mechanism for peptidergic mediators and their affiliated neurophysins. On the basis of single unit studies, these potentials have been shown to result from both orthodromic and antidromic stimuli. The synaptic events at somata and dendrites include a combination of stimulatory (cholinergic) and inhibitory (aminergic) input [21]. In addition, an inhibitory antidromic pathway via recurrent collaterals from neurosecretory axons has been demonstrated. The available morphological evidence concurs with these conclusions.

Two ultrastructural indications of nonconventional axo-axonal interaction have been observed. One is a synaptic contact between a conventional (afferent) and a neurosecretory neuron in the mammalian neurohypophysis [2], the other a junctional complex between a putative B fiber and a classical A-type neurosecretory fiber in the insect corpus cardiacum [10]. Even in the absence of accredited neurotransmitters, as in the second example, spatial intimacy, membrane thickenings, 'synaptic' vesicles and a postsynaptic web are indicative of the synaptic character of the structure.

B. Nonconventional Neural Signals to Regular Neurons

To go a step further, even conventional neurons may digress and become attuned to neurochemical signals other than those of the usual synaptic variety. For example, several studies carried out in insects have revealed the activities of certain neurons to be under the influence of neurohormones derived from the corpus cardiacum or other neurosecretory centers [18].

A highly interesting and promising recent finding in mammals concerns the effect of two hypothalamic neurohormones, the regulating factors MIF (MSH release inhibiting factor) and TRF (thyrotropin releasing factor) on dopaminergic neurons of the central nervous system. In their capacity to reduce parkinsonian-type tremor, these factors resemble L-dopa and even potentiate its effect. Since this action persists in the absence of the pituitary, the pathway to the neuronal effector cells must differ from that used by the same neurochemical mediators in their hypophysiotropic role [7, 9].

IV. Neurotrophic Activities

A class of neurochemical messengers, quite apart from those discussed thus far, is exemplified by a diffusible 'neurotrophic substance' which is released from sensory as well as motor fibers and, under certain conditions, even from some non-neural tissues [5]. Such trophic substances regulate tissue growth and maintenance. Their chemistry is still uncertain. They are thus not only the most unspecific but also the least understood among the known neurochemical principles.

Conclusions

In general conceptual terms, the result of the present survey points up the existence of an impressive range of neurochemical mechanisms by means of which signals reach a variety of effector cells. A large share of the evidence supporting this conclusion stems from an analysis of neural control over the endocrine system. A search for signs of equally diverse neural input to effectors that do not belong to either of the two integrative systems has revealed certain parallelisms with the neuroendocrine spectrum.

More surprising, perhaps, is the recent finding that not even all of the information chemically transmitted between neurons is of the conventional synaptic type. In all of these interactions, the parameters determining the type of signal as well as the type of response are as follows: (1) the chemical nature and amount of the neuronal mediator in use; (2) the extracellular pathway through which the chemical messenger reaches its destination; (3) the receptive properties of the effector cells, and (4) the duration of the signal, as determined by the method of inactivation of the messenger substance.

The spectrum of possibilities ranges from standard synaptic transmission, which affords high speed and greatest 'privacy', to classical neurohormonal messages that are channeled by way of the general circulation. Between these, there exist intermediate avenues for communication with various degrees of privacy and a choice of several, e.g. peptidergic, aminergic, purinergic, messenger substances.

Apparently, these multiple mechanisms have evolved in the course of specialization and, either by themselves or in combination, they serve to maintain the organism as a well functioning unit.

Summary

An analysis of neuroendocrine interactions reveals a remarkably broad spectrum of possible neural signals to endocrine effector cells. This information, and the detection of comparable modes of interaction between neurons and nonendocrine effectors, are indicative of a considerable capacity on the part of the nervous system to adapt itself to special demands. The modes of operation range from conventional information transfer by chemical synaptic transmission to neurohormonal signals, and encompass several intermediate nonconventional types in which the neurochemical mediator cannot be classified as either hormonal or neurohumoral.

Regular synaptic activity, by far the most common, is not restricted to interneuronal communication but is in use also in the control of non-neuronal effector cells, both endocrine and nonendocrine. Neurohormones affect endocrine (e.g., pituitary) and terminal 'targets' (e.g., kidney tubules, mammary gland). Furthermore, certain neurons may themselves respond to blood-borne neural mediators (e.g., effect of hypophysiotropic factors on dopaminergic neurons). Intermediate types of interaction, involving an extracellular stromal compartment that separates site of release and site of action of neurochemical messengers, occur in endocrine organs (e.g., adenohypophysis) and smooth musculature. Another specialization consists of neuroeffector junctions of peptidergic rather than cholinergic or aminergic character which are represented in all classes of effectors. Recognition of this wide range of possibilities adds new dimensions to our current concept of neurobiological activity.

References

1 BARER, R.: Speculations on the storage and release of hormones and transmitter substances. Bibl. anat., Vol. 8, pp. 72–75 (Karger, Basel 1967).

2 BARGMANN, W.: Über Synapsen im endokrinen System. Nova Acta Leopold. N.F. *30:* 199–206 (1965).

3 BARGMANN, W.; LINDNER, E. und ANDRES, K. H.: Über Synapsen an endokrinen Epithelzellen und die Definition sekretorischer Neurone. Untersuchungen am Zwischenlappen der Katzenhypophyse. Z. Zellforsch. *77:* 282–298 (1967).

4 FALCK, B. and OWMAN, C.: A detailed methodological description of the fluorescence method for the cellular demonstration of biogenic amines. Acta univ. Lund. II, *7:* 1–23 (1965).

5 GUTH, L.: 'Trophic' effects of vertebrate neurons. Neurosci. Res. Progr. Bull. *7:* 1–73 (1969).

6 HÖKFELT, T.: *In vitro* studies on central and peripheral monoamine neurons at the ultrastructural level. Z. Zellforsch. *91:* 1–74 (1968).

7 KASTIN, A. J. and BARBEAU, A.: Preliminary clinical studies with L-propyl-L-leucyl-glycine amide in Parkinson's disease. Canad. med. Ass. J. *107:* 1079–1081 (1972).

8 KÜHNEL, W.: On the innervation of the salt gland. Z. Zellforsch. *134:* 435–438 (1972).

9 PLOTNIKOFF, N. P.; KASTIN, A. J.; ANDERSON, M. S., and SCHALLY, A. V.: Oxotremorine antagonism by a hypothalamic hormone, melanocyte-stimulating hormone release-inhibiting factor (MIF). Proc. Soc. exp. Biol. Med. *140:* 811–814 (1972).

10 SCHARRER, B.: Neurosecretion. XIII. The ultrastructure of the corpus cardiacum of the insect *Leucophaea maderae.* Z. Zellforsch. *60:* 761–796 (1963).

11 SCHARRER, B.: Current concepts in the field of neurochemical mediation. Med. Coll. Virginia Quart. *5:* 27–31 (1969).

12 SCHARRER, B.: Neurohumors and neurohormones. Definitions and terminology. J. neuro-visc. relat. Suppl. 9, pp. 1–20 (1969).

13 SCHARRER, B.: General principles of neuroendocrine communication; in SCHMITT The neurosciences. Second study program, pp. 519–529 (Rockefeller University Press, New York 1970).

14 SCHARRER, B.: Concepts of neurochemical mediation. Neurocirugía, Santiago (Chile) *29:* 257–262 (1971).

15 SCHARRER, B.: Neuroendocrine communication (neurohormonal, neurohumoral, and intermediate); in ARIËNS KAPPERS and SCHADÉ Progress in brain research, vol. 38, pp. 7–18 (Elsevier, Amsterdam 1972).

16 SCHARRER, E.: The final common path in neuroendocrine integration. Arch. Anat. micr. Morph. exp. *54:* 359–370 (1965).

17 SCHARRER, E.: Principles of neuroendocrine integration, chap. I. In: Endocrines and the central nervous system. Res. Publ. Ass. nerv. ment. Dis. *43:* 1–35 (1966).

18 TRUMAN, J. W.: The eclosion hormone: Its role as a releaser and as a primer of insect behavior. Proc. 14th Int. Congr. Entomology, p. 163, Canberra 1972.

19 VOLLRATH, L.: Über die neurosekretorische Innervation der Adenohypophyse von Teleostiern, insbesondere von *Hippocampus cuda* und *Tinca tinca*. Z. Zellforsch. *78:* 234–260 (1967).

20 WEITZMAN, M. (ed.): Bibliographia neuroendocrinologica, vol. 1–10 (Albert Einstein College of Medicine, New York 1964–1974).

21 YEO, B. K.; LOCKETT, M. F., and HELLER, H.: Inhibition of the release of the neurohypophysial hormones of rats by catecholamines, *in vitro*. Austr. J. exp. Biol. med. Sci. *50:* 527–533 (1972).

Author's address: Dr. BERTA SCHARRER, Department of Anatomy, Albert Einstein College of Medicine, 1300 Morris Park Avenue, *Bronx, NY 10461* (USA)

Recent Studies of Hypothalamic Function
Int. Symp. Calgary 1973, pp. 17–25 (Karger, Basel 1974)

Structural and Functional Investigations of a Sexually Dimorphic Part of the Rat Preoptic Area

PAULINE M. FIELD and G. RAISMAN

Department of Human Anatomy, University of Oxford, Oxford

The discovery of the hypothalamo-hypophysial portal circulation was one of the major landmarks in establishing that the secretions of the pituitary gland are under the control of the central nervous system [7]. Through the integrative machinery of the central nervous system, the animal is able to modulate its reproductive functions so as to bring them in relation to its environment. It is by virtue of this responsiveness to environmental factors – whether they be seasonal, diurnal or associated with more immediate events such as the presence of a mate – that the peak of reproductive efficiency can be achieved. This communication will consider primarily the control of various reproductive phenomena in the female rat.

By use of selective lesion techniques, it is now established that the integrity of the mediobasal ('tuberal') region of the hypothalamus is essential for the maintenance of the secretion of the pituitary gonadotrophins, luteinising hormone (LH) and follicle-stimulating hormone (FSH) [5]. However, similar experiments have shown that a small region lying just in front of the hypothalamus – the preoptic area – is essential for the pre-ovulatory surge of gonadotrophins which precedes ovulation in the female rat. This region is sometimes referred to as the 'trigger' zone. A similar surge of gonadotrophins can be demonstrated in gonadectomised female rats which have been primed with oestrogen and then treated with progesterone; such a surge cannot be elicited in the castrated male rat [20]. This suggests that the function of the preoptic area is different in the two sexes. Furthermore, it has also been demonstrated that destruction of the preoptic area prevents the feminine type of receptive response (lordosis) to mounting by the male [10]. It has also been shown that normal genetic female rats when gonadectomised and treated with oestrogen and progesterone will show normal receptivity to a vigorous male, while castrated male rats similarly treated do not show this response [1].

So far, it is generally accepted that these functions depend upon the hypothalamus and preoptic area but it is far less clear what other parts of the central nervous system are involved in these responses, or even what are the pathways for the well-known effects of sensory stimuli (such as light) upon them. Because of our basically neuro-anatomical background, we have tried to approach some of these questions by a study of the fibre connections. In considering the afferent input to the mediobasal hypothalamus and medial preoptic area, it is at once apparent that two of the most prominent fibre bundles entering these regions both arise in the so-called 'limbic' structures of the forebrain. These connections are the medial cortico-hypothalamic tract from the hippocampus and the stria terminalis from the amygdala [for a review see ref. No. 13]. These are, of course, not the only efferent fibre projections arising in the amygdala and hippocampus, nor are they the sole known afferents to the medial hypothalamo-preoptic region. However, from the point of view of practical study the stria terminalis and the fimbria (which contains the medial cortico-hypothalamic tract fibres) have the great advantages that they form compact fibre bundles with a long extra-hypothalamic course through the brain. This means that they can be selectively and totally destroyed by lesions which do not encroach directly upon the hypothalamus itself.

In the present communication, we shall consider particularly the amygdala-stria terminalis system and its possible role in reproductive control. There is experimental evidence to suggest a role for the amygdala in the control of gonadotrophic hormone release and also in mating behaviour [for reviews see ref. No. 15, 19]. For example, in the experiments of VELASCO and TALEISNIK [21], it was demonstrated that various methods of stimulating the amygdala could induce ovulation in rats where spontaneous ovulation had been blocked by drugs such as atropine, reserpine or urethane. This effect of amygdaloid stimulation was prevented by section of the stria terminalis. While such experiments demonstrate that the amygdaloid projection fibres in the stria terminalis do have access to a mechanism for the initiation of ovulation, they only show them to be operative under the abnormal conditions of the particular experimental situation. This still leaves open the question of what normal functions can be demonstrated to use the fibre systems of the amygdala and stria terminalis.

Once again, a clue to this may be found by studying the fibre connections of the amygdala. It has recently been demonstrated that whereas the main olfactory bulb sends its efferent fibres to the pyriform cortex, the accessory olfactory bulb (which receives its input from the vomeronasal organ) projects to the cortico-medial group of amygdaloid nuclei [22]. The latter part of the amygdala is the source of those fibres of the stria terminalis which project to the medial preoptic area and to the medial hypothalamus [9, 12]. This special relationship between the amygdala, stria terminalis and the accessory olfactory bulb has been further emphasised by the demonstration that the cortico-medial group of amygdaloid nuclei also send efferent fibres through the stria terminalis to terminate in the granule cell layer of the accessory olfactory bulb [12, 14]. This type of neuronal circuitry strongly suggests that types of sensory stimuli which are detected by the vomeronasal organ may well use this route in order to have effects upon reproductive functions. The most likely effects in this regard are the so-called 'pheromonal' effects in which, for example, odours associated with the male may influence the secretion of gonadotrophins or prolactin by the female [3].

In an attempt to define the possible importance of the stria terminalis, we made bilateral lesions totally destroying the fimbria and stria terminalis in both male and female rats [2]. The results were in some ways disappointingly slight. In the female, this lesion did not interfere (after an initial period) with regular oestrous cycles and the animals could become pregnant, deliver and rear healthy litters of normal size. There was, however, some indication that while overall function was not grossly deranged, the set point for the secretion of gonatrophins had been altered in favour of higher levels of secretion in both sexes. This was seen in two ways. Firstly, in the males, there were elevated plasma levels of LH, FSH and testosterone (and increased seminal vesicle weights). Secondly, if the operation was carried out in pre-pubertal females, the onset of vaginal opening and of first ovulation were significantly advanced as compared with the controls [confirming the earlier observations of Critchlow and Bar-Sela, 4]. It is possible that this effect could also be related to a tonically increased output of gonadotrophins.

In a morphological study of the mode of termination of the fibres of the stria terminalis we were able to confirm at the electron-microscopic level that stria terminalis fibres establish synaptic contacts in a wide region of the medial preoptic area and medial hypothalamus [6]. The overall distribution of this tract has recently been described in a thorough anatomical study by DE OLMOS and INGRAM [12]. In our ultrastructural investigation it was possible to demonstrate terminal degeneration in the mid-dorsal part of the medial preoptic area (which we have called the 'strial part of the preoptic area') and further caudally around and also within the ventromedial hypothalamic nucleus. In each of these terminal regions we have made a quantitative study of the different types of synapses by taking advantage of the fact that axon terminals can be conveniently subdivided according to the part of the postsynaptic surface with which they make contact. In both the areas studied, only a small proportion (less than 3 %) of axon terminals were found to make direct contact with cell bodies (axosomatic contacts) and these will not be considered further. Most of the axon terminals appeared to make contact with dendrites or their appendages. It was found that a simple division of the dendritic post-synaptic surface could be made by distinguishing between the main dendritic trunks or shafts (which contain aligned microtubules) and the dendritic appendages or spines (which may vary considerably in structure, some being rather large expansions upon narrow necks and others consisting of much smaller dense knobs). This classification had the merit that figures could be collected rapidly and that samples from the same region in different animals gave results which were highly consistent. In the peripheral parts of the ventromedial nucleus (adjacent to the arcuate nucleus) the synapses were distributed roughly in the proportion of 3:1 on dendritic shafts and spines, respectively; a quite different ratio occurred in the samples from the strial part of the preoptic area, where spine synapses are far less common, so that the shaft synapses outnumber the spine synapses by a ratio of almost 13:1.

These observations implied that by the use of this rather simple classification of synapses it was possible to get a quantitative estimate of certain parameters by means of which the neuropil of different regions could be characterised. In functional terms, the significance of the relative distribution of synapses on dendritic spines and shafts is not clear. It is thought that in general spine synapses are of an excitatory type, whereas synapses upon dendritic shafts may be either excitatory or inhibitory. The observed ratios for the distribution of synapses may, therefore, have a relationship to the balance between excitatory and inhibitory inputs to a particular region. In the specific case of the preoptic area and ventromedial hypothalamus, it is not possible to relate differences to excitatory or inhibitory phenomena, although the different functional status of these two areas with regard to the control of gonadotrophin output has already been noted (see above).

In order to establish the mode of termination of the fibres in the stria terminalis, we have made use of the reaction of orthograde terminal degeneration. Two days after a lesion of the stria terminalis, the terminals belonging to the cut axons undergo a type of degeneration involving shrinkage and increased electron density although at this time they are mostly still in contact with their postsynaptic elements. By making such a lesion, therefore, it is possible to make a 4-way classification of synapses in both the preoptic area and the ventromedial nucleus. The 4 classes of synapses are degenerating synapses (i. e., those of strial origin) on dendritic shafts and spines, and non-degenerating (i. e., not of strial origin) on dendritic shafts and spines. By such an analysis we have established that the fibres of the stria terminalis tend to terminate preferentially upon dendritic spines in both of the areas studied. In both areas almost half the spine synapses were of strial origin, whereas usually less than 1 in 10 of the shaft synapses showed degeneration after the same lesions.

If it is argued that the regional differences in the neuropil of the hypothalamus and preoptic area are a reflection of their functional differences, then it could be supposed that the arrangement of the neuropil in the preoptic area of the female reflects in some way the presence there of the putative 'trigger' for the pre-ovulatory surge of gonadotrophins. In that case, one might expect that a quantitative analysis of the neuropil might reveal anatomical differences between the female preoptic area and that of the male, since the male does not possess the ability to initiate a surge of gonadotrophins even when gonadectomised and primed with oestrogen and progesterone. In a quantitative ultrastructural analysis of the neuropil of the preoptic area in a series of male and female rats we found that of all the 4 types of synapses described above, only the non-amygdaloid (i. e. non-degenerating) synapses upon dendritic spines differed in their incidence in males and females; the remaining 3 types had similar incidences in the two sexes. In the case of the non-amygdaloid synapses upon dendritic spines, however, it was clear that there were almost twice as many synapses in this category in the females as in the males [16].

This difference between the sexes may be of some interest because of its potential relationship to the presence in the female preoptic region of the putative 'trigger' zone which is necessary for the initiation of ovulation and also what may be a similar mechanism for the initiation of behavioural receptivity under the appropriate hormonal conditions. In order to strengthen this correlation we can take advantage of the facts known about the development of these reproductive functions [8]. It has been known for some time that the ability to show a cyclic surge of gonadotrophins in the presence of an ovary is not dependent on the genetic sex of the animal (at least for the rat). The crucial factor appears to be the presence or absence of certain sex steroid hormones during a critical stage of development. In the rat this is approximately the first 10 days after birth. In the female, where the central nervous system is not exposed to sex steroid hormones during this period, the adult functional pattern which develops is of the cyclic type (both with regard to gonadotrophin release and female mating behaviour). In the male the neonatal testis secretes enough androgen to cause differentiation of the brain away from the female type, and the adult is incapable of cyclic function. The crucial role of androgen (which it possibly exerts by being converted to oestrogen) [11] can be shown by the fact that a single injection of testosterone propionate into the female during the critical period of development is sufficient to prevent the development of cyclic adult function and, conversely, castration of the male within 12 h of birth (thus removing the source of its own endogenous androgen) results in an adult which, although genetically male, will show cyclic function in the presence of transplanted ovaries. Either of these manipulations carried out after the critical period is ineffective in preventing the development of the adult functional pattern appropriate to animals of that genetic sex.

This provides us with an ideal situation in which we can test whether the anatomical sexual differences we have observed in the strial part of the preoptic area are differentiated in the same way as the cyclic functions by the action of androgen during the neonatal period. Six groups of animals were prepared [18]. These consisted of three 'cyclic' groups: – (a) normal genetic females, (b) females treated with 1.25 mg of testosterone propionate on the 16th day of life (after the critical period), and (c) males castrated within 12 h of birth – and three 'non-cyclic' groups: – (a) normal genetic males, (b) males castrated on the 7th day of life, and (c) females treated with 1.25 mg of testosterone propionate on the 4th day of life (during the critical period). We have not tested the functional capacity of all these groups but did establish the vaginal cyclicity in the genetic females and the lordosis quotients in each group. All the functional observations were in accord with the hypothesis presented above. When the morphological assessments were carried out on the strial part of the preoptic area, it was found that the incidence of non-amygdaloid spine synapses was correlated not with the genetic sex of the animals but with the functional status. As will be seen from table I, the group of normal

Table I. Numbers of non-strial spine synapses (+ SEM) per 10 grid squares (1.8×10^4 μm^2) in the strial part of the preoptic area

Genetic sex	Treatment	Synapses	Functional status of adult
Female	none	53 ± 3	cyclic
	1.25 mg TP on day 4	35 ± 2	non-cyclic
	1.25 mg TP on day 16	54 ± 5	cyclic
Male	none	33 ± 2	non-cyclic
	castrated on day 0	50 ± 3	cyclic
	castrated on day 7	39 ± 4	non-cyclic

TP = testosterone propionate; day 0 = the day of birth. Data taken from a total count of 50,773 synapses in 64 rats.

females had incidences of 53 ± 3 synapses in 1.8×10^4 μm^2, the females treated with androgen on the 16th day of life and the males castrated within 12 h of birth had similar incidences. In the non-cyclic groups, the normal males had incidences of 33 ± 2, which were similar to those found in the females treated with androgen on the 4th day of life. The males castrated on the 7th day of life had higher incidences than the normal males but slightly lower incidences than the females; this group also turned out to be functionally somewhat heterogeneous when the lordosis quotient was measured [1, 18]. The most likely explanation for this heterogeneity is that by the 7th day of life only a proportion of the males are fully masculinised. Clearer results would almost certainly have been obtained if the castration had been delayed for another week. It is perhaps worth stressing that the anatomical differences were restricted to the non-amygdaloid spine synapses – none of the other 3 categories of synapse had a sexually dimorphic incidence. Furthermore, the difference was, as far as we could see, restricted to the strial part of the preoptic area; despite comparable counts of synapses, no such sexual difference was found in that part of the ventromedial hypothalamic nucleus receiving projections from the stria terminalis nor in other areas such as the septal nuclei [17].

In summary, therefore, we have very good circumstantial evidence to suggest that the anatomical difference found between the preoptic area of the male and that of the female is correlated in all 6 groups of animals with the ability to maintain the appropriately sexually dimorphic patterns of gonadotrophin release and mating behaviour. The correlation with the anatomical data rests on two facts – firstly, that the anatomical differences are found in the same general region (the preoptic area) which lesions and other studies indicate as essential for these functions and, secondly, that the establishment of the anatomical difference during normal ontogenetic development depends like the functional differences, not upon the genetic sex of the animal, but simply upon the presence or absence of androgen during the critical period.

It is perhaps best to end on a note of caution. We are studying here the end-point of a series of developmental processes which was influenced at a very early stage by the experimental manipulation of androgen. The location of the anatomical difference found in the adult, therefore, is not a direct indication that that region of the brain is in fact the primary target for the neonatal action of androgen. Furthermore, in attempting to establish the functional effects of small circumscribed lesions of the strial part of the preoptic area, we have found that total destruction of this specific region in the female rat is not incompatible with regular cyclic ovulation. Clearly, while the anatomical dimorphism is undeniably a clue to the functional difference between the male and the female, there are still many more stages in the puzzle which require to be elucidated.

References

1 Brown-Grant, K.: Recent studies on the sexual differentiation of the brain; in Comline, Cross, Dawes and Nathanielsz Foetal and neonatal physiology, pp. 527–545 (Cambridge University Press, London 1972).

2 Brown-Grant, K. and Raisman, G.: Reproductive functions in the rat following selective destruction of afferent fibres to the hypothalamus from the limbic system. Brain Res. *46:* 23–42 (1972).

3 Bruce, H. M.: Pheromones. Brit. med. Bull. *26:* 10–13 (1970).

4 Critchlow, B. V. and Bar-Sela, M. E.: Control of the onset of puberty; in Martini and Ganong Neuroendocrinology, vol. 2, pp. 101–162 (Academic Press, New York 1967).

5 Everett, J. W.: Central neural control of reproductive functions of the adenohypophysis. Physiol. Rev. *44:* 373–431 (1964).

6 Field, P. M.: A quantitative ultrastructural analysis of the distribution of amygdaloid fibres in the preoptic area and the ventromedial hypothalamic nucleus. Exp. Brain Res. *14:* 527–538 (1972).

7 Harris, G. W.: Central control of pituitary secretion; in Field, Magoun and Hall Handbook of physiology. I. Neurophysiology, vol. II, pp. 1007–1038 (American Physiological Society, Washington 1960).

8 HARRIS, G. W.: Hormonal differentiation of the developing central nervous system with respect to patterns of endocrine function. Philos. Trans. B *259:* 165–178 (1970).

9 LEONARD, C. M. and SCOTT, J. W.: Origin and distribution of the amygdalofugal pathways in the rat. An experimental neuroanatomical study. J. comp. Neurol. *141:* 313–330 (1971).

10 LISK, R. D.: Sexual behavior. Hormonal control; in MARTINI and GANONG Neuroendocrinology, vol. 2, pp. 197–240 (Academic Press, New York 1967).

11 NAFTOLIN, F.; RYAN, K. J., and PETRO, Z.: Aromatization of androstenedione by the anterior hypothalamus of adult male and female rats. Endocrinology *90:* 295–298 (1972).

12 OLMOS, J. S. DE and INGRAM, W. R.: The projection field of the stria terminalis in the rat brain. An experimental study. J. comp. Neurol. *146:* 303–334 (1972).

13 RAISMAN, G.: An evaluation of the basic pattern of connections between the limbic system and the hypothalamus. Amer. J. Anat. *129:* 197–202 (1970).

14 RAISMAN, G.: An experimental study of the projection of the amygdala to the accessory olfactory bulb and its relationship to the concept of a dual olfactory system. Exp. Brain Res. *14:* 395–408 (1972).

15 RAISMAN, G. and FIELD, P. M.: Anatomical considerations relevant to the interpretation of neuroendocrine experiments; in MARTINI and GANONG Frontiers in neuroendocrinology, pp. 3–44 (Oxford University Press, New York 1971).

16 RAISMAN, G. and FIELD, P. M.: Sexual dimorphism in the preoptic area of the rat. Science *173:* 731–733 (1971).

17 RAISMAN, G. and FIELD, P. M.: A quantitative investigation of the development of collateral reinnervation after partial deafferentation of the septal nuclei. Brain Res. *50:* 241–264 (1973).

18 RAISMAN, G. and FIELD, P. M.: Sexual dimorphism in the neuropil of the preoptic area of the rat and its dependence on neonatal androgen. Brain Res. *54:* 1–29 (1973).

19 SAWYER, C. H.: Functions of the amygdala related to the feedback actions of gonadal steroid hormones; in ELEFTHERIOU The neurobiology of the amygdala, pp. 745–762 (Plenum Press, New York 1972).

20 TALEISNIK, S.; CALIGARIS, L., and ASTRADA, J. J.: Feedback effects of gonadal steroids on the release of gonadotropins; in JAMES and MARTINI Hormonal steroids, pp. 699–707 (Excerpta Medica, Amsterdam 1971).

21 VELASCO, M. E. and TALEISNIK, S.: Release of gonadotropins induced by amygdaloid stimulation in the rat. Endocrinology *84:* 132–139 (1969).

22 WINANS, S. S. and SCALIA, F.: Amygdaloid nucleus: new afferent input from the vomeronasal organ. Science *170:* 330–331 (1970).

Authors' address: Dr. PAULINE M. FIELD and Dr. G. RAISMAN, Department of Human Anatomy, University of Oxford, South Parks Road, *Oxford OX1 3QX* (England)

Recent Studies of Hypothalamic Function
Int. Symp. Calgary 1973, pp. 26–38 (Karger, Basel 1974)

Temporal Organization of Neuroendocrine Function in Relation to the Sleep-Waking Cycle in Man

E. D. WEITZMAN

Department of Neurology, Montefiore Hospital and Medical Center, and the Albert Einstein College of Medicine, Bronx, N.Y.

During the past 8 years, our research group has been engaged in a series of studies emphasizing the temporal organization of certain hypothalamic-pituitary hormone systems as a function of the 24-hour sleep-waking cyclic pattern in man. The results of these studies have indicated that ACTH-cortisol, growth hormone, prolactin and the gonadotropins have important temporal relations with the sleep-waking cycles and in certain cases with sleep as a whole as well as with specific sleep stages. In addition, all the above hormones have been demonstrated to be secreted in an episodic manner throughout the 24-hour day such that a 'steady state' concentration cannot be recognized. The following sections describe our results to date for each of the above 4 hormonal systems.

I. ACTH-Cortisol Secretion

At the time of our first study it had been shown that all mammals including man had a characteristic temporal pattern of sleep stages such that the normal sleep period was composed of a series of recurrent short-term events consisting of a non rapid eye movement (REM) sleep stage cycle [10]. Man differed from nonprimate mammals, however, in that the early portion of his nightly sleep consisted largely of stages 3 and 4, whereas the latter portion of sleep consisted of two alternating stages, REM and stage 2 sleep. Since it was known that cortisol was secreted during the latter half of the night sleep, we developed a method which would enable us to define the temporal sequence of plasma concentration changes and compare these with polygraphically defined sleep stages in normal man. A small indwelling catheter was placed in an arm vein and plasma samples were obtained at frequent intervals with a minimum of disturbance to the sleeping subject. This method has been improved upon since 1964 and has allowed us subsequently to define the temporal sequence of events for 24- and 48-hour periods, not only for cortisol but for other hormones as well.

The results of our early study, with 30-min sampling, demonstrated that the rise in plasma 17-OHCS which occurred during the latter half of the night sleep period was characterized by a series of episodic peak

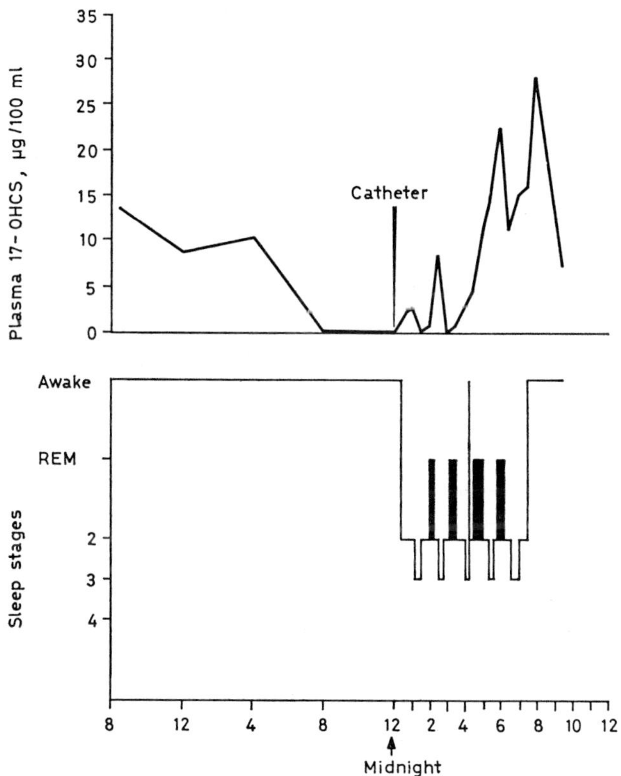

Fig. 1. Plasma 17-hydroxycorticosteroid (17-OHCS) concentrations, with 30-min sampling during the night and 4-hour sampling during the day in a normal young adult (D.R., total sleep time = 387 min). Sleep stages are depicted beneath hormonal graph. The rapid eye movement (REM) sleep periods (= 17 %) are shown in black [23].

elevations and not by a smooth gradual rise in hormonal concentration (fig. 1) [23]. A subsequent study in which ^{14}C-labeled cortisol was administered intravenously just prior to the initiation of the nocturnal secretory episodes, demonstrated that each episode of cortisol increment was accompanied by a proportionate decrease in cortisol specific activity [7]. This demonstrated that newly formed (unlabeled) cortisol was secreted in the episodic bursts during the early morning sleep period and that, between episodes, little or no cortisol was secreted. Although we speculated at the time that the intermittent release of cortisol might be related to recurrent changes in hypothalamic neuronal activity related to the REM-non-REM sleep cycle, future studies by our group did not support such a one-to-one relationship between these two recurrent cyclic physiologic events. That is, when subjects were sleep-deprived for 1 or 2 nights, the nocturnal episodic secretion of cortisol persisted, and when the sleep-wake cycle was acutely inverted (180°), a significant delay was present in the re-establishment of the circadian cortisol pattern, and the sleep stage patterns and plasma cortisol concentrations were dissociated [24].

We then extended the finding of episodic cortisol secretion to a study of the 24-hour pattern in normal adults with a normal sleep-wake cycle and 7–8 h of regular nocturnal sleep. The pattern of episodic secretion was found to be present throughout the entire 24-hour period. Indeed it was recognized that the 'circadian' or 24-hour pattern actually results from the temporal clustering of episodes of secretion (fig. 2) [25]. Utilizing a 20-min sampling technique, an average of 9 secretory episodes occurred in a 24-hour period. Approximately 25 % of the 24 h (350 min) was spent in active secretion, a total of approximately 16 mg of cortisol was secreted and a mean half-life of 66 min was measured. Four unequal temporal phases of episodic secretion for the 24-hour sleep-wake cycle were defined. Phase I was a 6-hour period of 'minimal secretory activity' (2 min of secretory activity per hour, with 0.28 mg of cortisol secreted per phase), this phase beginning 4 h before and ending 3 h after 'lights out' at night. Phase II was a 3-hour period called the 'preliminary nocturnal secretory episode' (16 min of secretory activity per hour, with 1.7 mg secreted); the phase occurring between the 3rd to 5th hours of sleep. Phase III was a 4-hour period called the 'main secretory' phase occurring from the 6 to the 8th hour of sleep and first hour after awakening (31 min of secretory activity per hour, 6.3 mg of cortisol secreted) and phase IV, the 11 h of 'intermittent waking secretory activity', occurring from 2 h after lights on in the morning to 5 h before sleep onset in the evening (15 min of secretory

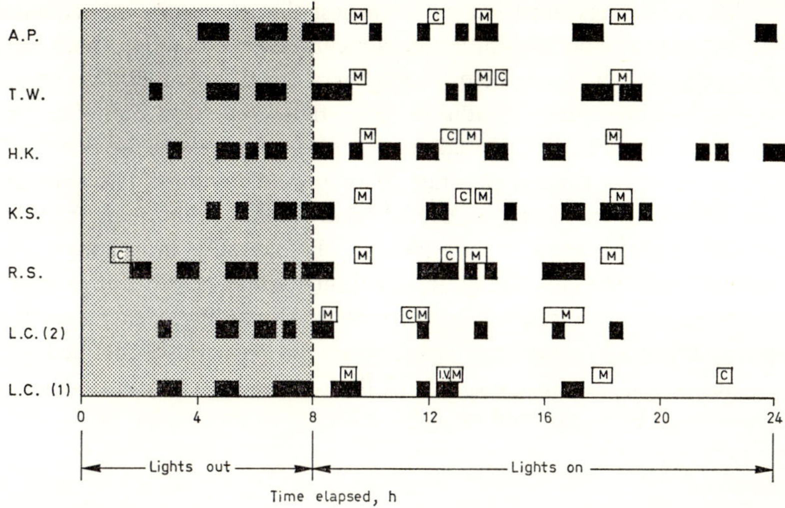

Fig. 2. Duration and time of secretory cortisol episodes for a 24-hour period for 7 separate studies. M is time of meals; C is time of insertion or reinsertion of the venous catheter; i.v. is time of intravenous injection of a tracer dose of ^{14}C-cortisol. Plasma samples were obtained every 20 min for each 24-hour period [25].

activity per hour, 7.2 mg secreted). A 'basal level' or 'steady state' cortisol concentration was not found for any extended time period of the 24-hour day. Indeed, only when the cortisol concentration fell to near zero (generally during 4 h at the time of sleep onset) was there any prolonged period of 'constancy'.

These studies were carried out with a 20-min sampling interval. We have recently extended these observations utilizing a 5-min sampling interval [6]. This very frequent sampling interval has demonstrated that smaller secretory episodes occur between the larger ones and that they may be missed with the longer sampling interval. In addition, the 5-min sampling demonstrated that longer half-life estimates are the result of several small intervening episodes of secretion. Concomitant with the cortisol measurement, we also measured ACTH in a series of 5-min plasma samples. This demonstrated that an appropriate temporal correlation was present for the major secretory episodes. In addition, we found that if several ACTH episodes of short duration occurred in close temporal juxtaposition, we could only detect one cortisol secretory episode having a smoothly rising slope.

These findings have led us to suggest that the temporal sequence of ACTH-cortisol episode initiation appears to be mainly under central nervous system (CNS) control as a 'programmed' sequence of events and, furthermore, that under conditions of a stable, repetitive daily life pattern the association of the ACTH-adrenal secretory events with the sleep-waking cycle is part of a general program of biological rhythms. The response of the ACTH-cortisol system to acute 'stress' stimuli (pain, fear, etc.) during the waking state cannot be considered to be operative in our studies as an explanation of the episodic pattern, although other possible environmental factors such as telephone calls, conversations, reading, visitors, meal time and changes in emotional state were not rigorously defined. Their contribution, if any, to the timing of the waking secretory episodes should be evaluated by future studies.

II. Growth Hormone (GH) Release; Relation to Sleep Stages

At the time of our first cortisol sleep study Dr. SEYMOUR GLICK measured GH in our nocturnal plasma samples and found that GH was released during the first hour after sleep onset, and not at the time cortisol was being secreted in large amounts in the latter third of the night [16]. We erroneously interpreted this sleep onset rise in GH as being related to a postprandial rise occurring 3–5 h after the evening meal. A series of subsequent investigations by other research groups confirmed our findings but demonstrated that the GH release was clearly a sleep onset event [8, 13, 20]. If the time of sleep onset was shifted several hours or inverted by 180° (12-hour shift), GH secretion clearly followed sleep onset and no release of GH took place at the expected time if sleep did not take place [17].

A number of studies have suggested that the sleep-related release of GH may be associated with the stages 3–4 sleep pattern. We recently carried out a study in which sequential plasma samples were obtained at 4-min intervals during and for about 90 min after the transition period from waking to sleep [15]. The results indicated that not until definitive sleep had occurred, when low frequency synchronous electro-cortical activity (stages 2–3) took place, was GH released (fig. 3). The consistent correlative temporal pattern suggests that a similar mechanism precipitates both the onset of 'slow wave' sleep and the release of GH. However, the presence of stages 3–4 sleep is not always associated with GH secretion since, during

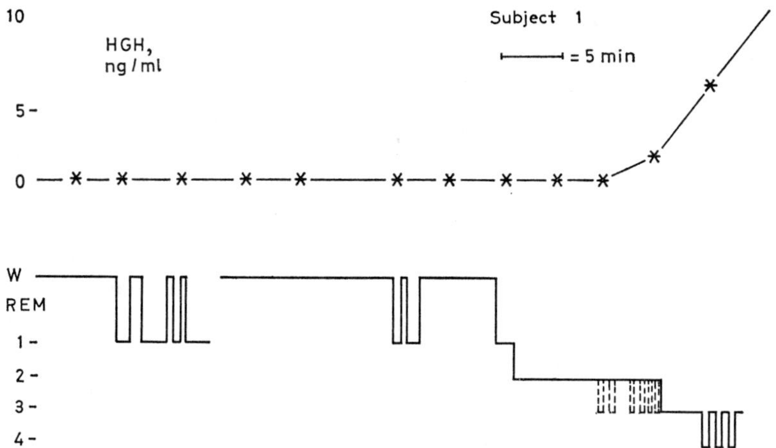

Fig. 3. Human growth hormone (HGH) plasma measurements during the transition from waking to sleep in a normal young adult subject. Sleep stages (REM, 1, 2, 3, 4) were scored in 30-sec epochs. A 10-sec epoch was used for scoring for the dotted lines in the transition period from stage 2 to stage 3 sleep [15].

the latter portion of the night stages 3–4 sleep has not always been associated with GH release.

During the waking day GH is characteristically secreted in transient episodic releases, some associated with acute stressful stimuli, but most appear without a clear related specific stimulus. It is of some interest that during the waking condition intravenous glucose will suppress the GH release to certain stimuli, whereas the sleep-related release cannot be prevented by intravenous infusion of glucose [14]. This suggests that the sleep-related release mechanism utilizes a different CNS pathway.

III. Prolactin Release in Man during Sleep

The recent development of an accurate radioimmunoassay method to measure specifically human prolactin has made it possible to study the 24-hour release patterns in normal young men and women [5, 9]. In a collaborative series of studies carried out with Dr. A. Frantz we measured prolactin in plasma at 20-min intervals for 24-hour periods [18]. In all subjects, a clear increase in the plasma concentration took place 60–90 min after sleep onset. This initial nocturnal peak was then followed by a series of increasingly larger secretory episodes during the rest of the night sleep period with peak concentrations occurring at approximately 0500–0700 h at the end of the night. During the hour after awakening, a rapid fall in concentration occurred with low plasma values reached between 1000 h and noon. Throughout the 24-hour period, the pattern of release is clearly episodic but differed from the GH pattern since the prolactin concentration never fell to undetectable values.

We recently studied the effects of shifts of the sleep-wake cycle on the temporal pattern of prolactin release and compared the changes with GH and cortisol concentrations [19]. We found that the sleep-related release of prolactin was shifted immediately in relation to the sleep pattern shift, as was the case with GH. However, with the exception of the sleep onset secretory episode, no consistent relation between the GH and prolactin episodes could be identified for the remaining 24-hour period. The ACTH-cortisol pattern differed considerably in that it took a period of 2–3 weeks before the 24-hour (including the sleep-related) episodic pattern shifted [21].

IV. The 24-Hour Pattern of Luteinizing Hormone (LH) Secretion

A number of years ago, our research group undertook a study of the 24-hour pattern of gonadotropic hormones in man utilizing plasma sampling and polygraphic sleep stage measurement techniques previously outlined. We have studied a group of normal young adult men, pubertal boys and girls, prepubertal children and normal women during different segments of the menstrual cycle.

We found that in normal young adult men, the 24-hour pattern was characterized by a sequence of LH secretory episodes, the initiation and cessation of these episodes taking place within a narrow, well-defined range of LH concentrations [4]. Approximately 12 secretory episodes occurred for a 24-hour period, with one-third present during the sleep period. We were unable to demonstrate a 24-hour rhythm either in concentrations or clustering of the episodes as a function of the time of day. However, a different pattern was found in pubertal boys and girls. During both early and late puberty, a major increment in LH concentration occurred during the entire sleep period, with a series of major augmentation of secretory episodes during this period (fig. 4) [3]. If the time of sleep onset was delayed, a delay of the sleep-related augmented LH

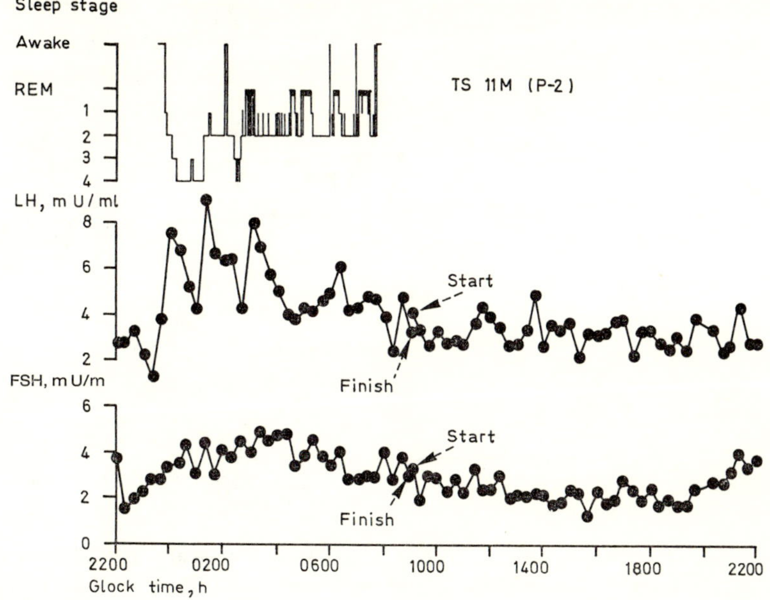

Fig. 4. The 24-hour pattern of LH and follicle-stimulating hormone (FSH) plasma concentrations in an 11-year-old pubertal boy. Sleep stages are depicted (REM, 1, 2, 3, 4) above the hormonal graphs. Samples were obtained every 20 min [3].

secretion also occurred. This indicates that the augmented LH release during puberty is directly related to sleep similar to that found for GH and prolactin in adults. In the late pubertal group, the daytime waking LH secretory episodes had a higher amplitude pattern than that found for the early pubertal group, resulting in a mean LH concentration approaching the adult range. In the prepubertal children, no such sleep-augmented release was found, rather, the pattern was one of a low range of plasma concentrations with evidence for a sequence of small secretory episodes present throughout the 24-hour day. In a recent study in our laboratories carried out on several children with idiopathic precocious puberty, patients with hypothalamic dysfunction due to a tumor and 2 patients with primary adrenal cortical abnormality (adreno-genital syndrome), we found that they too had an augmented LH secretion during sleep [2]. In addition, in several patients of pubertal age with gonadal dysgenesis, an augmented LH secretion pattern during sleep was also found [1].

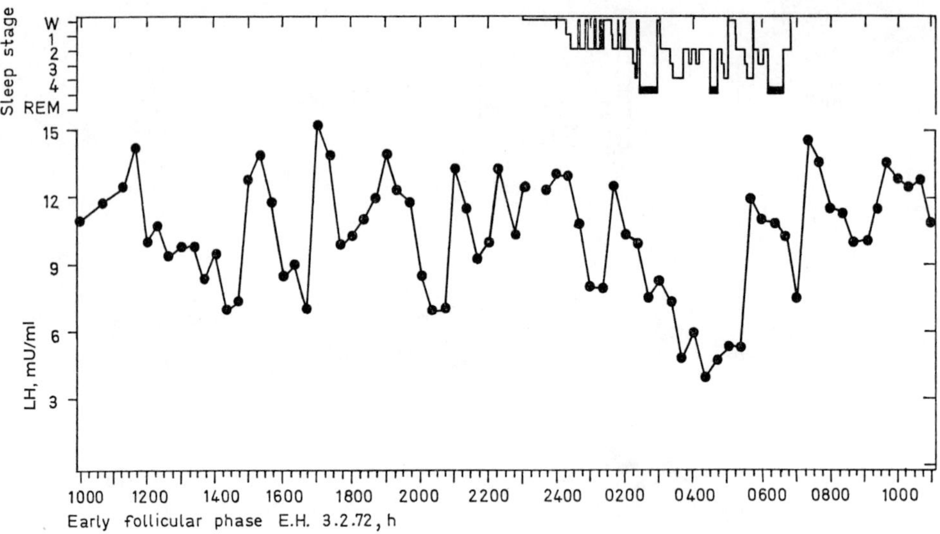

Fig. 5. The 24-hour plasma LH concentrations in a normal young adult woman, during the early follicular phase of the menstrual cycle. Samples were obtained every 20 min. Nocturnal sleep stages (REM, 1, 2, 3, 4) are depicted above the hormonal graph. W = wake [12].

Since the 24-hour pattern of LH secretion in adult women is complicated by changes occurring in relation to the monthly menstrual cycle, we have carried out such studies during selected portions of the cycle. During the early follicular phase (first 5 days), the pattern of LH secretion is characterized by a sequence of 10–15 episodes during the 24-hour period, similar to that found in adult men (fig. 5) [12]. No major increment in LH occurred during sleep such as found in puberty. However, the pattern differed from men in that there was a significant decrease in the plasma LH concentration during the first 3 h after sleep onset, a 33-percent decrease in the mean plasma LH taking place during the 3rd hour after the first onset of stage 2 sleep. This clear decrease in LH concentration during the first half of the night was followed by rise in the latter half of the sleep period (fig. 6). No such pattern could be recognized for men.

During the peri-ovulatory phase of the cycle, and especially at the time of the LH 'surge', larger secretory episodes took place during the 24-hour period superimposed on a progressive elevation of the 'baseline' concentration [11]. In several subjects, the presumptive onset of the LH surge could be recognized by an abrupt large increase in the plasma LH

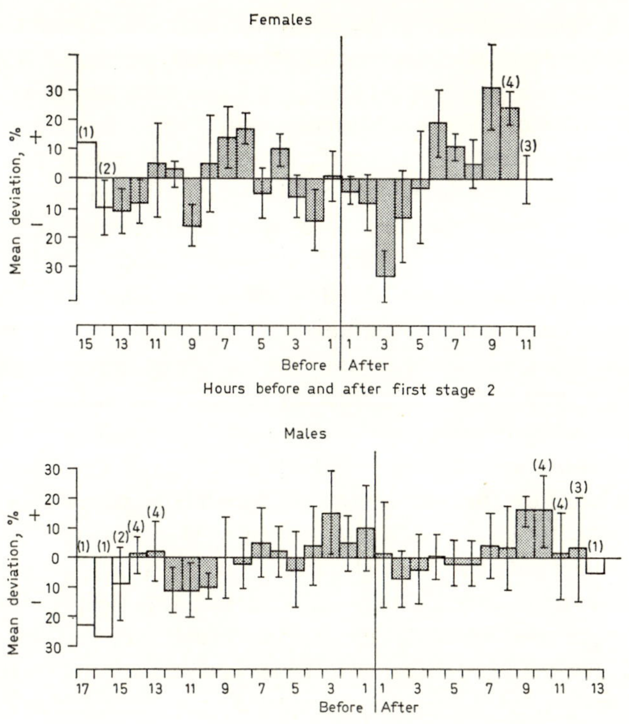

Fig. 6. Mean percent deviation of luteinizing hormone (LH) concentration for each hour from individual 24-hour mean values. The time scale is determined before and after the onset of stage 2 sleep for the night. Six women and 6 men compose the study groups except where specifically indicated above the hourly bar graphs [12].

concentration, occurring in close proximity to the end of the nocturnal sleep period [22]. However, further work must be done define the relationship of the initiation of the LH surge and the sleep-waking 24-hour cycle.

Conclusions

It is now clear that at least 4 hypothalamic-pituitary hormonal systems have temporal patterns of secretion that are closely linked to the 24-hour sleep-waking activity in man. No single principle or mechanism can explain these patterns, but rather each system has its own temporal organization and response to manipulations of the sleep-waking cycle. The ACTH-cortisol cycle is relatively resistent to change, and can be readily dissociated from sleep in spite of a highly correlative relationship under normal stable circadian conditions. Growth hormone, on the other hand, appears to be intimately associated with a specific sleep stage in relation to the period after sleep onset, and its release can be readily shifted by shifting the time of sleep. Following the sleep onset related GH release in adults there is usually none or only 1 or 2 small secretory episodes later in the night sleep period. Prolactin is released in large quantities throughout the night in an episodic manner with initiation at the time of sleep onset and increasing in concentration throughout the sleep period with a rather abrupt termination and fall of concentration just after awakening in the morning. Shift of sleep is associated with a shift of the release pattern. The gonadotropic hormone LH, has a complex pattern as expected because of the inter-related problems of maturation (prepuberty, puberty, sexual maturity and postmaturity), sex differences, menstrual cyclicity in females and the sleep-waking patterns. A most exciting finding is that in pubertal children there is an intimate association of LH release with sleep. Evidence suggests that at certain times during the menstrual cycle in sexually mature women, the release pattern of LH may also be related to sleep, with the question raised of a sleep ('critical period') associated LH episodic pattern during the 'LH surge' period. All of the above hormones studied, utilizing a sequential frequent sampling technique, have been shown to have an episodic secretory pattern throughout the 24-hour sleep-wake cycle.

The importance of the biological organization of neuroendocrine systems in relation to time is therefore clearly emphasized by these studies. The issues concerning the mechanisms maintaining normal temporal synchronization of hormonal release and the effect of disease upon these time-dependent processes should be pursued both from a diagnostic and pathological point of view.

References

1 BOYAR, R.; FINKELSTEIN, J.; ROFFWARG, H.; KAPEN, S.; WEITZMAN, E. D. and HELLMAN, L.: Augmentation of the 24-hour LH and FSH concentration during sleep in gonadal dysgenesis at the expected time of puberty. J. clin. Endocrin. *37:* 521 (1973).

2 BOYAR, R.; FINKELSTEIN, J. W.; DAVID, R.; ROFFWARG, H.; KAPEN, S.; WEITZMAN, E. D., and HELMANN, L.: Twenty-four hour luteinizing hormone and follicle stimulating hormone secretory pattern in sexual precocity. Clin. Res. *21:* 486 (1973).

3 BOYAR, R.; FINKELSTEIN, J.; ROFFWARG, H.; KAPEN, S.; WEITZMAN, E. D., and HELLMAN, L.: Synchronization of augmented LH secretion with sleep during puberty. New Engl. J. Med. *287:* 582–586 (1972).

4 BOYAR, R.; PERLOW, M.; HELLMAN, L.; KAPEN, S. and WEITZMAN, E. D.: Twenty-four hour pattern of luteinizing hormone secretion in normal men with sleep stage recording. J. clin. Endocrin. *35:* 73–81 (1972).

5 FRANTZ, A. G. and KLEINBERG, D. L.: Prolactin. Evidence that it is separate from growth hormone in human blood. Science *170:* 745–747 (1970).

6 GALLAGHER, T. F.; YOSHIDA, K.; ROFFWARG, H.; WEITZMAN, E. D.; FUKUSHIMA, D. K., and HELLMAN, L.: ACTH and cortisol secretory patterns in man. J. clin. Endocrin. *36:* 1058 (1973).

7 HELLMAN, L.; NAKADA, F.; CURTI, J.; WEITZMAN, E. D.; KREAM, J.; ROFFWARG, H.; ELLMAN, S.; FUKUSHIMA, D. K., and GALLAGHER, T. F.: Cortisol is secreted episodically by normal man. J. clin. Endocrin. *30:* 411–422 (1970).

8 HONDA, Y.; TAKAHASHI, K.; TAKAHASHI, S.; AZUMI, K.; IRIE, M.; SAKUMA, M.; TSUSHIMA, T., and SHIZUME, V.: Growth hormone secretion during nocturnal sleep in normal subjects. J. clin. Endocrin. *29:* 20–29 (1969).

9 HWANG, P.; GUYDA, H., and FRIESEN, H.: A radioimmunoassay for human prolactin. Proc. nat. Acad. Sci., Wash. *68:* 1902–1906 (1971).

10 KALES, A. (ed.): Sleep-physiology and pathology (Lippincott, Philadelphia 1969).

11 KAPEN, S.; BOYAR, R.; HELLMAN, L., and WEITZMAN, E. D.: Variations of plasma gonadotropin in normal subjects during the sleep-wake cycle. Psychophysiology *7:* 337 (1970).

12 KAPEN, S.; BOYAR, R.; HELLMAN, L., and WEITZMAN, E. D.: Episodic release of luteinizing hormone at mid-menstrual cycle in normal adult women. J. clin. Endocrin. *36:* 724–729 (1973).

13 PARKER, D. C.; SASSIN, J. F.; MACE, J. W.; GOTLIN, R. W., and ROSSMAN, L. G.: Human growth hormone release during sleep. Electroencephalographic correlation. J. clin. Endocrin. *29:* 871–874 (1969).

14 PARKER, D. C. and ROSSMAN, L. G.: Human growth hormone release in sleep. Nonsuppression by acute hyperglycemia. J. clin. Endocrin. *32:* 65–69 (1971).

15 PAWEL, M. A.; SASSIN, J. F., and WEITZMAN, E. D.: The temporal relation between HGH release and sleep stage changes at nocturnal sleep onset in man. Life Sci. *11:* 587–593 (1972).

16 ROTH, J.; GLICK, S. M.; CUATRECASAS, P., and HOLLANDER, C.: Acromegaly and other disorders of growth hormone secretion. Ann. intern. Med. *66:* 760–788 (1967).

17 SASSIN, J. F.; PARKER, D. C.; MACE, J. W.; GOTLIN, R. W.; JOHNSON, L. C., and ROSSMAN, L. G.: Human growth hormone release. Relation to slow wave sleep and sleep-waking cycles. Science *165:* 513–515 (1969).

18 SASSIN, J. F.; FRANTZ, A. G.; WEITZMAN, E. D., and KAPEN, S.: Human prolactin. 24-hour pattern with increased release during sleep. Science *177:* 1205–1207 (1972).

19 SASSIN, J. F.; FRANTZ, A.; KAPEN, S., and WEITZMAN, E. D.: The nocturnal rise of human prolactin is dependent on sleep. J. clin. Endocrin. *37:* 436 (1973).

20 TAKAHASHI, Y.; KIPNIS, D. M., and DAUGHADAY, W. H.: Growth hormone secretion during sleep. J. clin. Invest. *47:* 2079–2090 (1968).

21 WEITZMAN, E. D.: Unpublished observations (1970).

22 WEITZMAN, E. D.: Unpublished observations (1972).

23 WEITZMAN, E. D.; SCHAUMBURG, H., and FISHBEIN, W.: Plasma 17-hydroxycorticosteroid levels during sleep in man. J. clin. Endocrin. *26:* 121–127 (1966).

24 WEITZMAN, E. D.; GOLDMACHER, D.; KRIPKE, D.; MCGREGOR, P.; KREAM, J., and HELLMAN, L.: Reversal of sleep-waking cycle. Effect on sleep stage pattern and certain neuroendocrine rhythms. Trans. amer. Neurol. Ass. *93:* 153–157 (1968).

25 WEITZMAN, E. D.; FUKUSHIMA, D. K.; NOGEIRE, C.; ROFFWARG, H.; GALLAGHER, T. F. and HELLMAN, L.: The twenty-four hour pattern of the episodic secretion of cortisol in normal subjects. J. clin. Endocrin. *33:* 14–22 (1971).

Author's address: Dr. E. D. WEITZMAN, Department of Neurology, Montefiore Hospital and Medical Center, and the Albert Einstein College of Medicine, *Bronx, NY 10467* (USA)

Recent Studies of Hypothalamic Function
Int. Symp. Calgary 1973, pp. 39–49 (Karger, Basel 1974)

Functional Identification of Hypothalamic Neurones

B. A. CROSS

Department of Anatomy, University of Bristol, Bristol

I. Introduction

Anyone who aspires to unravel the complexities of the hypothalamus soon has to face the problem of neuronal specification: What are the differing roles of its constituent neurones? The outputs of the hypothalamus include vegetative responses like thermo-regulatory vasodilation and panting, endocrine effects involving both lobes of the pituitary and the adrenal medulla and behavioural reactions such as sexual receptivity or aggression. Obviously the various outputs utilise different neuronal pathways but we have little idea of their identity. Also our knowledge of the extent of sharing of interneurones between the various outputs is non-existent. Does a cell at one instant signal lordosis and at another ovulation, thirst or secretion of ADH; prolactin or oxytocin release; rage or adrenaline, and so on? Morphological studies only provide clues to major afferent and efferent pathways involved in different hypothalamic mechanisms. Other methods must be used to penetrate to the neuronal level of organisation, and microelectrode recording of signals from living neurones has taken an increasingly important part [see reviews by BEYER and SAWYER, 5; KOMISARUK, 25; FINDLAY, 19; CROSS, 7]. All this work has not yet gone very far towards settling outstanding questions of functional specificity of neurones. Indeed much of the evidence might imply a high degree of non-specificity in that many single cells respond to a wide variety of stimuli. But we must remember that the vast majority of recorded units are interneurones whose connections are unknown. In such circumstances one guess is as good as another in deciphering recorded signals and assigning a function to the cell. To make real progress it seems necessary to gather all the following items of information about recorded neurones: (A) the location of the cell body; (B) the distribution of its afferent synapses; (C) the neurotransmitters at the afferent synapses; (D) source of afferent fibres; (E) the site(s) of termination of its axon(s); (F) the neurotransmitter-neurohormone released at its axonal terminals, and (G) the relationship between action potential activity and output of transmitter-neurohormone.

In the mammalian hypothalamus there is probably only one instance where the majority of these requirements is met, if only tentatively, and that is the neurosecretory cell that releases oxytocin in the milk-ejection reflex.

II. The Oxytocin Neurone

A. Location

It is unnecessary to review the evidence that the hormones oxytocin and vasopressin are elaborated in separate neurosecretory cells in the paraventricular (PV) and supraoptic (SO) nuclei of the hypothalamus [see SCHARRER, this volume]. Still far from clear, however, is the proportion of oxytocin neurones in each nucleus. Since the work of OLIVECRONA [34] there has been increasing support for the view that the PV nucleus is the primary site of oxytocin cells and that vasopressin cells are concentrated in the SO nucleus. Nevertheless, there are equally strong indications that the segregation is not absolute, i.e. that the two neurosecretory cell types co-exist in both nuclei [6]. What has been lacking until recently has been a means of identifying an oxytocin neurone in life.

Experiments in lactating rabbits showed that neurosecretory cells in the PV nucleus could be identified by recording action potentials induced by antidromic stimulation of the neural lobe [33, 37]. We thought these might be oxytocin neurones, but could not prove the point because of difficulties in evoking a reflex oxytocin release in the anaesthetised rabbit. Then LINCOLN and WAKERLEY [27] discovered an exceptional phenomenon in anaesthetised lactating rats, i.e. that milk-ejection responses occurred at frequent intervals so long as the pups were left to suck the nipples. The recorded intra-mammary pressure changes were identical to those elicited by intravenous injection of about 1 mU oxytocin or a brief electrical stimulus to the neural lobe. The reflex response was blocked in a cannulated gland by a close arterial injection of the anti-oxytocin agent carbamylmethyloxytocin. Moreover, simultaneous measurements of urine output and milk ejection in lactating water-loaded rats established that less than 0.01 mU antidiuretic hormone accompanied the release of 1 mU oxytocin induced by suckling [42]. An essentially similar process of recurrent milk ejection occurs in unanaesthetised, unrestrained mother rats in their nesting cages [26].

By combining the antidromic technique for identifying PV neurosecretory neurones with the reflex milk-ejection responses just described, WAKERLEY and LINCOLN [43] were able to monitor the discharge activity of neurosecretory and non-neurosecretory cells during many cycles of reflex oxytocin release. The results of this manoeuvre exceeded all expectation. About half the neurosecretory cells advertised their oxytocic function by suddenly discharging a volley of high-frequency spikes, and then lapsed into silence for many seconds before resuming their previous slow rate of background firing. In every case, a reflex milk-ejection response followed some 15–18 sec after the explosion of unit activity. Such episodes of a brief rapid acceleration, from less than 2 spikes/sec to over 50 spikes/sec, followed by quiescence, are seen extremely rarely outside PV and SO nuclei and the constancy of the discharge, time locked to the subsequent milk-ejection response, makes it virtually certain that they are generating output of oxytocin from the neural lobe endings.

WAKERLEY and LINCOLN [44] have now transferred attention to the SO nucleus of the suckled lactating rat. They find, surprisingly, that exactly the same type of response can be seen in about half the antidromically identified neurosecretory cells in this nucleus also. It is fairly certain, therefore, that oxytocin cells are not confined to the PV nucleus and they may even be more numerous in the SO nucleus. This somewhat unexpected conclusion is supported by the discovery [18] that the SO nucleus of the rat contains more oxytocin than the PV nucleus.

B. Afferent Synapses

Despite the considerable interest that neurosecretory neurones have attracted there is still a paucity of morphological data about afferent terminations on these cells. Repeated attempts to demonstrate the terminal boutons with standard impregnation methods have been unavailing despite the profusion of such endings on non-neurosecretory neurones, e. g. in the ventromedial hypothalamic nucleus. There has also been little information about the number and form of the non-axonal processes of neurosecretory cells. Golgi impregnations of the cells so far have not been achieved. In electron-micrographs it is difficult to distinguish axonal from dendritic processes, for both appear to contain neurosecretory granules and microtubules. There is an absence of dendritic spines and no typical initial segment to aid in diagnosis. Synaptic endings are found, though not in high density, over the entire cell membrane and on the unmyelinated axon. Some of these synapses are almost certainly from axons coming from outside the hypothalamus since degenerating endings can be observed on paraventricular neurones in hypothalamic islands prepared 48 or 72 h previously in rats [7, 18]. As yet, however, there is a dearth of evidence to enable us to differentiate excitatory from inhibitory endings or to determine whether axons from different sources are separately located on the cell membrane. However, in preliminary studies J. F. MORRIS [unpublished observations] has found that 70 % of the synapses on the soma of PV neurosecretory cells contain dense cored vesicles, usually considered to store catecholamines. This is compatible with their being inhibitory endings.

C. Neurotransmitters

The histochemical observations of SHUTE and LEWIS [36] show the presence of cholinergic endings in the PV and SO nuclei, while fluorescent microscopy studies [20] demonstrate an even richer network of catecholaminergic (mostly noradrenergic) fibres within the nuclei. These findings are very suggestive of cholinergic and adrenergic control of neurosecretory cells and provide a structural basis for the earlier discoveries of DUKE and PICKFORD [13] which implied that adrenaline inhibited the release of ADH and of oxytocin and that acetylcholine was excitatory. This notion has recently been supported by experiments in which putative transmitters were injected micro-iontophoretically on to the cell bodies of antidromically identified neurosecretory cells. In the rat we found that the majority of neurosecretory cells in both nuclei were excited by iontophoretic injections of acetylcholine and none were inhibited, while noradrenaline inhibited more cells than it excited [29]. In a further study restricted to the PV nucleus in rabbits we found that 81 % were inhibited by noradrenaline [30]. Incidentally, this study underlined the importance of rigorously defining the antidromically activated neurones, for the non-neurosecretory cells reacted in the opposite fashion – being excited by noradrenaline and inhibited by acetylcholine. None of the other transmitters tested had this ability to differentiate the neurosecretory cells, e. g. glutamate excited and γ-aminobutyric acid (GABA) inhibited all the cells indiscriminately, while dopamine and serotonin had variable effects on both neurosecretory and non-neurosecretory cells. In the rat using a ventral approach to the SO nucleus, DREIFFUS and KELLY [11] also obtained evidence that acetylcholine was excitatory to neurosecretory neurones. Moreover, 3 laboratories concur that some at least of these excitatory responses are mediated by nicotinic receptors [3, 11, 30]. However, BARKER *et al.* [3] reported that in the cat SO nucleus most of antidromically identified cells were inhibited by acetylcholine. Since the inhibition was blocked by atropine they suggested that these effects were mediated by muscarinic receptors. Despite this discordant observation BARKER *et al.* [3] found that noradrenaline was inhibitory – in agreement with our own work in rats and rabbits – and also produced evidence that β-adrenergic receptors were involved.

So far, in what has been described, we have made no distinction between oxytocin and vasopressin cells. But at least for the rat, and probably the rabbit also, it seems that the majority of the oxytocin cells must be excited by acetylcholine and inhibited by noradrenaline. We are not yet fully entitled to conclude that orthodromic activation of oxytocin neurones in the physiological milk-ejection response is mediated by cholinergic synapses. But since we can now recognize oxytocin neurones *in vivo* it should not be long before this question is settled by the use of appropriate blocking agents applied iontophoretically during reflex activation, and by simulating the natural discharge by injections of transmitter. A similar analysis will be required to elucidate the physiological significance of noradrenaline as an inhibitory transmitter.

Here we should mention the curious fact that iontophoretic injection of oxytocin excites neurosecretory cells in the PV nucleus [31]. It is difficult to see the functional significance of this observation as there is no evidence that oxytocinergic endings occur on neurosecretory cells. Such cell to cell interactions as have been reported are predominantly inhibitory [24]. The weight of evidence is also against oxytocin exerting a stimulatory effect via a vascular or ventricular route [7, 15].

D. Source of Afferent Fibres

A good deal is known about the stimuli that excite or inhibit release of oxytocin [6, 35] but much less about the neural pathways involved [8]. There is agreement that the spinal afferent pathway for the milk-ejection reflex is ipsilateral but there is some conflict as to whether the fibres are in the lateral or dorsal funiculi. In the brain-stem, the ipsilateral situation remains but the precise location is disputed. On the basis of stimulation experiments TINDAL and his associates believe that the pathway is fairly compact in the region of the spinothalamic tract at the level of the midbrain tegmentum in guinea-pigs, rabbits and goats [23, 38, 39]. Using a similar technique in rabbits, we found positive stimulation points much more widely scattered through the midbrain tegmentum, suggesting a diffuse ascending pathway [40]. However, stimulation experiments do not demonstrate utilization of the pathways under physiological circumstances. A number of stimulation sites in scattered brain structures has been reported to elicit oxytocin release or its blockade [1, 2] but only lesion studies can show whether any of these mechanisms play an important physiological role. It is important, therefore, that the suckled lactating rat affords a means of monitoring oxytocin release together with the initiating discharge of oxytocin neurones, for this can be used in conjunction with lesion techniques to determine the essential neural pathways operating in the reflex release of oxytocin.

E. Axon Terminations

There is little doubt that the great majority of the axons from oxytocin cells terminate in the neural lobe. Some endings may occur in the median eminence, but the number in this category is uncertain. We know from antidromic stimulation studies that about a quarter of the cells in the PV nucleus do not project to the neural lobe [37] and it is assumed that these are non-neurosecretory cells which may include the neurones sending axons to the SO nucleus [46]. Electro-physiological evidence also points strongly to the existence of recurrent inhibitory collateral branches from neurosecretory cells [4, 12, 24, 32]. The location of the recurrent terminals is unknown, however, as is the identity of the inhibitory interneurone, though it is quite possible the latter is a noradrenergic cell (see above). Another difficulty is the nature of the chemical transmitter released from the collateral ending. Could this be oxytocin? The only transmitter-like action of oxytocin that has been reported is the excitation of neurosecretory cells which could scarcely be a basis for recurrent inhibition.

F. Substance Released from Axon Terminal

In the case of the oxytocin cell this presents little problem. The octapeptide oxytocin is released from the terminals in the neural lobe. It is becoming more and more likely, moreover, that the hormone is discharged by a process of exocytosis [10] in which the protein neurophysin II bound to oxytocin is also extruded from the axon [22]. The origin and nature of the 'synaptic' vesicles also seen in the terminals in electronmicrographs is still disputed but few workers any longer suppose them to contain synaptic transmitter substances.

G. Action Potentials and Hormone Release

The recent discoveries of WAKERLEY and LINCOLN [43, 44] throw light on this problem. Evidently, the reflex discharge of 1mU oxytocin in the suckled rat follows a near-synchronous discharge of about half the neurosecretory cells in the PV and SO nuclei for 2–4 sec at a peak rate of 30–80 spikes/sec. The resting discharge rate of 0–3/sec is associated with no measurable effect on milk ejection. Significantly, earlier work on the stimulation parameters for evoking milk ejection from the neural stalk indicated that shocks at frequencies of less than 20/sec were ineffective, while 50/sec were about optimal [21, 37]. Thus, it seems that only a crescendo of impulses arriving at the neural lobe terminals can release sufficient hormone to reach an adequate plasma concentration to contract the mammary myoepithelium. Of some interest is the after-inhibition that follows the brief barrage of spikes which is probably due to the inhibitory feedback loops referred to above. In effect, it allows an intermittent release of oxytocin by a re-distribution of action potentials in time – since the accelerated firing is balanced by the subsequent quiescence. Whether the action potential traffic affects transport of hormone down the axons as well as release from their endings is less certain. However, a reduction of firing in the PV neurosecretory cells occurs in hypothalamic islands and this is associated with increased storage of oxytocic activity in the nucleus [15, 18].

III. Concluding Remarks

Next to the oxytocin cell, the vasopressin cell most nearly satisfies our requirements for functional identification [14, 16, 41]. For other recorded hypothalamic neurones the evidence is fragmentary. For example, YAGI and SAWAKI [45] and MACARA *et al.* [28] have described a population of cells that were antidromically excited from the median eminence. These might include cells producing several different releasing factors, dopamine or serotonin. DYER and CROSS [17] have described preoptic cells excited antidromically from the ventromedial-arcuate region which may *inter alia* include interneurones to the releasing factor cells. We have also described fluctuations in firing rates of pre-optic-anterior hypothalamic neurones correlated with different days of the oestrous cycle in rats [9]. Many other cases of single neurones responding to particular stimuli, e. g. steroid hormones, anterior pituitary hormones, glucose, iontophoretically injected drugs and even behavioural cues have been described [7], but in default of the sort of information described above the physiological significance of these observations must remain in doubt.

Acknowledgements

Research in our laboratory described in this article was supported by grants from the Medical Research Council and the Population Council. I am grateful to my colleagues, R. E. J. DYBALL, R. G. DYER, D. W. LINCOLN, J. F. MORRIS and J. B. WAKERLEY, for helpful criticism and suggestions.

References

1 AULSEBROOK, L. H. and HOLLAND, R. C.: Central regulation of oxytocin release with and without vasopressin release. Amer. J. Physiol. *216:* 818–829 (1969).

2 AULSEBROOK, L. H. and HOLLAND, R. C.: Central inhibition of oxytocin release. Amer. J. Physiol. *216:* 830–842 (1969).

3 BARKER, J. L.; CRAYTON, J. W., and NICOLL, R. A.: Noradrenaline and acetylcholine responses of supraoptic neurosecretory cells. J. Physiol., Lond. *218:* 19–32 (1971).

4 BARKER, J. L.; CRAYTON, J. W., and NICOLL, R. A.: Antidromic and orthodromic responses of paraventricular and supraoptic neurosecretory cells. Brain Res. *33:* 353–366 (1971).

5 BEYER, C. and SAWYER, C. H.: Hypothalamic unit activity related to control of the pituitary gland; in GANONG and MARTINI Frontiers in neuroendocrinology, pp. 255–287 (Oxford Univ. Press, New York 1969).

6 CROSS, B. A.: The neural control of oxytocin secretion; in MARTINI and GANONG Neuroendocrinology, vol. 1, pp. 217–259 (Academic Press, New York 1966).

7 CROSS, B. A.: Unit responses in the hypothalamus; in GANONG and MARTINI Frontiers in neuroendocrinology, pp. 133–171 (Oxford Univ. Press, New York 1973).

8 CROSS, B. A. and DYBALL, R. E. J.: Central pathways for neurohypophysial hormone release. In: Handbook of physiology. Endocrinology (in press).

9 CROSS, B. A. and DYER, R. G.: Cyclic changes in neurons of the anterior hypothalamus during the oestrous cycle, and the effects of anaesthesia; in GORSKI and SAWYER Steroid hormones and brain function, pp. 95–102 (University of California Press, Los Angeles 1971).

10 DOUGLAS, W. W.; NAGASAWA, J., and SCHULZ, R.: Electron microscopic studies on the mechanism of secretion of posterior pituitary hormones and significance of microvesicles ('synaptic vesicles'). Evidence of secretion by exocytosis and formation of microvesicles as a by-product of this process; in HELLER and LEDERIS Memoirs of the Society for Endocrinology, No. 19, pp. 353–378 (Cambridge Univ. Press, London 1971).

11 DREIFFUS, J. J. and KELLY, J. S.: The activity of identified supraoptic neurones and their response to acetylcholine applied by iontophoresis. J. Physiol., Lond. *220:* 105–118, (1972).

12 DREIFFUS, J. J. and KELLY, J. S.: Recurrent inhibition of antidromically identified rat supraoptic neurones. J. Physiol., Lond. *220:* 87–103 (1972).

13 DUKE, H. N. and PICKFORD, M.: Observations on the action of acetylcholine and adrenaline on the hypothalamus. J. Physiol., Lond. *114:* 325–332 (1951).

14 DYBALL, R. E. J.: Oxytocin and ADH secretion in relation to electrical activity in antidromically identified supraoptic and paraventricular units. J. Physiol., Lond. *214:* 245–256 (1971).

15 DYBALL, R. E. J. and DYER, R. G.: Plasma oxytocin concentration and paraventricular neurone activity in rats with diencephalic islands and intact brains. J. Physiol., Lond. *216:* 227–235 (1971).

16 DYBALL, R. E. J. and POUNTNEY, P. S.: Discharge patterns of supraoptic and paraventricular neurones in rats given a 2 % NaCl solution instead of drinking water. J. Endocrin. *56:* 91–98 (1973).

17 DYER, R. G. and CROSS, B. A.: Antidromic identification of units in the preoptic and anterior hypothalamic areas projecting directly to ventromedial and arcuate nuclei. Brain Res. *43:* 254–258 (1972).

18 DYER, R. G.; DYBALL, R. E. J., and MORRIS, J. F.: The effect of hypothalamic deafferentation upon the ultrastructure and hormone content of the paraventricular nucleus. J. Endocrin. *57:* 509–516 (1973).

19 FINDLAY, A. L. R.: Hypothalamic inputs. Methods and five examples; in ARIEN, KAPPERS and SCHADÉ Progress in brain research, vol. 38, pp. 163–191 (Elsevier, Amsterdam 1972).

20 FUXE, K. and HÖKFELT, T.: The influence of central catecholamine neurons on the hormone secretion from the anterior and posterior pituitary; in STUTINSKY Neurosecretion, pp. 165–177 (Springer, Berlin 1967).

21 HARRIS, G. W.; MANABE, Y., and RUF, K. B.: A study of the parameters of electrical stimulation of unmyelinated fibres in the pituitary stalk. J. Physiol., Lond. *203:* 67–81 (1969).

22 JONES, C. W. and PICKERING, B. T.: Intra-axonal transport and turnover of neurohypophysial hormones in the rat. J. Physiol., Lond. *227:* 553–564 (1972).

23 KNAGGS, G. S.; MCNEILLY, A. S., and TINDAL, J. S.: The afferent pathway of the milk-ejection reflex in the mid-brain of the goat. J. Endocrin. *52:* 333–341 (1972).

24 KOIZUMI, K. and YAMASHITA, H.: Studies of antidromically identified neurosecretory cells of the hypothalamus by intracellular and extracellular recordings. J. Physiol., Lond. *221:* 683–705 (1972).

25 KOMISARUK, B. R.: Strategies in neuroendocrine neurophysiology. Amer. Zool. *2:* 741–754 (1971).

26 LINCOLN, D. W.; HILL, A., and WAKERLEY, J. B.: The milk-ejection reflex of the rat. An intermittent function not abolished by surgical levels of anaesthesia. J. Endocrin. *57:* 459–476 (1973).

27 LINCOLN, D. W. and WAKERLEY, J. B.: Intermittent release of oxytocin during suckling in the rat. Nature New Biol. *233:* 180–181 (1971).

28 MAKARA, G. B.; HARRIS, M. C., and SPYER, K. M.: Identification and distribution of tuberoinfundibular neurones. Brain Res. *40:* 283–290 (1972).

29 MOSS, R. L.; DYBALL, R. E. J., and CROSS, B. A.: Responses of antidromically identified supraoptic and paraventricular units to acetylcholine, noradrenaline and glutamate applied iontophoretically. Brain Res. *35:* 573–575 (1971).

30 MOSS, R. L.; DYBALL, R. E. J., and CROSS, B. A.: Excitation of antidromically identified neurosecretory cells of the paraventricular nucleus by oxytocin applied iontophoretically. Exp. Neurol. *34:* 95–102 (1972).

31 MOSS, R. L.; URBAN, I., and CROSS, B. A.: Microelectrophoresis of cholinergic and aminergic drugs on paraventricular neurons. Amer. J. Physiol. *223:* 310–318 (1972).

32 NEGORO, H. and HOLLAND, R. C.: Inhibition of unit activity in the hypothalamic paraventricular nucleus following antidromic stimulation. Brain Res. *42:* 385–402 (1972).

33 NOVIN, D.; SUNDSTEN, J. W., and CROSS, B. A.: Some properties of antidromically activated units in the paraventricular nucleus of the hypothalamus. Exp. Neurol. *26:* 330–341 (1970).

34 OLIVECRONA, H.: Paraventricular nucleus and pituitary gland. Acta physiol. scand. *40:* suppl. 136, pp. 1–78 (1957).

35 PICKFORD, M.: Control of secretion of the pars nervosa of the pituitary; in MARTINI, MOTTA and FRASCHINI The hypothalamus, pp. 499–514 (Academic Press, New York 1970).

36 SHUTE, C. C. D. and LEWIS, P. R.: Cholinergic and monoaminergic pathways in the hypothalamus. Brit. med. Bull. *22:* 221–226 (1966).

37 Sundsten, J. W.; Novin, D., and Cross, B. A.: Identification and distribution of paraventricular units excited by stimulation of the neural lobe of the hypophysis. Exp. Neurol. *26:* 316–329 (1970).
38 Tindal, J. S.; Knaggs, G. S., and Turvey, A.: The afferent path of the milk-ejection reflex in the brain of the guinea-pig. J. Endocrin. *38:* 337–349 (1967).
39 Tindal, J. S.; Knaggs, G. S., and Turvey, A.: The afferent path of the milk-ejection reflex in the brain of the rabbit. J. Endocrin. *43:* 663–671 (1969).
40 Urban, I.; Moss, R. L., and Cross, B. A.: Problems in electrical stimulation of afferent pathways for oxytocin release. J. Endocrin. *51:* 347–358 (1971).
41 Vincent, J. D.; Arnauld, E., and Nicolescu-Catargi, A.: Osmoreceptors and neurosecretory cells in the supraoptic complex of the unanaesthetised monkey. Brain Res. *45:* 278–281 (1972).
42 Wakerley, J. B.; Dyball, R. E. J., and Lincoln, D. W.: Milk ejection in the rat. The result of a selective release of oxytocin. J. Endocrin. *57:* 557–558 (1973).
43 Wakerley, J. B. and Lincoln, D. W.: The milk-ejection reflex of the rat. A 20- to 40-fold acceleration in the firing of paraventricular neurones during oxytocin release. J. Endocrin. *57:* 477–493 (1973).
44 Wakerley, J. B. and Lincoln, D. W.: Unit activity in the supraoptic nucleus during reflex milk ejection. J. Endocrin. *59:* xlvi–xlvii (1973).
45 Yagi, K. and Sawaki, Y.: On the localisation of neurosecretory cells controlling adenohypophysial function. J. physiol. Soc., Jap. *32:* 621–622 (1970).
46 Yamashita, H.; Koizumi, K., and Brooks, C. McC.: Electrophysiological studies of neurosecretory cells in the cat hypothalamus. Brain Res. *20:* 462–466 (1970).

Author's address: Dr. B. A. Cross, Department of Anatomy, University of Bristol, Medical School, *Bristol BS8 1TD* (England)

Recent Studies of Hypothalamic Function
Int. Symp. Calgary 1973, pp. 50–66 (Karger, Basel 1974)

Studies on the Hypothalamo-Neurohypophysial Complex in Organ Culture

H. SACHS, D. PEARSON, A. SHAINBERG, S. SHIN, G. BRYCE, S. MALAMED[1] and T. MOWLES

Roche Institute of Molecular Biology, Nutley, N.J.

Introduction

It is now well established that hypothalamic neurosecretory neurons are essential for the maintenance of the structure and function of the anterior and posterior pituitary glands [12]. These specialized neurons are found diffusely distributed throughout the medial basal region (hypophysiotrophic area) as well as in specific aggregates in the anterior portion (supraoptic and paraventricular nuclei) of the hypothalamus. In general, the hypophysiotrophic neurons terminate in the median eminence where they secrete specific peptide neurohormones (releasing factors) which reach anterior pituitary target cells via a portal vessel system. In the case of the supraoptic and paraventricular neurons, their axons end predominantly in the distal segment of the neurohypophysis where their respective secretory products, vasopressin and oxytocin, are stored and released into the peripheral circulation. Although remarkable progress has been made with respect to the chemical characterization [18, 21] and control of the secretory activity [21, 24] of both hypophysiotrophic and supraoptic neurosecretory neurons, there are relatively few studies concerned with the biochemistry and molecular biology of these neurons. This latter state of affairs stems largely from the fact that only intact animal preparations have been available for experimentation. However, many heretofore unavailable opportunities for such studies became possible after we were able to maintain either hypothalamic fragments [19] or the entire hypothalamo-neurohypophysial complex (HNC) [20] as viable organ cultures

1 Dr. MALAMED has been supported in this work by NIH Grant No. FR 05576.

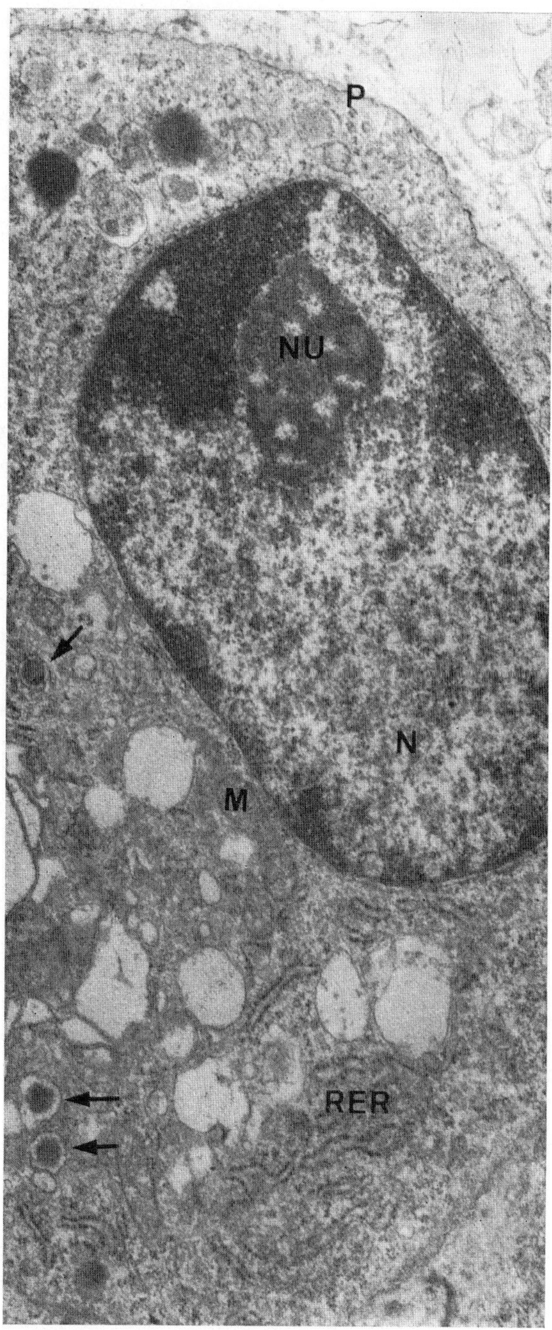

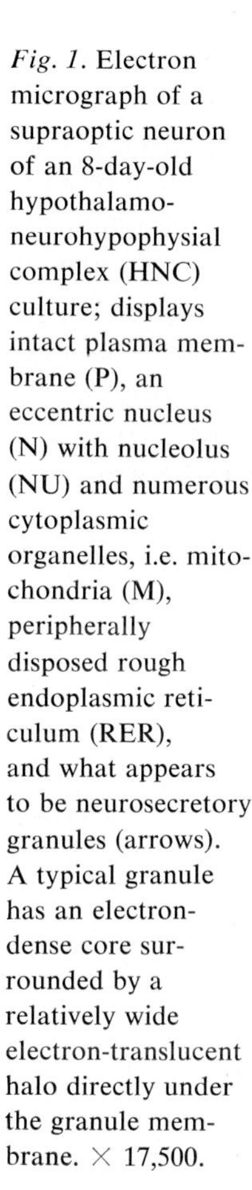
Fig. 1. Electron micrograph of a supraoptic neuron of an 8-day-old hypothalamo-neurohypophysial complex (HNC) culture; displays intact plasma membrane (P), an eccentric nucleus (N) with nucleolus (NU) and numerous cytoplasmic organelles, i.e. mitochondria (M), peripherally disposed rough endoplasmic reticulum (RER), and what appears to be neurosecretory granules (arrows). A typical granule has an electron-dense core surrounded by a relatively wide electron-translucent halo directly under the granule membrane. × 17,500.

for extended time-periods. We have examined many biochemical features of these hypothalamic cultures and in this manuscript we present the results of some of our studies on the biosynthesis of protein, nucleic acids, vasopressin, neurophysin and luteinizing hormone-releasing hormone (LHRH), as well as on axonal transport, estrogen binding and the action of a number of drugs and biologically important substances and possibly a fetal factor(s) on the modulation of the aforementioned processes.

Analytical Data

The two preparations employed consisted of either fragments of the young adult guinea pig anterior hypothalamus (hypothalamic-median eminence [HME]) severed from the neural lobe at the level of the neural stalk [19] or the entire HNC. Most of the studies in this report, except where noted, deal with the HNC cultures [which have already been described in preliminary form, 20]. Light and electron microscopy were used to identify supraoptic neurosecretory cells by their characteristic morphology and by their position immediately lateral to the optic tract. An electron-micrograph of a neurosecretory neuron of the supraoptic nucleus of an 8-day-old HNC culture is shown in figure 1. There is an abundance of cytoplasmic organelles, i.e. peripherally disposed RER (Nissl substance), mitochondria, dense bodies (possibly lysozomes) and what appear to be neurosecretory granules. The nucleus is in an eccentric position and contains a nucleolus; the plasma membrane is intact. The only abnormal feature of the cell that is apparent is the presence of clear vacuoles (possibly dilations of the Golgi apparatus). Even after 15–20 days in culture, despite widespread cellular degeneration in the surrounding hypothalamus and optic tract, the perikarya of many neurosecretory cells were observed to retain their ultrastructural integrity. However, the neural lobe portion which contained the axons of these cells showed an earlier central necrosis and deterioration as compared to the cell bodies in the hypothalamus.

Protein and Nucleic Acids

The nucleic acid, protein and vasopressin content of the hypothalamic section of the HNC for both freshly excised tissue and for identical tissues cultured 9 days are presented in figure 2. The protein determi-

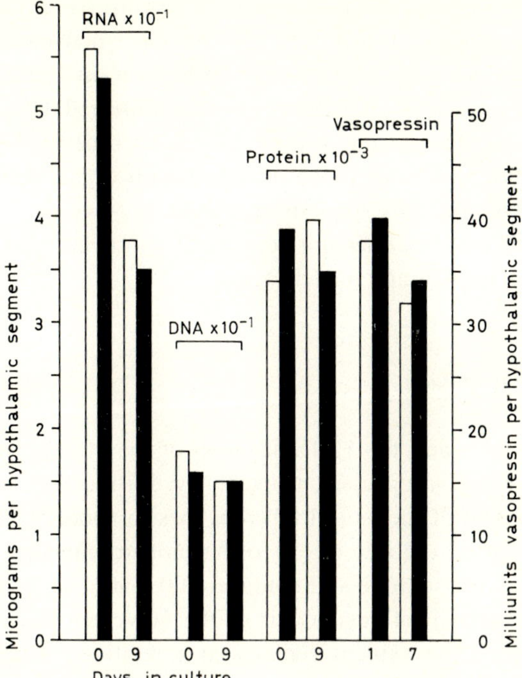

Fig. 2. Nucleic acids, protein and vasopressin content of the hypothalamic segment of the guinea pig HNC, either freshly excised (0 day) or after culture in the presence or absence of 5-fluoro-deoxyuridine (FUDR). Organ cultures were prepared as described [19, 20] and maintained in standard medium [20] or the same medium with FUDR (10^{-6} M). RNA and DNA were isolated [22] and determined by means of the orcinol [23] and diphenylamine reactions [4], respectively; protein was measured by the method of LOWRY *et al.* [10] and refers to the proteins insoluble in 0.2 N acetic acid-0.02 N HCl; vasopressin was extracted with the latter acid solution and estimated by radioimmunoassay [13] with rabbit antilysine vasopressin prepared according to SKOWSKY and FISHER [25].
▬ = FUDR, ▭ = control.

nations were performed on that portion of the hypothalamic fragments which was insoluble in the acid buffer (0.2 N acetic acid – 0.02 N HCl) used for the extraction of polypeptides and neurophysin. This acid-insoluble protein amounts to about 80 % of the total tissue proteins. After 9 days in culture there was about a 34- and 16-percent decline in the RNA and DNA content, respectively, whereas the protein remained relatively constant.

Since ^{3}H-thymidine incorporation increased over this same time-period (see below) due possibly to fibroblasts, glial or other, non-neuronal cell proliferation, analogous measurements were performed on HNC cultures maintained in the presence of 5-fluoro-deoxyuridine (FUDR), an inhibitor of DNA synthesis. The results of these latter experiments (fig. 2) were not significantly different from those observed with organ cultures in medium without FUDR and this suggests that over a 9-day period in culture an increase in cell number does not contribute significantly to any quantitative alterations in protein and nucleic acids.

Vasopressin and Neurophysin

Initially, the neural lobe and the HME segment of the HNC contain about 800 and 30–40 mU of vasopressin, respectively. By the 5th day in culture, the hormone content of the neural lobe declines to about 50 % of its initial value, but thereafter, the rate of loss of hormone diminishes and by the 9th to 10th day the neural lobe still contains 300–350 mU of vasopressin. By contrast, the content of vasopressin in the nerve cell bodies in the hypothalamic region of the explant remains at a relatively constant level of 30–40 mU (fig. 2) throughout the culture period. The corresponding value (hypothalamic section) for neurophysin after 8–9 days was about 1.2 μg (as determined by radioimmunoassay with purified guinea pig neurophysin as standard).

Incorporation Studies

Protein and Nucleic Acid Biosynthesis in Hypothalamic Segments of HNC Cultures

Although the absolute amount of RNA decreases with time in culture, the extent of incorporation of ^{3}H-uridine into total RNA actually increases markedly over the same time-period (fig. 2). Furthermore, we have previously shown [5] that this phenomenon is, in all likelihood, not due to changes in RNA pool size or changes in the specific activity of the precursor pools. At the first day, a 4-hour incubation period with ^{3}H-uridine results in a low degree and diffuse pattern of labeling which appears to be predominantly high molecular weight RNA. However, tissues cultured for 7 days showed a much higher level of ^{3}H-uridine incorporation, and also discrete radioactive peaks at 28, 18 and 4–7 S, indicative of renewed cytoplasmic RNA synthesis [20]. Labeling experiments with ^{3}H-thymidine showed an early onset of DNA synthesis, i.e. although zero-day cultures incorporate negligible quantities of ^{3}H-thymidine, after 2 days in culture measurable ^{3}H-thymidine incorporation can be demonstrated. That the observed increase in RNA synthesis may be largely due to non-neural cell proliferation was confirmed by experiments in which HNC cultures were maintained in the presence of FUDR. Under these latter conditions, it can be seen (fig. 3) that the increase in both DNA and RNA labeling is largely abolished. However, fractionation studies showed that FUDR inhibited predominately the incorporation of

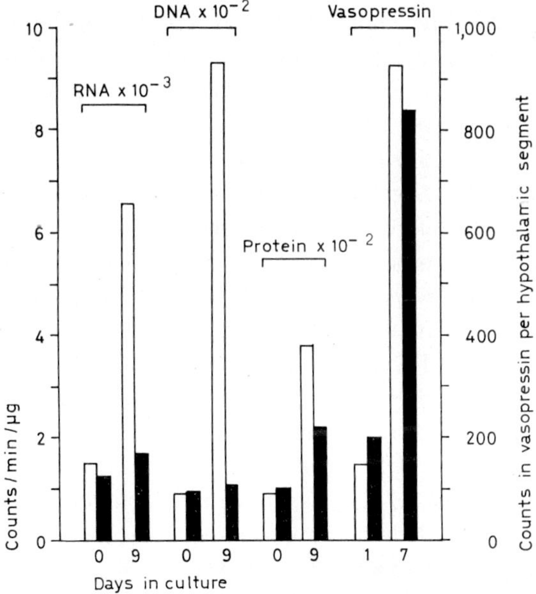

Fig. 3. Incorporation of ^{3}H-labeled precursors into nucleic acids, proteins and vasopressin of the hypothalamic segment of HNC organ cultures. Cultures were maintained for the indicated periods of time with or without FUDR (10^{-6} M) and then pulsed for 3 h with either ^{3}H-thymidine or ^{3}H-uridine (100 μCi/ml); DNA and RNA were isolated as for figure 2. Protein (acid-insoluble) and vasopressin labeling was determined after 22 h in the presence of a mixture of ^{3}H-labeled phenylalanine, tyrosine, proline and leucine (100 μCi/ml each). The protein-specific activities and the purification of ^{3}H-labeled vasopressin to constant specific activity were performed as previously described [19, 20]; see legend to figure 4.
■ = control, □ = FUDR.

^{3}H-uridine into the low molecular weight RNA species; 9-day-old cultures (with FUDR) pulsed with ^{3}H-uridine still displayed small but discreet peaks of labeled 28 and 18 S RNA. Furthermore, FUDR did not eliminate entirely or affect appreciably the enhanced incorporation of labeled amino acids into either protein or vasopressin, respectively, which occurs after several days in culture (fig. 3).

Vasopressin and Neurophysin Biosynthesis

Organ cultures of either the HNC or HME of guinea pigs were capable of incorporating ^{3}H- and ^{35}S-labeled amino acids into vasopressin, neurophysin and many other proteins and polypeptides. Because of the high input of labeled precursors (1–2 $\times$ 10^{8} cpm/ml medium) and the widespread labeling, it was necessary in each experiment to rigorously establish the isotopic purity of the newly synthetized vasopressin and neurophysin. This was done using slightly modified published procedures [19,20] (see legend to fig. 4) utilizing gel filtration, affinity, ion-exchange and thin-layer chromatography, gel electrophoresis and electrofocusing, and antibody interactions. In figure 4 are shown the final thin-layer chromatogram (top) and gel electrofocusing pattern (bottom) for ^{3}H-labeled vasopressin and neurophysin, respectively, each isolated from the hypothalamic segments of 8- to 9-day-old HNC cultures pulsed with ^{3}H amino acids for periods of 18–22 h. The constancy of the specific activities of the hormone and binding protein across the respective radioactive peaks was established by quantitative radioimmunoassays employing specific antibodies for each component. It is of some interest that organ cultures of the HNC incorporated about 1.7 times as much radioactivity into vasopressin as did cultures of hypothalamic fragments wherein the supraoptic neurons had been bisected essentially at the level of the neural stalk. Furthermore, vasopressin biosynthesis is consistently quite low and sometimes barely detectable in both the HNC and the HME during the first few days in culture. In both systems, with continued time in culture, the ability of the supraoptic neurons to incorporate labeled amino acids into vasopressin recovers dramatically (e.g., fig. 3), reaching a maximum between 1 and 2 weeks and then declines by the third or fourth week. The composition and specific activities of the free amino acid pools found in the hypothalamic sections were also determined and these data showed no significant changes with time in culture.

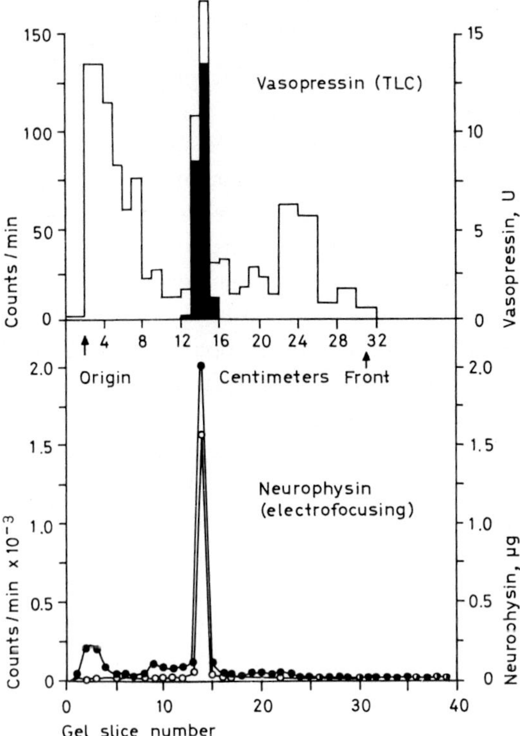

Fig. 4. Thin-layer chromatography (TLC) and isoelectric focusing of ^{3}H-labeled vasopressin (top) and ^{3}H-neurophysin (bottom), respectively, isolated from the hypothalamic segment of HNC cultures which had been pulsed with ^{3}H amino acids. Top panel, 8 HNC cultures maintained in the standard medium [20] for 9 days, pulse time, 22 h as in figure 3, TLC performed after Sephadex G-25 gel filtration, and affinity and CM-cellex chromatography [19, 20]. TLC chromatography was performed at room temperature with *n*-butanol:acetic acid:pyridine:water (15:3:10:6); the hormone was eluted from the silica with 0.5 M acetic acid and estimated by radioimmunoassay. Bottom panel, 10 HNC cultures (8 days), pulsed as above for 18 h; the hypothalamic segments were extracted at pH 1.0 [19, 20] and the extracts were passed successively through Sephadex G-25 and G-75; the neurophysin fraction was chromatographed on DEAE-Sephadex [3] prior to isoelectric focusing [6]. Neurophysin was estimated by radioimmunoassay [5] with rabbit antisera obtained from animals immunized with a mixture of bovine and porcine neurophysins II and I, respectively.

Axonal Transport

Axonal transport of the hormone and binding protein is an essential feature of the function of the neurosecretory cell. The perikarya of supraoptic neurosecretory neurons produce vasopressin and neurophysin and package hormone-protein complexes within neurosecretory granules which subsequently reach the axon terminals for storage and secretion. The ability of HNC cultures (7–9 days) to transport newly synthetized vasopressin and neurophysin was demonstrated by means of isotope experiments. After pulse periods of 18–24 h, from 7 to 15 % of the total radioactivity incorporated into peptide hormone and neurophysin were found in the neurohypophysial portion of the tissue complex, a result indicating that newly synthetized hormone and binding protein had been transported from perikarya into axonal compartments. In the same experiments, it was shown that neural lobes, bisected from hypothalami just prior to the labeling period, were unable to incorporate ^{3}H amino acids into either hormone or binding protein.

Inhibitors, Drugs, Effector Substances, Fetal Factor(s)

Puromycin, Cycloheximide, Actinomycin D, FUDR

Consistent with previous *in vivo* and *in vitro* studies on vasopressin biosynthesis [12], it was observed that inhibitors of protein biosynthesis such as puromycin (5×10^{-4} M) or cycloheximide (2×10^{-4} M) immediately blocked the ability of hypothalamic cultures to incorporate labeled amino acids into the octapeptide hormone as well as into protein. Actinomycin D, an inhibitor of RNA synthesis, also abolished vasopressin biosynthesis, but only after the cultures had been in contact with the drug for more extensive periods; the half-life of decay of hormonal biosynthetic activity in the presence of actinomycin D (10 μg/ml medium) was about 25 h.

The effects of FUDR on the HNC organ cultures have been described in previous sections.

Colchicine

Colchicine at a concentration of 1 μg/ml medium virtually elliminated the movement of labeled vasopressin into the neural lobe of HNC cultures pulsed with ^{3}H amino acids for periods of 18–24 h. Analysis of the incubation media at various time intervals and of the hormone content of the neural lobe at the termination of the experiment indicated that the drug did not affect vasopressin secretion. Somewhat surprisingly, it was observed that whereas colchicine inhibited the transport of labeled vasopressin to the nerve terminals, it apparently stimulated by 250–350 % the extent of incorporation of ^{3}H amino acids into the hormone in the neuronal perikarya or hypothalamic segment of the culture. It may be recalled that in the absence of colchicine, only 7–15 % of the radioactive hormone is transported.

Although preliminary studies with the electron microscope have indicated that HNC cultures incubated with colchicine show some diminution in the numerous microtubular elements found in the axons of control cultures [11], it is still premature (in view of the multiple actions of colchicine) to conclude that the inhibition of transport by colchicine is due solely to its ability to disaggregate microtubules [2].

Morphine

Morphine and a number of its derivatives have been shown to influence hypothalamic neuroendocrine function [17] and the HNC organ cultures offered a unique opportunity to study the long- and short-term effect of narcotic drugs on hypothalamic metabolism and specific neuronal function. Exposure of 7-day-old cultures to l-levorphanol (at 10^{-8} M) for 24 h led to a 50-percent inhibition of ribosomal RNA synthesis whereas the labeling of the higher molecular weight species was either unaffected or actually enhanced. The physiologically inactive d-isomer, by contrast, did not show these effects. In preliminary experiments, thus far, no effects of morphine on vasopressin biosynthesis by HNC cultures have been observed; i.e. with tissues either taken from morphine-treated animals or cultured in the presence of the drug.

Effector Substances, Fetal Factor(s), Nicotine, Estrogen, Dibutyryl Cyclic AMP

It has been a consistent finding that the incorporation of ^{3}H-labeled amino acids into vasopressin in organ cultures of the HNC of adult guinea pigs was stimulated 2- to 4-fold by media which had been preconditioned with guinea pig fetal hypothalamo-neurohypophysial explants. The fetal organ cultures used to condition the medium were taken from animals at 40–45 days of gestation; at this time the supraoptic neurosecretory cells have apparently differentiated [7] and have begun to produce vasopressin and neurophysin [8]. A representative experiment is shown in figure 5;

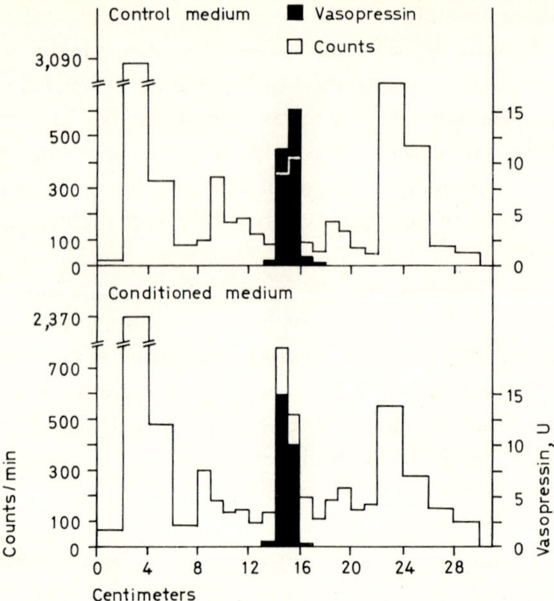

Fig. 5. Effect of medium conditioned with fetal guinea pig HNC cultures on the labeling of vasopressin in adult cultures. Groups of 10 adult HNC cultures were maintained in standard media for 6 days; they were then pulsed for 22 h with media containing ^{3}H amino acids (as in previous figures) which had been preincubated with either fetal HNC taken at 40–45 days of gestation (bottom, conditioned medium) or equivalent weights of adult tissue(top, control) for 2 days. Labeling experiments carried out simultaneously (not shown) with standard media gave results equivalent to those observed with adult HNC-conditioned medium (top).

analogous labeling experiments carried out with media similarly conditioned with organ cultures of fetal cortex, or liver, or the adult HNC did not show any stimulatory effects.

In a series of preliminary experiments, 6-day-old HNC cultures were further incubated for 3 days in the presence of either nicotine (10^{-4}M), dibutyryl cyclic AMP (10^{-4} M), or estradiol (10^{-10} M) and then pulsed with ^{3}H amino acids. None of these substances has thus far shown any pronounced effects on the rate of vasopressin labeling.

Releasing Factor Biosynthesis

Organ cultures of hypothalamic fragments or of the HNC contain not only vasopressin producing neurons but also many other morpho-

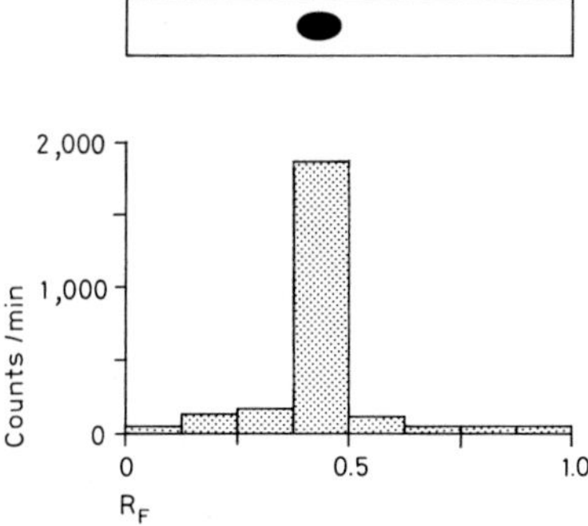

Fig. 6. Thin-layer chromatography (TLC) of ^{3}H-labeled luteinizing hormone-releasing hormone (LHRH) from hypothalamic fragments in culture for 7 days and pulsed with ^{3}H-leucine and ^{3}H-tyrosine for 20 h. Solvent system chloroform: methanol:ammonia, 60:45:20 on silica gel. The material was isolated from guinea pig hypothalamic organ cultures by extraction with 2 M acetic acid, CMC column chromatography and TLC in two different solvent systems. The panel at the top of the figure illustrates the appearance of the TLC plate under ultraviolet light. The fluorescent area is carrier LHRH.

logically intact neurosecretory neurons diffusely distributed throughout the basal-medial region of the hypothalamus. These latter cells have been implicated in the biosynthesis and secretion of a variety of polypeptide neurohormones (releasing factors) acting on specific adenohypophysial cells [12]. Incorporation studies with hypothalamic fragments (HME) cultured for 7 days demonstrated the existence of functional neurons which had retained their ability to synthetize both luteinizing and thyrotropic releasing hormones (LH-RH and TRH, respectively). In figure 6 is shown the final thin-layer chromatogram of the labeled LH-RH obtained from 7-day-old hypothalamic cultures which had been pulsed with a mixture of ^{3}H amino acids. Similar results (not shown) were obtained in studies on TRH biosynthesis.

Estrogen Binding

It has been repeatedly demonstrated in many species that hypothalamic sites serve as important feedback centers for gonadal steroid hormones [14] and the specific binding of ^{3}H-labeled steroids to target cells has been one of the means by which this phenomenon has been studied [26]. Hypothalamic fragments of adult female guinea pigs kept in culture for 7–9 days retain the ability to bind ^{3}H-estradiol; table I shows the

Table I. Nuclear uptake of ^{3}H-estradiol

Tissue	Days in culture	DPM/100 mg[1]
Cerebral cortex	1	7,570 ± 1,210
	6	8,545 ± 851
Hypothalamus	1	10,430 ± 1,010
	6	12,400 ± 806

Cultured hypothalamic fragments from female guinea pigs were weighed and homogenized in 3 ml 0.01 M tris-HCl buffer pH 7.4 (10^{-2} M KCl; 10^{-3} M EDTA) containing 10^{-9} M estradiol-17 β-6-7-^{3}H (SA 46.4 Ci/mM). After 10 min at 4 °C the homogenate was incubated at room temperature with shaking for 30 min. The nuclear pellet was collected by centrifugation at 800 g for 10 min and washed twice with 5 ml tris buffer pH 7.4. The nuclear pellet was dispersed in ice-cold water, the estradiol extracted with isoamyl alcohol-toluene (5:95) and the extracts counted in a liquid scintillation spectrometer.

1 Data is corrected for counting efficiency and is expressed as DPM/100 mg wet weight ± SEM; N = 6.

comparative binding of the steroid by nuclear fractions from 1- and 6-day-old organ cultures of either hypothalamus or cerebral cortex. The nuclear binding of steroid with the 6-day-old hypothalamic cultures was significantly greater ($p < 0.01$) than that observed with either cerebral cortex cultures of the same age or with hypothalami after 1 day in culture. On the other hand, specific cytosolic binding of the steroid (as measured by means of Sephadex gel filtration) usually declined with time in culture. At present, the precise nature and origin of the hypothalamic cells involved in this estrogen binding are unclear and require further investigation.

Secretion

The most important physiological function of the supraoptic neurosecretory neuron is to secrete vasopressin in response to appropriate stimuli, according to the needs of the organism to maintain the volume and composition of its body fluids [24]. According to electrophysiological investigations [1], acetylcholine is the stimulatory transmitter substance which interacts with nicotinic-like receptors on the perikaryon resulting in the generation of axon potentials. Varied lines of evidence suggest that the arrival of the axon potential at the nerve terminus leads to the sequence of events: (1) membrane depolarization; (2) the entrance or liberation of Ca^{++} into the axoplasm, and (3) ultimate emptying of the contents of the neurosecretory granule into the extracellular space via the process of exocytosis [15]. Thus far we have only examined the terminal secretory process in HNC organ cultures by using high concentrations of K^{+} ions as depolarizing agents. Under these circumstances, we have demonstrated that 7-day-old organ cultures of the HNC, in the presence of 56 mM K^{+}, will secrete in 20 min about 0.5–1.5 % of the vasopressin stored in the neural lobe [20]. Whether or not the supraoptic neurons of the cultured HNC have retained functional receptors or action potential capability is at present unknown.

Conclusions

We have demonstrated that the HNC of adult guinea pigs can be maintained in organ culture for several weeks. During this time many supraoptic and other hypothalamic neurosecretory neurons remain viable as judged by morphological criteria and by their ability to carry out major steps involved in their neurosecretory function, i.e. the biosynthesis, transport and secretion of vasopressin and neurophysin, the biosynthesis of hypothalamic releasing factor hormones, and possibly estrogen binding. Within the first few days in culture, the supraoptic neurons undergo what appears to be a chromatolytic response similar to that observed after nerve section, with noticeable morphological recovery at about the fourth or fifth days [22]. This recovery is paralleled by an enhanced ability of the cultures to incorporate labeled precursors into both cytoplasmic RNA and protein which appears to be independent of new cell division (FUDR-insensitive) and the specific activities of 'precursor pools'. More specifically, the incorporation of ^{3}H amino acids into vasopressin also increased several-fold after the first few days. It thus appears that after the initial insult on placing the tissue in culture, neurosecretory neurons first recover their cellular integrity prior to synthetizing neurohormones for extracellular export. The finding that medium, conditioned by incubation with the fetal HNC, stimulates vasopressin biosynthesis 2- to 4-fold (whereas fetal cortex, fetal liver and the adult HNC are inneffective) suggests the interesting possibility that the fetal HNC synthetizes a factor(s) which stimulates the functional activity of the supraoptic neurosecretory neuron, and this is presently under investigation. The inability, thus far, to demonstrate any pronounced effects on hormone synthesis by HNC cultures in the presence of a number of drugs or 'effector molecules' (e.g., morphine, nicotine, dibutyryl cyclic AMP, estrogen) may be due to several alternative factors; these are: (1) the concentration of material employed; (2) the period of the contact with the tissue; (3) the loss of responsiveness because the neurons have lost receptor or other essential functions during culture, and (4) a combination of the above. The stimulation of vasopressin labeling by colchicine is somewhat surprising and could not be explained on the basis of the action of the alkaloid to inhibit transport; this result may be related to the report [16] on studies in which colchicine given to intact rats caused marked ultrastructural alterations in the perikarya of supraoptic neurons. The inhibition by colchicine of the movement of newly synthetized hormone and binding protein from cell body into the axonal elements in the neural lobe is, however, consistent with previous *in vivo* studies [9] implicating microtubule structures in the transport process. It should be noted that the transport studies have sometimes given variable results which is presumably due to the more rapid onset of the degeneration of the axon as compared to the nerve cell body. The precise reasons for this latter phenomenon are presently unknown.

It has also been demonstrated that organ cultures of the adult guinea pig HNC are capable of specific estrogen binding and contain yet another class of functional neurosecretory neurons which can produce in culture a class of biologically important peptide neurohormones concerned with the regulation of anterior pituitary function. In the light of the results of the studies described in this report, it is apparent that the successful organ culture of the guinea pig HNC offers inumerable opportunities for studying at a biochemical level many hypothalamic cellular processes and their control.

References

1 BARKER, J. L.; CRAYTON, J. W., and NICOLL, R. A.: Supraoptic neurosecretory cells. Adrenergic and cholinergic sensitivity. Science *171:* 208–210 (1971).

2 BORISY, G. G. and TAYLOR, E. W.: The mechanism of action of colchicine. Binding of colchicine ^{3}H to cellular protein. J. Cell Biol. *34:* 525–533 (1967).

3 BRESLOW, E.; AANNING, H. L.; ABRASH, L., and SCHMIR, M.: Physical and chemical properties of the bovine neurophysins. J. biol. Chem. *246:* 5179–5188 (1971).

4 BURTON, K.: Determination of DNA concentration with diphenylamine; in GROSSMAN and MOLDAVE Methods in enzymology, vol. 12, part B, pp. 163–166 (Academic Press, New York 1968).

5 DESBUQUOIS, B. and AURBACH, G. D.: Use of polyethylene glycol to separate free and antibody-bound peptide hormones in radioimmunoassays. J. clin. Endocrin. *33:* 732–738 (1971).

6 DOERR, P. and CHRAMBACH, A.: Anti-estradiol antibodies. Isoelectric focusing in polyacrylamide gel. Analyt. Biochem. *42:* 96–107 (1971).

7 DONEV, S.: Occurrence of the neurosecretory substance during the embryonic development of the guinea pig. Z. Zellforsch. *104:* 517–529 (1970).

8 GOODMAN, R.; OSINCHAK, J., and SACHS, H.: Developmental and biosynthetic studies on the guinea pig hypothalamo-neurohypophysical complex, *in vivo* and in organ culture (abstract). 2nd Ann. Meeting Society for Neuroscience, 1972.

9 GRIFFIN, J.; KEEN, P., and LIVINGSTON, A.: Effects of vinblastine and colchicine on oxytocin levels of the rat after saline treatment. J. Endocrin. *52:* 407–408 (1972).

10 LOWRY, O. H.; ROSEBROUGH, N. J.; FARR, A. L., and RANDALL, R. J.: Protein measurement with the folin phenol reagent. J. biol. Chem. *193:* 265–275 (1951).

11 MALAMED, S.: Personal commun. (1973).

12 MARTINI, L.; MOTTA, M., and FRASCHINI, F.: The hypothalamus (Academic Press, New York 1970).

13 MILLER, M. and MOSES, A. N.: Radioimmunoassay of vasopressin with a comparison of immunological and biological activity in the rat posterior pituitary. Endocrinology *84:* 557–562 (1969).

14 MOTTA, M.; PIVA, F., and MARTINI, L.: The hypothalamus as the center of endocrine feedback mechanisms; in MARTINI, MOTTA and FRASCHINI The hypothalamus, pp. 463–489 (Academic Press, New York 1970).

15 NAGASAWA, J.; DOUGLAS, W. W., and SCHULZ, R. A.: Ultrastructural evidence of secretion by exocytosis and of 'synaptic vesicle' formation in posterior pituitary glands. Nature, Lond. *227:* 407–409 (1970).

16 NORSTRÖM, A.; HANSSON, A., and SJÖSTRAND, J.: Effects of colchicine on axonal transport and ultrastructure of the hypothalamo-neurohypophyseal system of the rat. Z. Zellforsch. *113:* 271–293 (1971).

17 RAY, A. K. and GHOSH, J. J.: Changes in the hypothalamo-neurohypophysial neurosecretory materials of rats during different phases of morphine administration. J. Neurochem. *16:* 1–5 (1969).

18 SACHS, H.; FAWCETT, P.; TAKABATAKE, Y., and PORTANOVA, R.: Biosynthesis and release of vasopressin and neurophysin; in ASTWOOD Rec. Prog. Hormone Res., vol. 25, pp. 447–484 (Academic Press, New York 1969).

19 SACHS, H.; GOODMAN, R.; OSINCHAK, J., and MCKELVY, J.: Supraoptic neurosecretory neurons of the guinea pig in organ culture. Proc. nat. Acad. Sci., Wash. *68:* 2782–2786 (1971).

20 SACHS, H.; GOODMAN, R.; SHIN, S.; SHAINBERG, A., and PEARSON, D.: Vasopressin and neurophysin biosynthesis in hypothalamic organ cultures. Proc. 4th Int. Congr. of Endocrinology, Washington 1972 (in press).

21 SCHALLY, A. V.; ARIMURA, A., and KASTIN, A. J.: Hypothalamic regulatory hormones. Science *179:* 341–350 (1973).

22 SCHMIDT, G. and THANNHAUSER, S. J.: A method for the determination of deoxyribonucleic acid, ribonucleic acid, and phosphoproteins in animal tissues. J. biol. Chem. *161:* 83–89 (1945).

23 SCHNEIDER, W. C.: Phosphorous compounds in animal tissues. I. Extraction and estimation of DNA and RNA. J. biol. Chem. *161:* 293–303 (1945).

24 SHARE, L. and CLAYBAUGH, J. R.: Regulation of body fluids. Annu. Rev. Physiol. *34:* 235–260 (1972).

25 SKOWSKY, W. R. and FISHER, D. A.: The use of thyroglobulin to induce antigenicity to small molecules. J. Lab. clin. Med. *80:* 134–144 (1972).

26 STUMPF, W. E.: Hypophyseotropic neurons in the periventricular brain. Topography of estradiol concentrating neurons; in SAWYER and GORSKI Steroid hormones and brain function, pp. 215–227 (University of California Press, Los Angeles 1972).

Author's address: Dr. H. SACHS, Roche Institute of Molecular Biology, *Nutley, NJ 07110* (USA)

Recent Studies of Hypothalamic Function
Int. Symp. Calgary 1973, pp. 67–79 (Karger, Basel 1974)

Substances Affecting the Release of Neurohypophysial Hormones[1]

H.-J. RUOFF, J. L. GOSBEE and K. LEDERIS

Division of Pharmacology and Therapeutics, Faculty of Medicine, University of Calgary, Calgary, Alberta

Introduction

Studies of the release of neurohypophysial hormones from isolated neural lobes have shown that the presence of acetylcholine (Ach) in the incubation medium does not stimulate the release of the hormones [3, 5, 6, 18]. In contrast, the findings of PICKFORD and her colleagues [1, 2, 19, 20] that Ach stimulates release of vasopressin and oxytocin *in vivo* when acting upon the supraoptic (SO) and paraventricular (PV) nuclei of the hypothalamus have been confirmed in *in vitro* experiments in which the isolated hypothalamo-hypophysial system has been incubated in the presence of Ach [3, 18]. These findings, as well as those of LEDERIS and LIVINGSTON [16] indicating the occurrence of Ach in 'non-neurosecretory' nerve terminals of the neurohypophysis, speak against the contention that Ach, supposedly present in the hormone-containing neurosecretory nerve terminals of the neurohypophysis, is involved in the release of the neurohypophysial hormones by virtue of its 'dual neurohumoral role' [7, 9, 10].

The presence of Ach and of related enzymes in the neural lobe has been established in a number of mammalian species [11–15]. At present no evidence is available in support of a direct local involvement of Ach in the release of hormones from the neural lobe.

Much of the work, from which the presently held concepts concerning local events associated with hormone release have been arrived at, has been done on the isolated neural lobe. The isolated neural lobe con-

1 Supported by Medical Research Council of Canada Grant No. MT 3911.

sists of severed nerve terminals and is, therefore, not an 'intact' organ in that sense. Any findings on hormone release, obtained by the use of the severed nerve terminals (i.e., the neural lobe *in vitro*), may have only a limited relationship, or no bearing at all, to the actual hormone release mechanisms operative in the intact neuron (i.e., either in the intact animal *in vivo* or on the 'intact' isolated hypothalamo-neurohypophysial system *in vitro*). Such findings have elucidated some of the local membrane (neuronal and/or granular) phenomena. Confirmation of these findings using the complete neuronal unit(s) and elucidation of the events between the point of arrival of a secretory signal at the cell body of the neurosecretory neuron and the final changes at, or in, the nerve terminal seems to be strongly indicated.

Recent experiments employing the anesthetized water-loaded rat and by applying acetylcholine direct to the neural lobe by microinfusion via the transauricular route [8] showed that, under these conditions, a dose-related release of vasopressin occurs. The findings are at variance with the results obtained on the isolated neural lobe. While the mechanism of action of Ach under these experimental conditions has not been elucidated, a retrograde depolarization of the entire neuron has been suggested [2a].

To study hormone-release phenomena acting locally on the nerve terminals of the neural lobe and to investigate the significance of the involvement of Ach at the level of the nerve terminals, two different approaches were used.

1. Comparison was made of the effects on hormone release of Ach and of other locally occurring bioamines (noradrenaline, dopamine, 5-hydroxytryptamine), prostaglandins (PGE_1, $PGF_{2\alpha}$), c-AMP and angiotensin II.

2. Rates of synthesis of Ach in the neural lobe *in vivo* and *in vitro* and in the SO and PV nuclei *in vitro* were measured in relation to neurohypophysial function.

II. Release of Vasopressin from the Rat Neural Lobe

A. Experiments on Vasopressin Release *in vivo*

The procedure described by Gosbee and Lederis [8] was used. An ethanol anesthetized water-loaded rat (5 % body weight), in which the urinary bladder and a femoral vein were cannulated, is held in a Hoffman-Reiter hypophysectomy instrument. Test substances are infused via a 26-gauge needle through a hollow ear bar directly into the neural lobe at the rate of 2 μl/min, for 5 min. The urine flow (drops/min) is measured with a drop counter and the urine is passed through a recording conductivity meter. Decrease in urine flow accompanied by an increase in conductivity is interpreted to indicate an increased release of vasopressin. Quantitation of vasopressin release is obtained by intravenous injections of matched doses of exogenous vasopressin.

It has already been reported [8] that this *in vivo* preparation made possible the measurement of vasopressin release at, or near to, physiological magnitudes, in contrast to the spontaneous or stimulated release from isolated neural lobes *in vitro* of large quantities of hormones

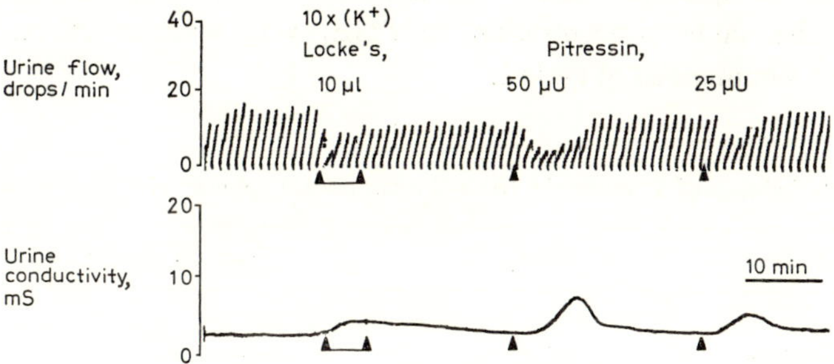

Fig. 1. Local effect of excess potassium on the release of vasopressin from the neural lobe *in vivo*. Upper tracing: urine flow (drops per minute). Lower tracing: urine conductivity (measured in milli-Siemens, mS). The effect of 10 × [K^+] approximates the antidiuretic effect of 25 μU of vasopressin (Pitressin, Parke-Davis).

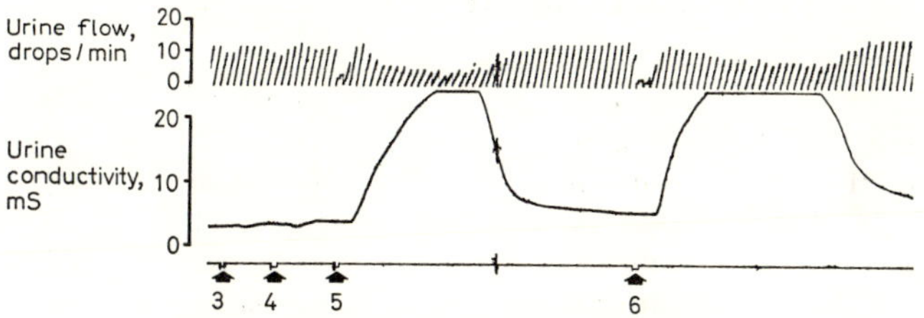

Fig. 2. Antidiuretic effect of electrical stimulation of the rat neural lobe *in vivo*. Upper tracing: urine flow (drops per minute). Lower tracing: urine conductivity (mS). All stimuli were 2 msec duration at a frequency of 60 cps for 1 min. Voltage was increased until a response was obtained. No. 3, 2 V; No. 4, 4 V; No. 5 and 6, 8 V. Arrows: application of electrical stimuli.

amounting up to 10–20 % of the total neural lobe content over brief observation periods, ranging from 10–40 min.

Release of vasopressin obtained with Ach or carbachol, the latter producing a long-lasting release, could be reproduced either by the infusion of excess potassium or by a direct electrical stimulation of the neurohypophysis (fig. 1, 2). Infusion of noradrenaline, dopamine or 5-hydroxytryptamine, at doses ranging from 10 ng to 1 μg, did not affect the release of vasopressin. Equally, dibutyryl c-AMP (with or without theophylline) did not affect vasopressin release. Prostaglandins PGE_1 and $PGF_{2\alpha}$ were without effect at 10^{-9}–10^{-7} M. However, the infusion of 10 μl of PGE_1 at 10^{-6} M caused a decrease in urine flow and an increase in conductivity suggesting a stimulation of vasopressin release. Angiotensin II, at dose ranges of 10^{-9}–10^{-7} M, caused an increase in urine flow indicating a further inhibition of vasopressin release over and above that caused by ethanol. $PGF_{2\alpha}$ at the same concentration as PGE_1 was without effect. The findings are summarized in table I.

Table I. Release of vasopressin from the rat neural lobe *in vivo* and *in vitro*

Test substance	Dose	Effect on vasopressin release	
In vivo			
Acetylcholine	10–100 ng	*increased*	
Carbachol	5–20 ng	*increased*	
Noradrenaline	10–1,000 ng	no change	
Dopamine	10–1,000 ng	no change	
5-HT	100–1,000 ng	no change	
diBcAMP	5 μg	no change	
Angiotensin II	1–5 ng	decreased	
PGE_1	10 U/l 10^{-6} M	*increased*	
In vitro			
diBcAMP	10 U/M	no change	(3)
Theophylline	1 m/m		
Angiotensin II	10^{-7} M	decreased	(5)
	10^{-9} M	decreased	(3)
$PGF_{2\alpha}$	10^{-7} M	no change	(7)
	10^{-9} M	no change	(6)
PGE_1	10^{-7} M	no change	(5)
	10^{-9} M	no change	(4)

5-HT = 5-hydroxytryptamine; diBcAMP = dibutyryl cAMP; PGE_1 = prostaglandin E_1; $PGF_{2\alpha}$ = prostaglandin $F_{2\alpha}$.

B. Release of Vasopressin from Rat Neural Lobes *in vitro*

In view of the apparent further inhibition of vasopressin release by angiotensin and a stimulation of release by PGE_1 in the *in vivo* experiments, a comparison was made using isolated rat neural lobes. Halved neural lobes (5 neural lobes in each experiment) were incubated in phosphate buffer as used earlier [4, 6]. Angiotensin II, at similar concentrations to those used in the *in vivo* experiments, caused a decreased release of vasopressin as compared with spontaneous (control) release in the phosphate buffer alone. Inhibition of release was observed in all experiments using either 10^{-9} or 10^{-7} M concentrations of angiotensin but the inhibition was more pronounced after incubation with higher concentrations of angiotensin (fig. 3).

In the same series of *in vitro* experiments, dibutyryl c-AMP, and varying concentrations of prostaglandins PGE_1 or $PGF_{2\alpha}$ did not affect the release of vasopressin (table I).

In an experiment in which the halved neural lobes were incubated in the presence of c-AMP (10 μM dibutyryl c-AMP and 1 mM theophylline), an indication of a possible stimulation of vasopressin release by c-AMP was obtained when the nucleotide was added to the incubation medium

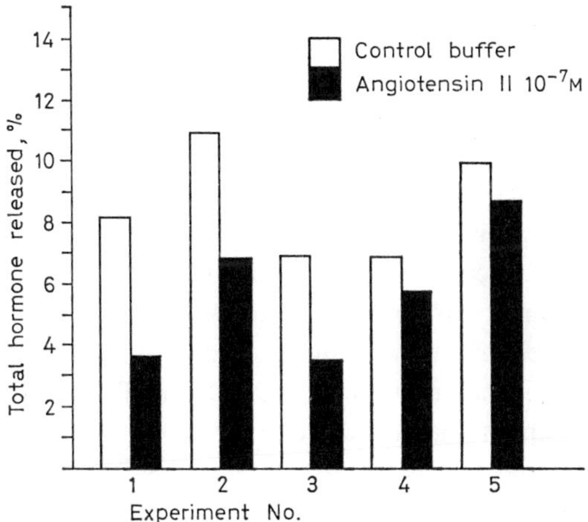

Fig. 3. Effects of angiotensin II (10^{-7} M in phosphate buffer) on the release of vasopressin from the isolated rat neural lobe. Five experiments (1–5) are shown. Note: inhibition of release amounting to approximately 50 % of control values in 3 of 5 experiments.

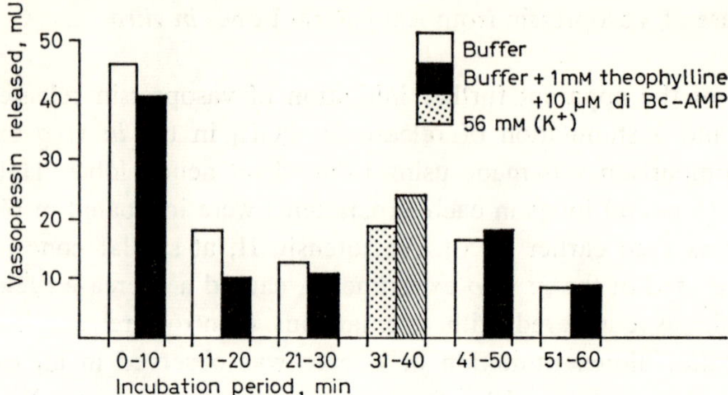

Fig. 4. Release of vasopressin from rat neural lobes in the presence of dibutyryl c-AMP (diBc-AMP; 10 μM) and theophylline (1 mM). Stimulation of the release of vasopressin was indicated only when c-AMP and theophylline were added to buffer containing excess $K+$ (56 mM; incubation period at 31–40 min).

containing excess potassium (fig. 4). Further studies (*in vivo* as well as *in vitro*) are required to decide if c-AMP is involved in the release of vasopressin when the neurohypophysial nerve terminals are in the depolarized state.

III. Biosynthesis of Ach in the Neural Lobe and in Hypothalamic Nuclei

Cholinergic secretomotor innervation of the hypothalamo-neurohypophysial system at the hypothalamic perikarya is generally accepted [see Introduction and B.A. CROSS, this volume]. Nervous control of hormone release may also be studied by considering the turnover or the rates of synthesis of a neurotransmitter at different loci in the hypothalamo-neurohypophysial system and in relation to varying rates of hormone release. The significance of cholinergic innervation at the SO and PV perikarya and in the neural lobe was investigated indirectly by measuring rates of synthesis of Ach from isotope-labeled precursors.

A. Synthesis of Ach in the Neural Lobe *in vivo*

The *in vivo* preparation as described earlier in this report was used. Urethane anesthetized rats were held in the hypophysectomy instrument and a solution containing ^{3}H-choline chloride (2 × 10^{-4} M, 1 Ci/mM), acetyl-coenzyme A (CoA) (2 × 10^{-4} M), eserine sulphate (10^{-4} M) and NaCl (10^{-1} M) was infused into the neural lobe at the rate of 0.4 μl/min for 5–60 min. After infusion, the animals were decapitated, the neural lobes removed, homogenized in 0.2 N perchloric acid, homogenates centrifuged at 1200 g for 10 min and the supernatants (extracts) collected. Choline and Ach were separated by thin-layer chromatography (TLC) on cellulose in butanol-water-acetic acid (4 : 5 : 1). Labeled Ach and choline were measured by liquid scintillation counting in dioxane-toluene. Newly synthetized radioactive Ach was identified (1) by comparing R_f-values of cold Ach and choline with the radioactive TLC peaks of the extracts, and (2) when aliquots of the extracts (after infusion of ^{3}H-choline) were treated with NaOh at pH 10 and heated for 5 min, no radioactivity was found in the Ach region of the chromatogram.

The total radioactivity in the neural lobe increased progressively during a 60 min (fig. 5) infusion, the bulk of radioactivity being due to the uptake of ^{3}H-choline at undetermined cellular sites at the rate of 2–4 % of the infused choline. Most of the infused choline appears to have

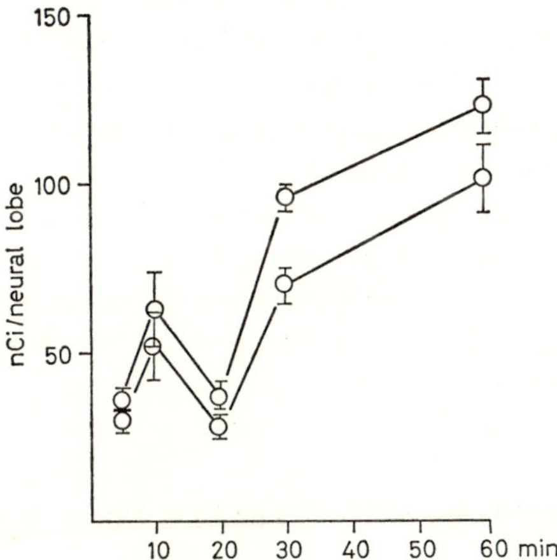

Fig. 5. Total radioactivity in neural lobe homogenates (upper tracing) and radioactivity due to ^{3}H-choline in TLC eluates (lower tracing) after infusion of ^{3}H-choline chloride into the rat neural lobe for varying periods of time (5–60 min). The differences between the two tracings can be accounted for by presence of ^{3}H-choline and ^{3}H-acetylcholine (Ach) in the upper tracing and by incomplete recovery of ^{3}H-choline after chromatography (in lower tracing).

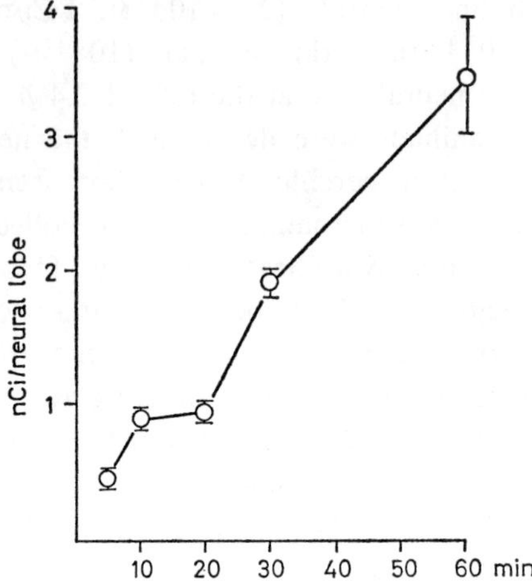

Fig. 6. Incorporation of ^{3}H-choline (nCi ≡ pM) into Ach, after varying periods of infusion (10–60 min) of ^{3}H-choline chloride into the rat neural lobe.

been removed into the general circulation. Of the ^{3}H-choline taken up by the neural lobe (about 100 pM/neural lobe after an infusion for 60 min), approximately 3 % were incorporated into newly formed radioactive Ach (i.e., of the 100 pM of ^{3}H-choline taken up after 1-hour infusion, 3.4 pM/neural lobe incorporated into Ach; fig. 6).

The amount of radioactive Ach formed represents only an undetermined proportion of total Ach formed in the neural lobe, assuming that formation of cold Ach from endogenous precursors proceeds simultaneously. Since total Ach content of the neural lobes was not estimated, Ach turnover has not been calculated.

B. Synthesis of Ach in Neural Lobe and in SO and PV Nuclei *in vitro*

Rates of synthesis of Ach were investigated in the hypothalamic nuclei and in neural lobes of lactating rats (15 days *post partum*), dehydrated rats (2 % NaCl for 12–72 h) and in controls.

Rats were killed by decapitation; the neural lobes and hypothalamic tissue blocks containing the SO or the PV nuclei were removed. Reaction

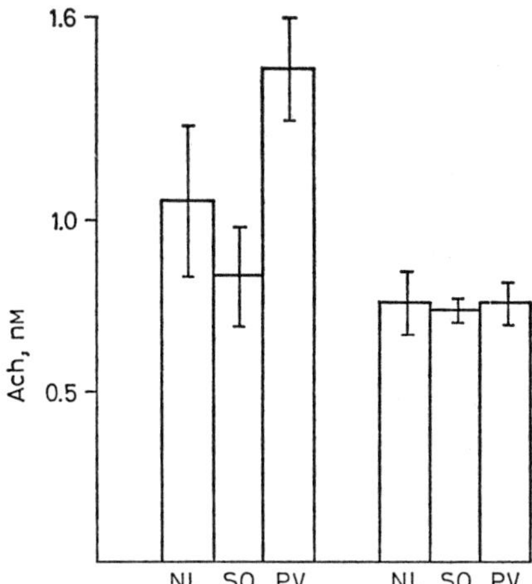

Fig. 7. Synthesis of ^{14}C-acetylcholine (nM of Ach/mg tissue/20 min) by homogenates of neural lobes (NL), supraoptic (SO) and paraventricular (PV) nuclei of lactating rats (NL, SO, PV in left-hand columns) and in controls (right-hand columns). Rates of synthesis of Ach in NL and SO not significantly different between lactating and control animals. Significantly higher rate of synthesis of Ach in PV nuclei of lactating rats ($p < 0.05$).

was started (incubation for 20 min at 30 °C) by adding squashed tissue samples to 200 μl of incubation medium consisting of acetyl-^{14}C-CoA (1.5×10^{-4} M, 1.4 mCi/mM), choline chloride (5×10^{-3} M), eserine sulphate (2.5×10^{-4} M), n-butanol (0.5 %), NaCl (3×10^{-1} M), potassium phosphate buffer (1×10^{-1} M, pH 6.8). The newly synthetized ^{14}C-Ach and acetyl-^{14}C-CoA were separated on a Dowex-1 anion-exchange column and estimated by liquid scintillation counting in toluene-triton X-100 according to the procedure of SCHRIER and SHUSTER [21] as modified by SHUBERT [22].

In the experiments with lactating rats (fig. 7) at 15 days *post partum*, where rats at the end of the lactation period (28–30 days *post partum*) were used as controls, the rate of synthesis of Ach in the neural lobes did not differ significantly between the lactating rats and controls (1.06 nM Ach/mg tissue/20 min, and 0.76 nM Ach/mg tissue/20min, respectively). Equally, the rates of synthesis of Ach by the SO nuclei were similar (0.84 nM and 0.74 nM Ach/mg tissue/20 min in lactating rats and in controls, respectively). A significantly higher rate of synthesis of Ach occurred in the PV nuclei of the lactating rats as compared with the controls (1.45 nM and 0.76 nM/mg tissue/20 min, respectively).

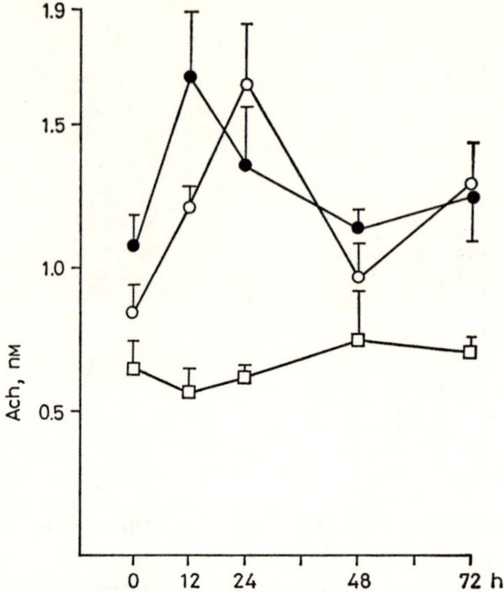

Fig. 8. Synthesis of Ach by homogenates of neural lobes (□—□) and of SO (o—o) and PV nuclei (●—●) in control rats (0 h) and animals given to drink 2 % NaCl for 12–72 h. *Note:* no change in the rate of synthesis of Ach in the neural lobes of experimental animals and controls; significantly increased rate of synthesis of Ach in SO ($p < 0.01$) and PV nuclei ($p < 0.05$) at 24 and 12 h of NaCl treatment respectively.

In the NaCl-dehydrated rats, the rate of synthesis of Ach by the neural lobes did not differ from controls at all treatment periods studied (12, 24, 48 and 72 h). A significant increase in the rate of synthesis of Ach was found in both, the SO and PV nuclear areas ($p < 0.01$ and < 0.05, respectively) at 12–24 h of NaCl treatment (fig. 8).

Summary

Release of Vasopressin in vivo and in vitro

The findings that Ach and cholinomimetic substances (e.g., carbachol) when applied directly to neural lobe tissue *in vivo,* stimulate the release of vasopressin, are at variance with the *in vitro* observations from numerous laboratories. The implications and the significance of the local action of Ach on vasopressin release remain to be analyzed. When these findings are considered in the light of local synthesis (with the possible implication of local turnover) of Ach, one conclusion presents itself as likely. Namely, the increased release of vasopressin that is observed, may be unspecific and, possibly, unphysiological, representing either a local depolarization of nerve terminal membranes which may not normally be accessible to sufficient concentrations of Ach or, alternatively, a retrograde ('antidromic') depolarization which initiates a vasopressin release process by triggering a releasing signal from the hypothalamic perikarya.

The release-inhibiting effects of angiotensin II observed in the present experiments, *in vivo* and *in vitro,* may again be unspecific and may have no bearing on the proposed centrally mediated action of angiotensin on the release of vasopressin [17]. Interestingly, and in contrast to the experiments with Ach or PGE_1, comparable effects were obtained with angiotensin under the *in vivo* and *in vitro* conditions.

Biosynthesis of ACH

The study of biosynthesis of Ach in the neural lobe and in the hypothalamic nuclei, by showing significantly increased rates of Ach synthesis in 'stimulated' animals (e.g., lactating and NaCl-treated rats) in the SO or PV nuclei, or both, but not in the neural lobe, confirms the view that sites for the neural control in the release of vasopressin and oxytocin are located at the SO and/or PV perikarya (e.g., cholinergic, adrenergic synapses), and that an explanation has to be found for the presence of neurotransmitters (and corresponding nerve terminals) in the neural lobe.

The findings presented here (e.g., vasopressin release from the neural lobe *in vivo* and *in vitro,* rates of biosynthesis of Ach in the neural lobe and in the hypothalamic nuclei in relation to the function of the hypothalamo-neurohypophysial system) can be used to argue that: (1) meaningful studies (and conclusions) to elucidate mechanisms of secretion of vasopressin and oxytocin, other than local membrane phenomena, can only be carried out by considering the neurosecretory neuron as a whole, and (2) any findings obtained on the nerve terminals (i.e., the neural lobe – *in vivo* or isolated) can only be applied to interpret local nerve terminal phenomena; application of findings obtained from the study of the neural lobe may have little bearing on the function of the hypothalamo-neurohypophysial system as a whole as far as the control of hormone release is concerned.

References

1 Abrahams, V. C. and Pickford, M.: Simultaneous observations on the rate of urine flow and spontaneous uterine movements in the dog and their relationship to posterior lobe activity. J. Physiol., Lond. *126:* 329–346 (1954).

2 Abrahams, V. C. and Pickford, M.: The effect of anticholinesterase injected into the supraoptic nuclei of the chloralosed dog on the release of the oxytocic factor of the posterior pituitary. J. Physiol., Lond. *133:* 330–333 (1956).

2a Abrahams, V.: Personal commun. (1973).

3 Daniel, A. R. and Lederis, K.: Effects of acetylcholine on the release of neurohypophysial hormones *in vitro.* J. Endocrin. *34:* x–xi (1966).

4 DANIEL, A. R. and LEDERIS, K.: Release of neurohypophysial hormones *in vitro*. J. Physiol., Lond. *190:* 171–187 (1967).

5 DICKER, S. E.: Release of vasopressin and oxytocin from isolated pituitary gland of adult and new-born rats. J. Physiol., Lond. *185:* 429–444 (1966).

6 DOUGLAS, W. W. and POISNER, A. M.: Stimulus-secretion coupling in a neurosecretory organ and the role of calcium in the release of vasopressin from the neurohypophysis. J. Physiol., Lond. *172:* 1–18 (1964).

7 GERSCHENFELD, H. M.; TRAMEZZANI, J. H., and ROBERTIS, E. DE: Ultrastructure and function of the neurohypophysis of the toad. Endocrinology *66:* 741–762 (1960).

8 GOSBEE, J. L. and LEDERIS, K.: *In vivo* release of antidiuretic hormone by direct application of acetylcholine or carbachol to the rat neurohypophysis. Canad. J. Physiol. *50:* 618–620 (1972).

9 KOELLE, G. B.: A proposed dual neurohumoral role of acetylcholine. Its function at the pre- and post-synaptic sites. Nature, Lond. *190:* 208–211 (1961).

10 KOELLE, G. B. and GEESEY, C.: Localization of acetylcholinesterase in the neurohypophysis and its functional implications. Proc. Soc. exp. Biol. Med. *106:* 625–628 (1961).

11 LABELLA, F. S.: Storage and secretion of neurohypophysial hormones. Canad. J. Physiol. *201:* 695–709 (1969).

12 LABELLA, F. S.; BINDLER, E.; MINNICH, J., and SHEN, S.: Acetylcholine, acetylcholinesterase and choline acetylase in the bovine posterior pituitary. Fed. Proc. *26:* 225 (1967).

13 LABELLA, F. S. and SHIN, S.: Estimation of cholinesterase and choline acetyltransferase in bovine anterior pituitary, posterior pituitary, and pineal body. J. Neurochem. *15:* 335–342 (1968).

14 LEDERIS, K. and LIVINGSTON, A.: Acetylcholine content of the rabbit neurohypophysis. J. Physiol., Lond. *185:* 37–38 (1966).

15 LEDERIS, K. and LIVINGSTON, A.: Acetylcholine and related enzymes in the neural lobe and anterior hypothalamus of the rabbit. J. Physiol., Lond. *201:* 695–709 (1969).

16 LEDERIS, K. and LIVINGSTON, A.: Neuronal and subcellular localisation of acetylcholine in the posterior pituitary of the rabbit. J. Physiol., Lond. *210:* 187–204 (1970).

17 MOUW, D.; BONJOUR, J. P.; MALVIN, R. L., and VANDER, A. J.: Central action of angiotensin in stimulating ADH release. Amer. J. Physiol. *220:* 239–242 (1971).

18 NORDMANN, J. J.; BIANCHI, R.; DREIFUSS, J. J., and RUF, K. B.: Release of posterior pituitary hormones from the entire hypothalamo-neurohypophysial system *in vitro*. Brain Res. *25:* 669–671 (1971).

19 PICKFORD, M.: The inhibitory effect of acetylcholine in water diuresis in the dog, and its pituitary transmission. J. Physiol., Lond. *95:* 226–238 (1939).

20 PICKFORD, M. and WATT, J. A.: A comparison of the effect of the intravenous and intracarotid injections of acetylcholine in the dog. J. Physiol., Lond. *114:* 333–335 (1951).

21 SCHRIER, B. K. and SHUSTER, L.: A simplified radiochemical assay for choline acetyltransferase. J. Neurochem. *14:* 977–985 (1967).
22 SCHUBERT, J.: Measurement of choline acetylase; in GLICK Methods of biochemical analysis. Suppl. vol., Analysis of biogenic amines and their related enzymes, pp. 275–296 (Interscience Publishers, New York 1971).

Authors' addresses: Dr. H.-J. RUOFF, Department of Pharmacology, Faculty of Medicine, University of Tübingen, *D-74 Tübingen* (FRG); Dr. J. L. GOSBEE and Dr. K. LEDERIS, Division of Pharmacology and Therapeutics, Faculty of Medicine, University of Calgary, *Calgary, Alberta T2N 1N4* (Canada)

Recent Studies of Hypothalamic Function
Int. Symp. Calgary 1973, pp. 80–89 (Karger, Basel 1974)

Control of Gonadotropin and Prolactin Secretion by Hypothalamic Releasing and Inhibiting Hormones

S. M. McCann, K. Cooper, P. G. Harms, C. Libertun, A. Negro-Vilar, S. R. Ojeda, R. Orias, R. L. Moss, C. P. Fawcett and L. Krulich

Department of Physiology, University of Texas, Southwestern Medical School, Dallas, Tex.

Introduction

In this brief communication we would like to review recent work from our laboratory on the complex control system which regulates the secretion of gonadotropins and prolactin in the rat. The recent introduction of radioimmunoassay of pituitary hormones has greatly facilitated the unraveling of these complex systems and a rather clear picture has now emerged.

Effects of Electrical or Electrochemical Stimulation of the Hypothalamus on Gonadotropin Release

It has long been known that ovulation can be induced by hypothalamic stimulation. With the availability of radioimmunoassay, it has been possible to quantitate the release of luteinizing hormone (LH) which has occurred and to also measure follicle-stimulating hormone (FSH) release. Stimulation of a band of tissue extending from the medial preoptic region through the anterior hypothalamus to the median eminence led to an increase in LH release in proestrous rats in which the spontaneous release of LH was blocked by Nembutal. On the other hand, FSH release resulted from stimulation of a somewhat more caudally located region extending from the anterior hypothalamic area to the median eminence region. Stimulation of the preoptic region was followed by increases in LH even though FSH secretion was unmodified, whereas in two animals in which stimulations were carried out in the anterior hypothalamic area, FSH release was enhanced while LH release was unaltered [20]. Thus, it appears that the LH-controlling region extends more rostrally than the overlapping FSH-controlling area.

The Effects of Hypothalamic Lesions on the Release of Gonadotropins and Prolactin

Lesions in the median eminence region have resulted in a dramatic decrease in the release of both FSH and LH; however, the decrease in plasma LH of approximately 95 % in ovariectomized animals was greater than the decrease in FSH of 75 %. This could mean that the pituitary has some autonomous capability to release FSH; alternatively, the lesions may not have been complete even though no intact tissue in the median eminence region was seen. In striking contrast to the effects on gonadotropins were the effects on prolactin which, instead of being decreased, was markedly elevated. Median eminence lesions resulted in an immediate discharge of all 3 hormones which is thought to be due to release of stored releasing factors from the injured tissue [5]. Thus, in agreement with earlier work, it appears that the hypothalamus exerts a net stimulatory effect on the release of FSH and LH, on the one hand, and a dramatic inhibitory effect on the release of prolactin, on the other.

Lesions in the suprachiasmatic region are capable of blocking ovulation and inducing a state of constant vaginal cornification. On examination of the ovaries in these animals, they are filled with large follicles. The animal appears to be poised on the brink of ovulation and is unable to release an ovulatory quota of gonadotropins. It has been hypothesized that these lesions destroy the region of the brain which is involved in the stimulatory feedback of gonadal steroids to induce ovulation [2]. Recent studies by radioimmunoassay further support this hypothesis. These lesions prevented the release of LH in response to progesterone treatment, whereas the release of FSH could still be produced [6]. In these experiments the animals are first primed with estrogen and then given a single subcutaneous injection of progesterone. In normal animals this provokes a discharge of both gonadotropins mimicking the normal preovulatory surge. The results with these lesions in which the response of LH but not FSH was blocked by destruction of the suprachiasmatic region agree with the stimulation experiments cited above and suggest that the hypothalamic centers controlling FSH release are located slightly more caudally than those controlling LH release.

Hypothalamic Gonadotropin-Releasing Factors

Hypothalamic control over gonadotropin release is mediated neurohumoraly via gonadotropin-releasing factors. Crude or partially purified hypothalamic extracts of murine, ovine or bovine origin have a much more pronounced effect on the release of LH than FSH *in vivo,* whereas *in vitro* considerable FSH release also takes place [17, 24, 38, 45]. Matsuo *et al.* determined the structure of a peptide isolated from porcine hypothalami and the synthesis of this decapeptide was accomplished [27]. Subsequently, Burgus *et al.* reported that ovine LRF had a similar structure [7]. Single injections of the synthetic decapeptide LH-releasing factor (LRF) or purified LRF promote LH release almost exclusively in normal rats, castrate rats or castrate rats primed with estrogen and/or estrogen plus progesterone [9, 15, 24]. Infusions of the decapeptide enhance not only LH but also FSH release as reported by Arimura *et al.* [1], and we have confirmed this finding [34]. Apparently, when the gonadotrophs are given long exposure to the decapeptide, FSH as well as LH release is promoted; however, it should be pointed out that the relative increase in FSH release is always less than that of LH.

In earlier work utilizing bioassay, a separate FSH-releasing factor (FRF) was demonstrated in our laboratory [11, 12] and in that of a number of other investigators [19, 41]. Utilizing radioimmunoassay, a number of workers have had difficulty in separating a discrete FRF from LRF in hypothalamic extracts. We have similarly had difficulty separating the two activities but have encountered suggestive evidence for such separation following chromatography of hypothalamic extracts [14]. Bowers, Johansson *et al.* [18] have reported recently a partial separation of the two activities. In addition, the LH- to FSH-releasing ratio of hypothalamic extracts appears to be altered at certain stages of the estrous cycle and after estrogen treatment of ovariectomized females [15, 24]. Furthermore, on the basis of the lesion and stimulation studies noted above, plus many dissociations of LH and FSH release which either occur naturally or can be induced, the conclusion is almost inescapable that a distinct FRF will ultimately be isolated.

Effects of Steroids on the Pituitary Responsiveness to LRF

Recent studies have shown that single intravenous injections of estradiol can acutely inhibit the response of the pituitary to the decapeptide *in vivo* [31]. Furthermore, ovariectomy is followed by enhanced sensitivity to purified LRF [8]. In chronically ovariectomized animals, the sensitivity to LRF is further enhanced by treatment with estradiol 72 h prior to the measurements. In these experiments, the sensitization to LRF was observed both *in vivo* and *in vitro* [24]. These changes in sensitivity to the releasing factor at the pituitary level play a role during the normal estrous cycle since an enhanced sensitivity to purified LRF has been observed in rats prior to the proestrous discharge of gonadotropins [9].

Localization of the Gonadotropin-Releasing Factors in Hypothalamus

Gonadotropin-releasing activity can be extracted from a region extending from the medial preoptic area caudally through the anterior hypothalamic area to the median eminence arcuate region, where the bulk of the activity is concentrated. This has been revealed by making frozen sections of the tissue along 3 planes at right angles to each other and assaying extracts of the sections by their ability to alter release of gonadotropins from pituitaries incubated *in vitro* [10, 36]. Since lesions in the suprachiasmatic area caused a decline in the content of LRF stored in the median eminence, we have postulated that some LRF-secreting neurons have cell bodies in the suprachiasmatic region and long axons extending down to the median eminence region where the secretory products are emptied into the hypophysial portal vessels. Since some LRF remained in the median eminence region after these lesions, we also postulate that other LRF neurons have short axons extending to the median eminence and cell bodies located near it, perhaps in the arcuate nuclei [42]. The largest amount of activity is stored in the median eminence region. There is also LRF activity in the hypophysial stalk, but none in the neural lobe proper [10]. Thus, the median eminence serves an analagous function to that of the neural lobe which functions as a reservoir for neurohypophyseal hormones. In these studies, small amounts of FSH- as well as LH-releasing activity were found as far forward as the preoptic region [36], whereas in the lesion and stimulation experiments, the FSH-controlling center appeared to be located slightly more caudally than the LH center. We believe that the detection of FSH-releasing activity in this rostral site may be related to the potentiation of the FSH-releasing response by the *in vitro* system as indicated above.

Putative Synaptic Transmitters Altering Release of Gonadotropin-Releasing Factors and Prolactin-Inhibiting Factor

Extensive investigations have now been performed in an attempt to determine the possible synaptic transmitters which may mediate the influence of the rest of the central nervous system (CNS) on the neurosecretory cells which secrete the releasing factors. As a result of studies both *in vitro* and *in vivo,* it appears that dopamine and, to lesser extent, norepinephrine, can stimulate the release of gonadotropin-releasing factors and prolactin-inhibiting factor [28, 35]. Initially dopamine was incubated with pituitaries alone. In this situation, it failed to alter FSH and LH release except at high doses which destroyed the hormones, but it inhibited prolactin release from the gland at relatively low doses as has also been shown by other workers [4, 37]. When anterior pituitaries were incubated in a co-incubation system with ventral hypothalamic fragments, dopamine was capable of stimulating a release of FRF and LRF and prolactin-inhibiting factor (PIF) [28, 37]. In the *in vivo* experiments the catecholamines were injected into third ventricular cannulae and it was possible to show that dopamine could stimulate the release of LRF, FRF and PIF [28, 35]. The stimulatory effect on LRF release could be blocked by estrogen, which suggests that this may be one site for the negative feedback of the steroid [43].

Another approach was to inject drugs which would either block adrenergic receptors or alter biosynthesis of catecholamines. This approach lent further support for a role of dopamine as a synaptic transmitter to release PIF. On the other hand, in the case of the gonadotropins, most of the evidence points to norepinephrine as the probable transmitter involved. Thus, drugs which block norepinephrine synthesis impair the postcastration rise of gonadotropins, interfere with the stimulation of gonadotropin release brought about by either estrogen or progesterone and block the preovulatory surge of gonadotropins [13, 21, 32]. The noradrenergic synapse may lie in the suprachiasmatic region, since inhibitors of norepinephrine synthesis blocked LH release from preoptic but not median eminence stimulation [22].

There is a possibility that cholinergic synapses may also be involved in controlling gonadotropin and prolactin release since atropine given either subcutaneously or into the third ventricle can block gonadotropin and prolactin release. For example, the drug blocks the preovulatory discharge of FSH, LH and prolactin in the rat and also inhibits the post-castration rise of gonadotropins [25, 26].

Serotonin has been shown to inhibit gonadotropin and stimulate prolactin release in several studies [28, 35]. In our laboratory, no effects have been observed with the inhibitor of serotonin synthesis, p-chlorophenylalanine, casting doubt on the physiological significance of serotonin in the control of these hormones; however, in a recent abstract, KORDON *et al.* [23] have reported that p-chlorophenylalanine can block the suckling-induced rise in prolactin which suggests that it may be involved in mediating suckling-induced increases.

The pineal hormone melatonin has effects similar to serotonin, that is, it stimulates prolactin and inhibits gonadotropin release. We have recently observed that pinealectomy blocks the early-morning discharge of prolactin which occurs in male rats and in general leads to lower levels of the hormone than those found in normal animals, which suggests a role for the pineal in the control of prolactin release [39]. The effects of pinealectomy on gonadotropin release by contrast have been quite small [40].

The Possible Role of Prostaglandins in Mediating Gonadotropin and Prolactin Release

Inhibitors of prostaglandin synthesis can block ovulation, but there has been controversy as to the locus of action [3, 44]. In recent studies from our laboratory, it has been possible to show that injection of prostaglandins either into the third ventricle or in larger doses peripherally can alter gonadotropin and prolactin release [16]. Prostaglandin E_1 elevated prolactin release, perhaps by stimulating prolactin-releasing factor discharge or, alternatively, by inhibiting the release of prolactin-inhibiting factor. On the other hand, prostaglandin E_2 caused the discharge of LH apparently by stimulating the release of LRF. Prostaglandin $F_{1\alpha}$ and $F_{2\alpha}$ were without effect. No effects have been demonstrated from intrapituitary injection of the prostaglandins.

Possible Role of Cyclic Nucleotides in Control of PIF Release

In other studies, third ventricular injections of cyclic AMP or dibutyryl cyclic AMP have been shown to result in a lowering of plasma prolactin levels, whereas similar injections into the pituitary were without effect [33]. It appears that cyclic AMP may be involved as a second messenger to control the release of PIF, perhaps mediating the response to dopaminergic input.

Relationship of the LH Controlling Center to Mating Behavior

In the normal animal the preovulatory surge of gonadotropins is followed several hours later by the onset of mating behavior. It occurred to us that the discharge of LRF into the portal vessels might be accompanied by a local discharge of the hormone from its storage sites in the preoptic area. The LRF might then have an influence on the mating center which lies in juxtaposition to the presumed cell bodies of the LRF neurons in the preoptic area. To test this hypothesis, we injected LRF subcutaneously in a dose of 500 ng into ovariectomized, estrogen-primed rats. The subcutaneous route was chosen because of the ease of administration and lack of disturbance to the animal. The dose was one previously shown to be adequate to cause a discharge of LH no larger than the preovulatory release [46]. The animals were primed with a dose of estrone which did not by itself produce mating behavior. Within 2 h of administration of LRF, nearly all of the animals came into behavioral receptivity and exhibited lordosis behavior. The effect appeared to be specific for LRF since ovulating doses of LH, large doses of FSH or a large dose of thyrotropin-releasing factor were ineffective in inducing mating behavior [29]. Recently, it has been shown that injection of LRF in some males will shorten the latency to mounting and ejaculation [30]. If these effects carry over to the human, they will obviously have profound significance.

References

1 Arimura, A.; Debeljuk, L., and Schally, A. V.: Stimulation of FSH release *in vivo* by prolonged infusion of synthetic LH-RH. Endocrinology *91:* 529–532 (1972).

2 BARRACLOUGH, C. A.: Modifications in the CNS regulation of reproduction after exposure of prepubertal rats to steroid hormones. Recent Progr. Hormone Res. *22:* 503–539 (1966).

3 BEHRMAN, H. R.; ORCZYK, G. P., and GREEP, R. O.: Effect of synthetic gonadotropin-releasing hormone (GN-RH) on ovulation blockade by aspirin and indomethacin. Prostaglandins *1:* 245–258 (1972).

4 BIRGE, C. A.; JACOBS, L. S.; HAMMEO, C. T., and DAUGHADAY, W. H.: Catecholamine inhibition of prolactin secretion by isolated rat adenohypophyses. Endocrinology *86:* 120–130 (1970).

5 BISHOP, W.; FAWCETT, C. P.; KRULICH, L., and MCCANN, S. M.: Acute and chronic effects of hypothalamic lesions on the release of FSH, LH and prolactin in intact and castrated rats. Endocrinology *91:* 643–656 (1972).

6 BISHOP, W.; KALRA, P. S.; FAWCETT, C. P.; KRULICH, L., and MCCANN, S. M.: The effects of hypothalamic lesions on the release of gonadotropins and prolactin in response to estrogen and progesterone treatment in female rats. Endocrinology *91:* 1404–1410 (1972).

7 BURGUS, R.; BUTCHER, M.; LING, N.; MONAHAN, M.; RIVIER, J.; FELLOWS, R.; AMOSS, M.; BLACKWELL, R.; VALE, W. et GUILLEMIN, R.: Structure moléculaire du facteur hypothalamique (LRF) d'origine ovine contrôlant la sécretion de l'hormone gonadotrope hypophysaire de luteinisation (LH). C. R. Acad. Sci. *273:* 1611–1613 (1971).

8 COOPER, K. J.: Personal commun. (1973).

9 COOPER, K. J.; FAWCETT, C. P., and MCCANN, S. M.: Variations in pituitary responsiveness to LH-releasing factor (LRF) during the rat oestrous cycle. J. Endocrin. *57:* 187–188 (1973).

10 CRIGHTON, D. B.; SCHNEIDER, H. P. G., and MCCANN, S. M.: Localization of LH-releasing factor in the hypothalamus and neurohypophysis as determined by *in vitro* assay. Endocrinology *87:* 323–329 (1970).

11 DHARIWAL, A. P. S.; NALLAR, R.; BATT, M., and MCCANN, S. M.: Separation of FSH-releasing factor from LH-releasing factor. Endocrinology *76:* 290–294 (1965).

12 DHARIWAL, A. P. S.; WATANABE, S.; ANTUNUES-RODRIGUES, J., and MCCANN, S. M.: Chromotographic behavior of follicle stimulating hormone-releasing factor on Sephadex and carboxy methyl cellulose. Neuroendocrinology *2:* 294–303 (1967).

13 DONOSO, A. O.; BISHOP, W.; FAWCETT, C. P.; KRULICH, L., and MCCANN, S. M.: Effects of drugs that modify brain monoamine concentrations on plasma gonadotropin and prolactin levels in the rat. Endocrinology *89:* 774–784 (1971).

14 FAWCETT, C. P.: Discussion of paper No. 24; in Hypothalamic hypophysiotropic hormones, p. 111 (Excerpta Medica, Amsterdam 1973).

15 FAWCETT, C. P.: Personal commun. (1973).

16 HARMS, P. G.; OJEDA, S. R., and MCCANN, S. M.: Prostaglandin involvement in hypothalamic control of gonadotropin and prolactin release. Science *181:* 760–761 (1973).

17 IGARASHI, M.; NALLAR, R., and MCCANN, S. M.: Further studies on the follicle stimulating hormone-releasing action of hypothalamic extracts. Endocrinology *75:* 901–907 (1964).

18 Johansson, K. N.; Currie, B. L.; Folkers, K., and Bowers, C. Y.: Biosynthesis and evidence for the existence of the follicle stimulating hormone releasing hormone. Biochem. biophys. Res. Commun. *50:* 8–26 (1973).

19 Jutisz, M.: Purification and chemistry of gonadotropin releasing factors. The human testis, p. 207 (Plenum Press, New York 1970).

20 Kalra, S. P.; Ajika, K.; Krulich, L.; Fawcett, C. P.; Quijada, M., and McCann, S. M.: Effect of hypothalamic and preoptic electrochemical stimulation on gonadotropin and prolactin release in proestrous rats. Endocrinology *88:* 1150–1158 (1971).

21 Kalra, P. S.; Kalra, S. P.; Krulich, L.; Fawcett, C. P., and McCann, S. M.: Involvement of norepinephrine in transmission of the stimulatory influence of progesterone on gonadotropin release. Endocrinology *90:* 1168–1176 (1972).

22 Kalra, S. P. and McCann, S. M.: Effect of drugs modifying catecholamine synthesis on LH release from preoptic stimulation in the rat. Endocrinology *93:* 356–362 (1973).

23 Kordon, C. A.; Blake, C. A., and Sawyer, C. H.: Participation of serotonin-containing neurons in suckling-induced rise in plasma prolactin. Proc. 4th Int. Congr. of Endocrinology, Washington, D.C. 1972.

24 Libertun, C.; Cooper, K. J.; Fawcett, C. P., and McCann, S. M.: Effects of ovariectomy and steroid treatment of hypophyseal sensitivity to purified LRF Endocrinology *94:* 518–525 (1974).

25 Libertun, C. and McCann, S. M.: Blockade of preovulatory release of FSH and LH by subcutaneous or intraventricular injection of atropine. Biol. Reprod. *7:* 110 (1972).

26 Libertun, C. and McCann, S. M.: Blockade of the release of gonadotropins and prolactin by subcutaneous or intraventricular injection of atropine in male and female rats. Endocrinology *92:* 1714 (1973).

27 Matsuo, H.; Baba, Y.; Nair, R. M. G.; Arimura, A., and Schally, A. V.: Structure of the porcine LH- and FSH-releasing hormone. I. The proposed amino acid sequence. Biochem. biophys. Res. Commun. *43:* 1334–1339 (1971).

28 McCann, S. M.; Kalra, S. P.; Kalra, P. S.; Bishop, W.; Donoso, A. O.; Schneider, H. P. G.; Fawcett, C. P., and Krulich, L.: The role of monoamines in the control of gonadotropin and prolactin secretion. Brain-endocrine interaction. Median eminence. Structure and function. Int. Symp., Munich 1971, pp. 224–235 (Karger, Basel 1972).

29 Moss, R. L. and McCann, S. M.: Induction of mating behavior in rats by luteinizing hormone releasing factor. Science *181:* 177–179 (1973).

30 Moss, R. L.: Personal commun. (1973).

31 Negro-Vilar, A.; Orias, R., and McCann, S. M.: Evidence for a pituitary site of action for the acute inhibition of LH release by estrogen in the rat. Endocrinology *92:* 1680–1684 (1973).

32 Ojeda, S. R. and McCann, S. M.: Evidence for participation of a catecholaminergic mechanism in the postcastration rise in plasma gonadotropins. Neuroendocrinology *12:* 295–315 (1973).

33 OJEDA, S. R.; KRULICH, L., and MCCANN, S. M.: Modification of plasma prolactin and LH levels of intraventricular injection of cyclic AMP. Endocrinology (submitted for publication).

34 ORIAS, R. and LIBERTUN, C.: Effects of synthetic LH-releasing factor (LRF) infusion on gonadotropin release and inhibition of these effects by estrogen (E_2) and progesterone (P). Progr. 55th Meet. of the Endocrine Society, p. 197 (1973).

35 PORTER, J. C.; KAMBERI, I. A., and ONDO, J. G.: Role of biogenic amines and cerebrospinal fluid in the neurovascular transmittal of hypophysiotropic substances. Brain-endocrine interaction. Median eminence. Structure and function. Int. Symp., Munich 1971, pp. 245–253 (Karger, Basel 1972).

36 QUIJADA, M.; KRULICH, L.; FAWCETT, C. P.; SUNDBERG, D., and MCCANN, S. M.: Localization of TSH-releasing factor (TRF), LH-RF and FSH-RF rat hypothalamus. Fed. Proc. *30:* 197 (1971).

37 QUIJADA, M.; ILLNER, P.; KRULICH, L., and MCCANN, S. M.: The effect of catecholamines on hormone release from anterior pituitaries and ventral hypothalami incubated *in vitro*. Neuroendocrinology *13:* 151–163 (1973).

38 RAMIREZ, V. D. and MCCANN, S. M.: A highly sensitive test for LH-releasing activity. The ovariectomized, estrogen progesterone-blocked rat. Endocrinology *73:* 193–198 (1963).

39 RONNEKLEIV, O.; KRULICH, L., and MCCANN, S. M.: An early morning surge of prolactin in the male rat and its abolition by pinealectomy. Endocrinology *92:* 1339–1342 (1973).

40 RONNEKLEIV, O. and MCCANN, S. M.: The effect of pinealectomy on plasma gonadotropins and prolactin in the rat. Physiologist *16:* 436 (1973).

41 SCHALLY, A. V.; SAITO, T.; ARIMURA, A.; MULLER, E. E.; BOWERS, C. Y., and WHITE, W. F.: Purification of follicle-stimulating hormone-releasing factor (FSH-RF) from bovine hypothalamus. Endocrinology *79:* 1087–1094 (1966).

42 SCHNEIDER, H. P. G.; CRIGHTON, D. B., and MCCANN, S. M.: Suprachiasmatic LH-releasing factor. Neuroendocrinology *5:* 271–280 (1969).

43 SCHNEIDER, H. P. G. and MCCANN, S. M.: Release of LRF into the peripheral circulation of hypophysectomized rats by dopamine and its blockade by estradiol. Endocrinology *87:* 249–253 (1970).

44 TSAFRIRI, A.; LINDNER, H. R.; ZOR, U., and LAMPRECHT, S. A.: Physiological role of prostaglandins in the induction of ovulation. Prostaglandins *2:* 1–10 (1972).

45 WATSON, J. T.; KRULICH, L., and MCCANN, S. M.: Effect of crude rat hypothalamic extract on serum gonadotropin and prolactin levels in normal and orchidectomized male rats. Endocrinology *89:* 1412–1417 (1971).

46 ZEBALLOS, G. and MCCANN, S. M.: The effect of subcutaneous administration of synthetic LH-releasing factor (LRF) on plasma gonadotropins and prolactin in the rat. Proc. Soc. exp. Biol., N.Y. (in press).

Author's address: Prof. S. M. MCCANN, Department of Physiology, The University of Texas, Southwestern Medical School, *Dallas, TX 75235* (USA)

Recent Studies of Hypothalamic Function
Int. Symp. Calgary 1973, pp. 90–99 (Karger, Basel 1974)

The Influence of Hyperthyroidism on the Hypothalamo-Hypophysial-Thyroid Axis[1]

J. M. McKenzie[2], P. D'Amour and M. Zakarija

Department of Medicine, McGill University Clinic, Royal Victoria Hospital, Montreal, Quebec

Introduction

Involvement of the hypothalamus and pituitary in the hyperthyroidism of Graves' disease was long suspected on two distinct bases [10]. First, a syndrome similar to the human state was produced experimentally by injection of pituitary extract rich in thyrotropin. Secondly, a possible psychic 'trigger' for the start of hyperthyroidism is a common anamnestic finding; thus, although the evidence remains largely anecdotal, an attractive postulate of emotion activating the hypothalamo-hypophysial-thyroid axis was readily embraced. An obvious test of this hypothesis was to measure thyrotropin in the blood of patients with hyperthyroidism, but by bioassay criteria [10] little confirmation was obtained. Even by more sensitive techniques of radioimmunoassay, thyrotropin was usually reported as unmeasurable [13].

Although many points of circumstantial data weigh in favor of the conclusion reached by these experiences, i.e. that hyperthyroidism is indeed not due to an excess of thyrotropin [10], an objective assessment of the findings leads to a 'not disproven' verdict. In most reports, the immunoassay of thyrotropin gives normal values that range from zero, indicating they do not have the sensitivity to measure the hormone in all euthyroid persons; thus, in theory, hyperthyroidism could be associated with an increase in thyrotropin in the blood and it still might be unmeasurable. If this were the case, it would be reasonable to expect that the

1 Financially supported by grants from the Medical Research Council of Canada (MT884) and the US Public Health Service (AMO4121).

2 Research Associate, Medical Research Council of Canada.

pituitary in Graves' disease was responsive to thyrotropin-releasing hormone (TRH). This has been shown not to occur in frank hyperthyroidism [2] but we decided to examine the hypothesis in patients with Graves' disease who were euthyroid at the time of testing with TRH.

Clinical Studies

We identified 3 categories of euthyroid patients to study, two with Graves' disease and one with autonomous functioning adenoma of the thyroid. The first Graves' disease group consisted of patients who had ended a course of treatment with antithyroid drugs and were euthyroid at the time of study. Secondly, we had patients who had ophthalmopathy characteristic of Graves' disease but no overt evidence of thyroid disease. The third group comprised 3 patients with thyroid adenoma functioning autonomously to the degree that there was suppression of pituitary thyrotropin, resulting in nonfunction of the para-adenomatous tissue, as evidenced by 'scanning' with radioiodine [9]. All of these patients were clinically euthyroid and had a normal serum concentration of thyroxine. They were tested by having blood taken at timed intervals before and after the intravenous injection of 400 μg TRH (Abbott Laboratories, North Chicago, Ill.). Thyroid function, including measurement of serum thyroxine, triiodothyronine and thyrotropin, the latter two by radioimmunoassay, was assessed using the first blood (–15 min); subsequent bloods (at 0, 5, 10, 15, 30, 45, 60 and 90 min after the injection of TRH) were used for radioimmunoassay of triiodothyronine, thyrotropin and of prolactin, that is known also to be responsive to TRH [2].

The Graves' disease patients were further tested for 'suppressibility' of thyroid function. The normal subject, taking 100 μg triiodothyronine daily, will have prompt cessation of the thyroid release of hormone [6] and, by 8–10 days, will have only low uptake of orally administered radioiodine [19]; by one or other of these tests our patients were classed as 'suppressible' or 'nonsuppressible'. Precise details of the various procedures used in these studies will be published elsewhere [5].

The data obtained with the Graves' disease patients are shown in figure 1–4 and for those patients with autonomous adenoma in figure 5. It may be seen that all patients responded to injection of TRH by an increase in prolactin; yet, of the 22 whose data are illustrated, 16 showed no increase in serum thyrotropin.

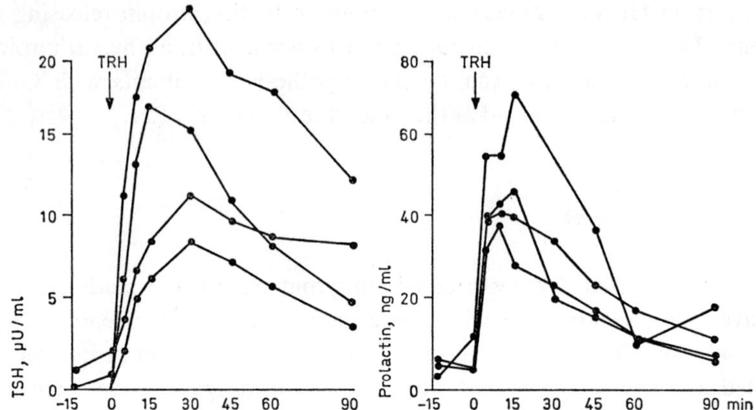

Fig. 1. Effect of thyrotropin-releasing hormone (TRH) on the concentration of thyrotropin (TSH) and of prolactin in the blood of euthyroid patients after a course of propylthiouracil for hyperthyroidism (group I, table I). 400 µg TRH was injected intravenously at time 0 and blood was taken at timed intervals as shown on the abscissa.

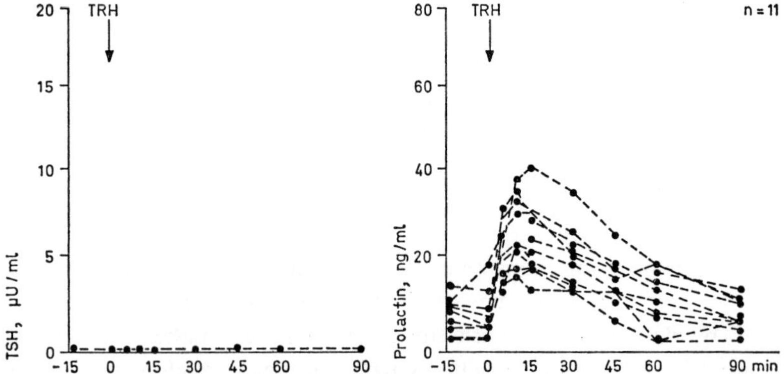

Fig. 2. Effect of TRH on the concentration of thyrotropin (TSH) and of prolactin in the blood of euthyroid patients after a course of propylthiouracil for hyperthyroidism (group II, table I). Other details as in figure 1.

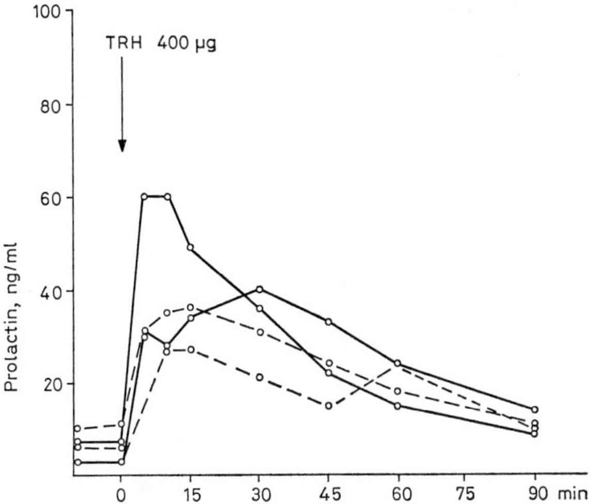

Fig. 3. Effect of TRH on the concentration of prolactin in the blood of euthyroid patients with ophthalmopathy of Graves' disease (group III, table I). Prolactin is plotted on the ordinate as nanograms per milliliter serum; other details as in figure 1. ------ = TSH increase, ——— = no TSH increase.

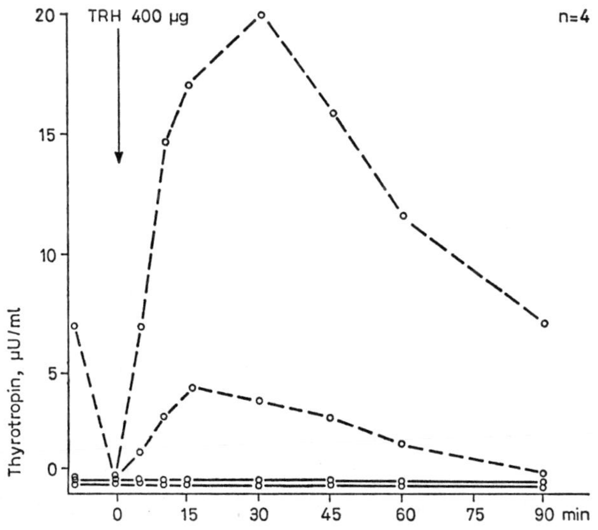

Fig. 4. Effect of TRH on the concentration of thyrotropin in the blood of euthyroid patients with ophthalmopathy of Graves' disease (group III, table I). Thyrotropin is plotted on the ordinate as microunits per milliliter serum; other details as in figure 1.

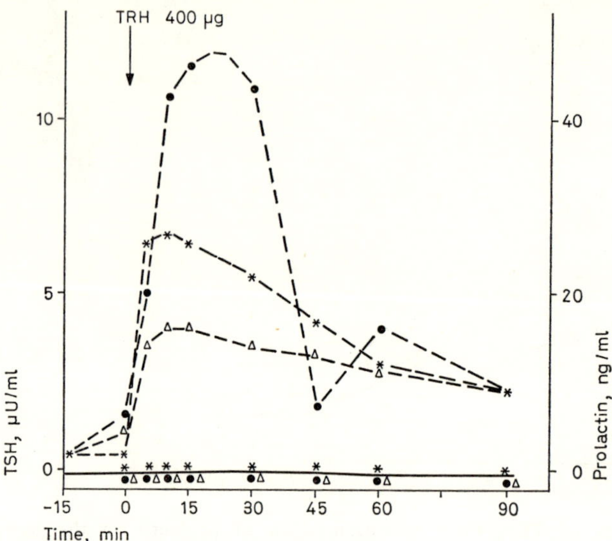

Fig. 5. Effect of TRH on the concentration of thyrotropin (TSH, ——) and of prolactin (– – – –) in the blood of euthyroid patients with autonomous adenoma of the thyroid gland. Details as in figure 1.

When suppressibility of thyroid function was considered, there was absolute concordance of the data; those patients who had a thyrotropin response to TRH had suppressible thyroid function, and those with no thyrotropin response had nonsuppressible thyroid function. It may be that those in the post-treatment group who had suppressible function may be considered as no longer having Graves' disease, although rapid relapse or recurrence certainly may occur in such subjects [11].

Another correlative aspect of these studies is shown in table I. Here it is clear that the majority of patients with Graves' disease who had non-suppressible thyroid function, and no thyrotropin response to TRH, had a supranormal concentration of triiodothyronine in blood. However, two had unequivocally normal levels of serum triiodothyronine (142 and 160 mg/100 ml) and had no thyrotropin response to TRH; in addition, one patient had supranormal serum concentrations of triiodothyronine and responded to TRH.

From these data no firm conclusion may be achieved regarding pituitary responsiveness to TRH in Graves' disease. Tentatively, however, one might surmise that the gland is as readily made unresponsive as is the

Table I. Correlation of serum triiodothyronine, thyroid gland suppressibility and thyrotropin response to thyrotropin-releasing hormone (TRH) in euthyroid patients with Graves' disease

Group	Number of patients	Thyroid function	Thyrotropin response to TRH	Serum triiodothyronine
I a	1	suppressible	positive	high
I b	3	suppressible	positive	normal
II a	9	nonsuppressible	negative	high
II b	2	nonsuppressible	negative	normal
III a	2	suppressible	positive	normal
III b	2	nonsuppressible	negative	normal

Patients are grouped as: I, post-treatment for hyperthyroidism, suppressible thyroid function; II, post-treatment for hyperthyroidism, nonsuppressible thyroid function; III, ophthalmopathy of Graves' disease. Normal values for serum triiodothyronine ranged from 96 to 230 ng/100 ml and those defined as 'high' from 240 to 900 ng/100 ml.

normal. Presumably the unresponsiveness is related to the concentration of circulating thyroid hormone, especially triiodothyronine [15, 16, 18], and in most of our patients this was supranormal. In those in whom it was normal the conjecture may be proffered that perhaps these patients normally maintained serum thyroid hormones in the lower range of what we consider normal, or it is possible that the sensitivity of the pituitary is more related to rates of disposal or production of triiodothyronine or thyroxin, than to the absolute concentration. In any case, from these studies, a pituitary gland readily secreting thyrotropin in response to hypothalamic command does not appear at all likely as an explanation for hyperthyroidism in Graves' disease.

Mechanism of Pituitary Unresponsiveness to TRH in Hyperthyroidism

Even before TRH was available as a synthetic tripeptide, there was some understanding of the action of thyroid hormones in the system. Table II presents data illustrating points previously established by others

Table II. Effects of triiodothyronine (T_3) and of puromycin on the *in vivo* bioassay of TRH

Group	Pretreatment of mice	TRH	Response (mean ± SD), %	p
1	nil	nil	103 ± 14	–
2	nil	1 μg	884 ± 94	vs. 1 < 0.001
3	10 μg T_3 s.c. at –1 h	nil	119 ± 9	–
4	10 μg T_3 s.c. at –1 h	1 μg	274 ± 126	vs. 2 < 0.01
5	2 mg Puromycin i.p. at –2 h	nil	113 ± 14	–
6	2 mg Puromycin i.p. at –2 h	1 μg	781 ± 144	vs. 2 NS
7	+10 μg T_3 s.c. at –1 h	1 μg	644 ± 44	vs. 6 NS vs. 4 < 0.01

The bioassay for TRH was in mice with thyroid iodine labeled with ^{125}I as described by Bowers *et al.* [4]; the response in blood ^{125}I, 2 h after the injection of test material, expressed as a percent of ^{125}I in blood taken immediately before the i.v. injection of TRH or control solution. Probability of significance of difference in responses was calculated by Student's t-test; n = 5 in each group. NS = not significant.

[3, 17]; these are that *in vivo* an injection of triiodothyronine readily inhibits a thyrotropin response to TRH and this, in turn, is prevented by prior injection of Puromycin. The reasonable interpretation of these findings is that triiodothyronine rapidly stimulates the synthesis of a protein antagonistic to the thyrotropin releasing action of TRH, and an inhibitor of protein synthesis, such as Puromycin, prevents this occurrence [3, 17]. However, the mechanism whereby an inhibitory protein might act to prevent thyrotropin release has been little explored.

Although confirmatory data are not extensive, it is likely that TRH acts at the cell membrane of 'thyrotropes' within the adenohypophysis, enhances the activity of adenyl cyclase and thereby stimulates the release of thyrotropin that is normally available in storage granules within these cells [7, 8, 12]. Other studies tend to corroborate the involvement of 3′,5′-adenosine monophosphate (cyclic AMP), the product of adenyl cyclase action, in the action of TRH [1, 14], although Bowers, in discussing an extensive series of experiments relevant to this question, could only summarize that [1]: 'it is not possible to conclude from our results that cyclic AMP does not mediate the release of TSH...'

As might be expected, considering the fact that only a minority of

Table III. Effect of TRH *in vitro* on the concentration of cyclic AMP in rat anterior pituitary

Incubation duration, min	Concentration of cyclic AMP, mean ± SD (test lobe : control lobe × 100)	p
3	92 ± 34	NS
10	122 ± 11	<0.01
60	109 ± 20	NS

Cyclic AMP was measured by radioimmunoassay as pM/mg wet weight of tissue. Each pituitary was divided into its two lobes that were incubated for 3–60 min, as shown in Krebs-Ringer bicarbonate buffer that contained 0.2 % glucose and 0.2 % human serum albumen; the buffer for the test lobe had, in addition, 20 ng TRH/ml. For each pair of lobes the cyclic AMP in the test lobe at the end of incubation was calculated as a percentage of that in the corresponding control lobe. NS = not significant.

adenohypophyseal cells secrete thyrotropin, the effect of TRH on the total gland content of cyclic AMP is minimal but, as shown in table III, it can be identified. BOWERS [1] reported that 5 weeks after thyroidectomy, the concentration of cyclic AMP in the rat pituitary is markedly increased and this was acutely reduced by injection of triiodothyronine. Although BOWERS' data were poorly reproducible, we have attempted to combine these experiences to design an experiment to test the effect of TRH on the concentration of cyclic AMP in the pituitary of the rat treated chronically with propylthiouracil and injected with triiodothyronine shortly before removal of the gland. As shown in table III pituitaries from such animals, whether or not they had been injected with triiodothyronine, had after incubation *in vitro* a concentration of cyclic AMP that differed little from that in control glands, and there was the normal small response to TRH.

It seems therefore that the triiodothyronine-induced inhibitor of TRH action does not act at the level of TRH stimulation of adenyl cyclase. Firstly, if it did, the 10-min period of *in vitro* incubation ought to have led to a decrease in the concentration of cyclic AMP compared with that in control glands, because of a reduced rate of synthesis of the nucleotide; secondly, we found the small increase in pituitary cyclic AMP, brought about by TRH, to be unimpaired by the injections of triiodothyronine. Therefore, the putative inhibitor of TRH action presumably acts to modify the subsequent steps [8] in the release of thyrotropin.

References

1 Bowers, C. Y.: Studies on the role of cyclic AMP in the release of anterior pituitary hormones. Ann. N.Y. Acad. Sci. *185:* 263–290 (1971).

2 Bowers, C. Y.; Friesen, H. G.; Hwang, P.; Guyda, H. J., and Folkers, K.: Prolactin and thyrotropin release in man by synthetic pyroglutamyl-histidyl-prolinamide. Biochem. biophys. Res. Commun. *45:* 1033–1041 (1971).

3 Bowers, C. Y.; Lee, K. L., and Schally, A. V.: A study on the interaction of the thyrotropin-releasing factor and l-triiodothyronine. Effects of Puromycin and cycloheximide. Endocrinology *82:* 75–82 (1968).

4 Bowers, C. Y.; Schally, A. V.; Reynolds, G. A., and Hawley, W. D.: Interactions of L-thyroxine or L-triiodothyronine and thyrotropin-releasing factor on the release and synthesis of thyrotropin from the anterior pituitary gland of mice. Endocrinology *81:* 741–747 (1967).

5 D'Amour, P.; Banovac, K.; Salisbury-Murphy, S.; Friesen, H. G., and McKenzie, J. M.: Thyrotropin and prolactin secretion in euthyroid patients with non-suppressible thyroid function (submitted for publication).

6 Ecklund, R. and Ryan, R.: Suppression of release of radioactive iodine as a test of thyroid function. J. clin. Endocrin. *22:* 26–30 (1962).

7 Labrie, F.; Barden, N.; Poirier, G., and Lean, A. de: Binding of thyrotropin-releasing hormone to plasma membranes of bovine anterior pituitary gland. Proc. nat. Acad. Sci., Wash. *69:* 283–287 (1972).

8 Labrie, F.; Lemaire, S.; Poirier, G.; Pelletier, G., and Boucher, R.: Adenohypophyseal secretory granules. I. Their phosphorylation and association with protein kinase. J. biol. Chem. *246:* 7311–7317 (1971).

9 McKenzie, J. M.: Hyperthyroidism caused by thyroid adenomata. J. clin. Endocrin. *26:* 779–781 (1966).

10 McKenzie, J. M.: Humoral factors in the pathogenesis of Graves' disease. Physiol. Rev. *48:* 252–310 (1968).

11 McKenzie, J. M.: Does LATS cause hyperthyroidism in Graves' disease? (A review biased toward the affirmative.) Metabolism *21:* 883–894 (1972).

12 McKenzie, J. M.; Adiga, P. R., and Solomon, S. H.: Hypothalamic control of thyrotropin; in Martini, Motta and Fraschini The hypothalamus, pp. 335–350 (Academic Press, New York 1970).

13 Odell, W. D.; Utiger, R. D., and Wilber, J.: Studies of thyrotropin physiology by means of radioimmunoassay. Recent Progr. Hormone Res. *23:* 47–56 (1967).

14 Poirier, G.; Labrie, F.; Barden, N., and Lemaire, S.: Thyrotropin-releasing hormone receptor. Its partial purification from bovine anterior pituitary gland and its close association with adenyl cyclase. FEBS Letters *20:* 283–286 (1972).

15 Shenkman, L.; Mitsuma, T., and Hollander, C. S.: Modulation of pituitary responsiveness to thyrotropin-releasing hormone by triiodothyronine. J. clin. Invest. *52:* 205–209 (1973).

16 Snyder, P. J. and Utiger, R. D.: Inhibition of thyrotropin response to thyrotropin-releasing hormone by small quantities of thyroid hormones. J. clin. Invest. *51:* 2077–2084 (1972).

17 Vale, W.; Burgus, R., and Guillemin, R.: On the mechanism of action of TRF. Effects of cycloheximide and actinomycin on the release of TSH stimulated *in vitro* by TRF and its inhibition by thyroxine. Neuroendocrinology *3:* 34–46 (1968).

18 Vigneri, R.; Papalia, D.; Pezzino, V.; Squatrito, S.; Motta, L., and Polosa, P.: Triiodothyronine and thyrotropin-releasing hormone interaction on TSH release in man. Hormones *3:* 250–256 (1972).

19 Werner, S. C. and Spooner, M.: New and simple test for hyperthyroidism employing l-triiodothyronine and the twenty-four hour I-131 uptake method. Bull. N.Y. Acad. Med. *31:* 137–145 (1955).

Authors' address: Dr. J. M. McKenzie, Dr. P. D'Amour and Dr. M. Zakarija, Department of Medicine, McGill University Clinic, Royal Victoria Hospital, *Montreal, Quebec* (Canada)

Recent Studies of Hypothalamic Function
Int. Symp. Calgary 1973, pp. 100–113 (Karger, Basel 1974)

Brain Catecholamines in the Regulation of ACTH Secretion

G. R. Van Loon

Departments of Medicine and Physiology, University of Toronto, Toronto, Ontario

I. Introduction

The role of the central nervous system in the regulation of secretion of adrenocorticotropic hormone (ACTH) has been well documented [34, 52]. The neuroanatomical substrate for this regulation appears to involve pathways from several brain areas including brain stem, amygdala, hippocampus, etc., which converge in the hypothalamus probably on the neurons secreting corticotropin releasing factor (CRF). Neurophysiological studies have provided evidence for both stimulatory and inhibitory mechanisms in this regulation, but the neurotransmitters in these pathways remain to be defined. The catecholamines, norepinephrine and dopamine, appear to function as neurotransmitters at central synapses, and the possible role of catecholamines in neural pathways regulating ACTH secretion will be discussed in this chapter.

A considerable body of experimental data has been described relating studies of catecholamine metabolism and ACTH secretion. Until recently, however, there have been no unifying hypotheses presented concerning such a relationship. The catecholamines are apparently not, in themselves, corticotropin releasing factors [21] but there is abundant evidence that they are involved in the control of ACTH secretion [57].

Central catecholamine-containing neuronal pathways have been described [17].

II. Correlation between Catecholamines and ACTH Secretion

A. Effect of Drugs which Alter Brain Catecholamine Metabolism on ACTH Secretion

Early attempts to elucidate the role played by biogenic amines in the central regulation of the hypothalamic-pituitary-adrenal axis involved the use of pharmacological agents known to alter amine metabolism; reserpine was the most commonly employed drug. Reserpine stimulates ACTH secretion in several species [5, 26, 33] when the dose is sufficient to decrease brain monoamine content by more than 50 %. The effects of reserpine and other stresses on ACTH secretion can be prevented by monoamine oxidase inhibitors [24, 44]. The literature relating the effects of chlorpromazine on ACTH secretion is confusing and has been reviewed by DE WIED [60].

B. Effect of Stress on the Metabolism of Brain Catecholamines

Many stressful stimuli that increase ACTH secretion are associated also with a decrease in brain norepinephrine content [21]. Stressful agents and situations vary considerably in their effectiveness to produce brain norepinephrine depletion [8, 36, 48]. Increased turnover of brain norepinephrine has been reported in response to stressful stimuli [9, 12, 25]. The effects of stress on central dopaminergic neurons have been less well documented. Both change [9] and failure to find change [12] in dopamine depletion after α-methyl-p-tyrosine have been reported in animals exposed to stress.

C. Effect of Experimental Endocrine Manipulation on Brain Catecholamine Metabolism (Table I)

Various changes in brain catecholamine metabolism have been reported following experimental manipulation of the hypothalamic-pituitary-adrenal axis. Adrenalectomy has been associated with increased brain norepinephrine turnover [18, 28], lack of change in brain norepinephrine concentration [28, 41, 50], and no change in brain dopamine metabolism [16]. Hypophysectomy is followed by a decrease in brain

Table I. Effect of endocrine manipulation on brain catecholamine metabolism

Endocrine manipulation	Brain area	Norepinephrine		Dopamine		References
		concentration	turnover	concentration	turnover	
Adrenalectomy	Median eminence			↑		AKMAYEV and DONATH (1966)
				N.C.	N.C.	FUXE and HOKFELT (1967)
	Hypothalamus		↑			FUXE *et al.* (1973a)
	Midbrain	loss of diurnal rhythm				FRIEDMAN and WALKER (1968)
	Neostriatum				N.C.	FUXE *et al.* (1973b)
	Cerebral cortex		↑			FUXE *et al.* (1973a)
	Whole brain	N.C.	↑			JAVOY *et al.* (1968)
		N.C.				PFEIFER *et al.* (1963)
		N.C.				UTEVSKII *et al.* (1971)
		variable effect				DESCHAEPDRYVER *et al.* (1969)
Hypophysectomy	Median eminence				↓	FUXE *et al.* (1973b)
	Hypothalamus		↓			FUXE *et al.* (1973a)
	Neostriatum				N.C.	FUXE *et al.* (1973a)
	Cerebral cortex		↓			FUXE *et al.* (1973b)
	Whole brain		↓			FUXE *et al.* (1973b)
Corticosteroid administration	Infundibulum			N.C.		BARRY (1968)
	Median eminence			N.C.	N.C.	FUXE and HOKFELT (1967)
					↑	FUXE *et al.* (1973b)
	Hypothalamus		slight ↓			FUXE *et al.* (1973a)
			↑			FUXE *et al.* (1973b)
	Neostriatum				N.C.	FUXE *et al.* (1973a)
					↑	FUXE *et al.* (1973b)
	Cerebral cortex		slight ↓			FUXE *et al.* (1973a)
			↑			FUXE *et al.* (1973b)

	Whole brain	variable effect		variable effect		DeSchaepdryver *et al.* (1969)
		N.C.				Petrovic and Davidovic (1970)
			↑			Fuxe *et al.* (1973b)
ACTH administration	Median eminence				↑	Fuxe *et al.* (1973b)
	Hypothalamus		slight ↑			Fuxe *et al.* (1973a)
	Neostriatum				↑	Fuxe *et al.* (1973b)
	Cerebral cortex		slight ↑			Fuxe *et al.* (1973a)
	Whole brain	↑				Petrovic and Davidovic (1970)
			slight ↑			Fuxe *et al.* (1973b)
Metyrapone administration	Infundibulum			N.C.		Barry (1968)
Corticosteroid administration after adrenalectomy	Hypothalamus		↑ blocked			Fuxe *et al.* (1973a)
	Neostriatum				slight ↑	Fuxe *et al.* (1973a)
					N.C.	Fuxe *et al.* (1973b)
	Mesolimbic pathway				slight ↑	Fuxe *et al.* (1973a)
Corticosteroid administration after hypophysectomy	Hypothalamus		↓ not altered			Fuxe *et al.* (1973b)
	Neostriatum				N.C.	Fuxe *et al.* (1973b)
	Cerebral cortex		↓ not altered			Fuxe *et al.* (1973b)
ACTH administration after hypophysectomy	Hypothalamus		↓ not altered			Fuxe *et al.* (1973a)
	Neostriatum				slight ↓	Fuxe *et al.* (1973a)
	Limbic forebrain				slight ↓	Fuxe *et al.* (1973a)
	Whole brain		further ↓			Fuxe *et al.* (1973b)

↑ Increased. ↓ Decreased.
N.C. No change.

norepinephrine turnover [18, 19]. Administration of ACTH produces increased turnover of brain norepinephrine and dopamine without change in concentrations [18, 19]; however, ACTH fails to prevent the decreased norepinephrine turnover in hypophysectomized animals [18]. Corticosteroid administration is not associated with consistent changes in brain catecholamine administration [2, 18, 19, 40, 44].

Thus, a correlation between change in brain catecholamine activity and change in ACTH secretion has been demonstrated in numerous studies. Other studies have failed to demonstrate a correlation between brain catecholamines and ACTH secretion [7, 11, 27, 46].

III. Excitatory Role of Catecholamines in the Regulation of ACTH Secretion

Most investigators have sought a role for catecholamines as excitatory neurotransmitters in central mechanisms regulating ACTH secretion. For example, release of ACTH following injection of catecholamines directly into the brain has been reported by several workers [4, 14].

Several possible mechanisms by which catecholamines stimulate ACTH secretion have been suggested. Neurohumors liberated into the hypothalamic-hypophyseal portal blood might affect anterior pituitary secretion by altering the caliber of portal vessels, thus changing adenohypophyseal blood flow [21, 57]. The possibility that catecholamines may stimulate ACTH secretion via a peripheral neural mechanism has been discussed by several authors [38, 39, 57, 58]. Although exogenous catecholamines do stimulate ACTH secretion, these substances do not seem to be necessary mediators of the stimulation of ACTH secretion [57].

IV. Inhibitory Role of Catecholamines in the Regulation of ACTH Secretion

A. Re-Examination of Pharmacological Studies

Examination of the literature relating catecholamines to ACTH secretion has suggested the possibility of an inverse relationship [57]. Reserpine, a drug that decreases brain catecholamine activity, increases

ACTH secretion. Chlorpromazine may act in the brain to decrease catecholamine activity by blocking catecholamine receptor sites and is also associated with stimulation of ACTH secretion. The monoamine oxidase inhibitor, iproniazid, potentiated the effect of dexamethasone in blocking the plasma corticosterone response to stress [13, 20].

Depletion of median eminence CRF in association with increased plasma and adrenal corticosterone and decreased pituitary ACTH has been reported in rats treated with reserpine or chlorpromazine [5].

The systemic administration of pargyline and amphetamine, drugs which increase central catecholaminergic activity, produced decreased resting levels of plasma corticosterone and decreased response to stress [6]. In other studies from the same laboratory, the intraventricular administration of norepinephrine or dopamine caused a depression of reserpine-induced increases in adrenal corticosterone [35]. Using a microelectrophoretic technique, it has been demonstrated that the action potentials of some hypothalamic and midbrain neurons are depressed by both dexamethasone and norepinephrine [47].

B. Hypothesis of a Central Adrenergic System Inhibiting ACTH Secretion

A review of the literature relating brain catecholamines and ACTH secretion clearly suggests the hypothesis that brain catecholamines inhibit ACTH secretion, probably by inhibiting the secretion of corticotropin releasing factor.

The catecholamine precursor, L-dopa, administered intravenously, inhibited the adrenal venous 17-hydroxycorticosteroid (17-OHCS) response to laparotomy in dogs [53]. On the other hand, the catecholamines, norepinephrine and dopamine, which do not cross the blood-brain barrier to any extent and do not alter brain catecholamine concentration, were ineffective in preventing the 17-OHCS response to surgical stress. Furthermore, the time-course of inhibition of the 17-OHCS response to surgical stress produced by L-dopa in dogs [53] coincides with both the alteration in motor activity produced by L-dopa in mice [10] and the increase in brain catecholamine content following administration of L-dopa [3].

The minimum effective dose of L-dopa necessary to inhibit the 17-OHCS response was increased by a drug (α-methyl-p-tyrosine) that decreases adrenergic activity and decreased by a drug (pargyline) that increases adrenergic activity [53]. These data provide further support for the concept that L-dopa inhibits ACTH secretion via an increase in adrenergic activity.

The adrenal venous 17-OHCS response to surgical stress was studied in response to the administration of large doses of several adrenergic agents directly into the third ventricle in dogs [54]. L-dopa, L-norepinephrine, L-isoproterenol, dopamine, tyramine and α-ethyl-tryptamine inhibited the 17-OHCS response to stress whereas D-norepinephrine and vanillymandelic acid failed to inhibit the 17-OHCS response.

Adrenergic inhibition was studied with regard to different parameters of ACTH secretion in the rat. The effects of alterations in brain catecholamine metabolism on plasma corticosterone resting level and circadian variation rather than stress response were investigated. The soluble methyl ester of α-methyl-*p*-tyrosine administered intraperitoneally increased plasma corticosterone, and this increase was inhibited by the simultaneous administration of L-dopa but not by norepinephrine [55]. Thus, the catecholamine precursor which freely crosses the blood-brain barrier prevented the effect of α-methyl-*p*-tyrosine, whereas norepinephrine, which fails to penetrate the blood-brain barrier, also failed to prevent the effect of α-methyl-*p*-tyrosine on plasma corticosterone.

Alpha-methyl-*p*-tyrosine did not produce its effect on plasma corticosterone by acting directly on the adrenals since it did not increase plasma corticosterone concentration in the acutely hypophysectomized rat [VAN LOON, NICHOLSON and LIDDLE, unpublished observations]. Additional evidence that the effect of α-methyl-*p*-tyrosine on plasma corticosterone is mediated through the brain is provided by the finding that the intraventricular administration of a systemically ineffective dose of α-methyl-*p*-tyrosine increased plasma corticosterone concentration [55]. Also, guanethidine, a drug that depletes catecholamines from neurons but fails to cross the blood-brain barrier to any significant extent, increased plasma corticosterone concentration after intraventricular administration [42].

It has been reported that norepinephrine is the catecholamine mediating inhibition of ACTH secretion. The dopamine-β-oxidase inhibitor, FLA-63, produced decreased hypothalamic norepinephrine concentration and increased plasma corticosterone with no change in hypothalamic dopamine concentration [42]. Also, dihydroxyphenylserine was found to prevent both the depletion of hypothalamic norepinephrine and the increase in plasma corticosterone induced by α-methyl-p-tyrosine without altering the depletion of hypothalamic dopamine [23].

Alpha-adrenergic mediation of catecholamine inhibition of ACTH secretion is suggested by several studies. Firstly, intraventricular administration in the rat of a systemically ineffective dose of phentolamine, an α-adrenergic blocking agent, increased plasma corticosterone [43]. Secondly, intraventricular administration in dogs of phenoxybenzamine, another α-adrenergic blocking agent, counteracted the inhibition of ACTH secretion produced by intravenous L-dopa [22, 23]. However, neither phentolamine nor the β-adrenergic blocking agent, propranolol, affected this inhibition. Thirdly, potentiation by iproniazid of the suppressive effect of dexamethasone on plasma corticosterone response to stress in rats is blocked by phentolamine [43].

The effect of catecholamines on ACTH secretion does not appear to be mediated by a direct action on the pituitary, since dopamine failed to alter the release of ACTH from rat anterior pituitaries incubated *in vitro* [51].

Some support for the hypothesis of central adrenergic neural inhibition of ACTH secretion has been provided in man. In 3 of 5 patients studied, chronic treatment with L-dopa for 1 month decreased the plasma 11-hydroxycorticosteroid (11-OHCS) response to insulin-induced hypoglycemia; one of these patients failed completely to show a rise in 11-OHCS after treatment with L-dopa [59]. Resting levels of plasma 11-OHCS and urinary 17-OHCS and 17-ketosteroids were unaffected by treatment with L-dopa. It is possible that some patients treated with L-dopa may not be able to respond adequately to stress with an appropriate rise in plasma corticosteroids.

C. An Animal Model and Speculation Concerning Clinical Correlation

Evidence has been presented in support of a hypothesis of a central adrenergic system that inhibits ACTH secretion. If such an inhibitory system regulating ACTH secretion were absent, increased secretion of ACTH might be expected. Increased secretion of ACTH in man manifests as Cushing's disease. Could Cushing's disease be somewhat analagous to Parkinson's disease with functional loss of a catecholaminergic system in a different brain area? In an attempt to provide a possible model for further investigation of the pathophysiology of human Cushing's disease, the effect of catecholamine depletion with α-methyl-*p*-tyrosine on various parameters of ACTH secretion was studied in the rat [56, 57]. Acute treatment with α-methyl-*p*-tyrosine is associated with increased plasma corticosterone levels, abolishment of diurnal rhythm of plasma corticosterone, resistance to suppression of plasma corticosterone with dexamethasone, enhanced response of plasma corticosterone to ACTH and enhanced response of plasma deoxycorticosterone to metyrapone. All of these abnormalities in pituitary-adrenal function are present in patients with Cushing's disease. Furthermore, in the rat the simultaneous administration of L-dopa with α-methyl-*p*-tyrosine returns the abnormal function tests toward normal. It is possible that the hyperadrenocortisolism induced in animals may not be relevant to human disease, and these studies may simply represent the pituitary-adrenal response to stress. Preliminary studies with administration of L-dopa to patients with Cushing's disease have not provided consistent correction of pituitary-adrenal hyperfunction [31, VAN LOON, unpublished observations]. This mode of treatment may not provide a suitable means of increasing norepinephrine concentration in the relevant brain areas, or alternatively this animal model may not be a valid model of the human disease.

D. Recent Studies Challenging the Inhibitory Hypothesis

Recently, other authors have investigated the effects of α-methyl-*p*-tyrosine on plasma corticosterone in rats. KAPLANSKI *et al.* [29] have confirmed the finding [55] that α-methyl-*p*-tyrosine depletes brain norepinephrine concentration and increases plasma corticosterone. However, when the former authors altered the dosage regimen they were able to decrease brain norepinephrine without increasing plasma corticosterone concentrations. Thus, they concluded that the effects of α-methyl-*p*-tyrosine cannot be used as evidence for the existence of central adrenergic systems inhibiting ACTH secretion. A significant difference in the effects of these two drug regimens on brain norepinephrine concentration may account for this discrepancy. The dosage of α-methyl-*p*-tyrosine which increased plasma corticosterone in the hands of both sets of authors depleted brain norepinephrine to less than 40% of control, whereas the dosage which failed to increase plasma corticosterone depleted brain norepinephrine to only 56% of control. The possibility that this may be a significant difference is suggested by the findings of MAICKEL *et al.* [33]

that reserpine increases plasma corticosterone only when brain norepinephrine has been depleted by more than 50 %.

McKinney *et al.* [37] found that α-methyl-*p*-tyrosine in a dosage of 80 mg/kg (about one-third of the dosage used by Van Loon *et al.* [55]) depleted brain norepinephrine concentration to 46 % of controls without altering plasma corticosterone. However, the morning control values which they report for plasma corticosterone of 30 μg/100 ml are in the range considered by most authors to demonstrate a stress response.

The effects of administration of 6-hydroxydopamine on plasma corticosterone have been reported by several authors [49]. Five to 7 days after intracisternal injections of 6-hydroxydopamine no significant alteration in resting levels of plasma corticosterone was found at 0900 or 2100 h [49]. It is of interest that whole brain norepinephrine and hypothalamic norepinephrine concentrations were decreased by the drug protocol to 20 and 38 % of controls; whole brain dopamine was decreased to 53 % of control but hypothalamic dopamine remained unaltered. At present it is possible only to speculate on the significance of these differences in brain catecholamines.

Decreased mid-day resting levels of plasma corticosterone were found 3 days after intraventricular injection of 6-hydroxydopamine, and normal plasma corticosterone levels 11 days after injection [32]. Decreased plasma corticosterone response to ketamine hydrochloride 28 days after 6-hydroxydopamine administration was also observed. These results were used to argue against an inhibitory role for catecholamines in the regulation of ACTH secretion.

Neither the intracisternal nor the intraventricular administration of 6-hydroxydopamine, which depleted whole brain norepinephrine to 56 and 21 % and whole brain dopamine to 72 and 33 %, affected the resting level of plasma corticosterone 1 week later [30]. It is possible that in situations of chronic catecholamine depletion compensatory neural mechanisms may normalize the control of the pituitary-adrenal system.

Thus, much of the evidence relating a possible role of catecholamines in the regulation of ACTH secretion is conflicting, and the data could support several different hypotheses regarding such a possible interaction. Considerable experimentation will be necessary before the role of brain catecholamines in the regulation of ACTH secretion is clearly understood.

References

1 AKMAYEV, I. G. und DONATH, I.: Die Katecholamine der Zona palisadica der Eminentia mediana des Hypothalamus bei Adrenalektomie, Hydrocortisonverabreichung und Stress. Z. mikr.-anat. Forsch *74:* 83–91 (1966).

2 BARRY, J.: Etude comparée des neurones et des fibres monoaminergiques de l'hypothalamus et de la région infundibulaire chez le cobaye mâle normal ou soumis à des injections de cortisone ou de métopirone. C. R. Acad. Sci. *266:* 2230–2233 (1968).

3 BERTLER, A. and ROSENGREN, E.: Occurrence and distribution of dopamine in brain and other tissues. Experientia *15:* 10–11 (1959).

4 BHARGAVA, K. P.; BHARGAVA, R., and GUPTA, M. D.: Central adrenoceptors concerned in the release of adrenocorticotrophic hormone. Brit. J. Pharmacol. *45:* 682–683 (1972).

5 BHATTACHARYA, A. N. and MARKS, B. H.: Reserpine and chlorpromazine-induced changes in hypothalamo-hypophyseal-adrenal system in rats in the presence and absence of hypothermia. J. Pharmacol. exp. Ther *165:* 108–116 (1969).

6 BHATTACHARYA, A. N. and MARKS, B. H.: Effects of pargyline and amphetamine upon acute stress response in rats. Proc. Soc. exp. Biol. Med. *130:* 1194–1198 (1969).

7 BHATTACHARYA, A. N. and MARKS, B. H.: Effects of α-methyltyrosine and *p*-chlorophenylalanine on the regulation of ACTH secretion. Neuroendocrinology *6:* 49–55 (1970).

8 BLISS, E. L. and ZWANZIGER, J.: Brain amines and emotional stress. J. psychiat. Res. *4:* 189–198 (1966).

9 BLISS, E. L.; AILION, J. and ZWANZIGER, J.: Metabolism of norepinephrine, serotonin and dopamine in rat brain with stress. J. Pharmacol. exp. Ther. *164:* 122–134 (1968).

10 BOISSIER, J. R. et SIMON, P.: De la potentialisation des effets de la DOPA par les inhibiteurs de la monoamineoxydase. Psychopharmacologia *8:* 428–436 (1966).

11 CARR, L. A. and MOORE, K. E.: Effects of reserpine and α-methyltyrosine on brain catecholamines and the pituitary-adrenal response to stress. Neuroendocrinology *3:* 285–302 (1968).

12 CORRODI, H.; FUXE, K., and HÖKFELT, T.: The effect of immobilization stress on the activity of central monoamine neurons. Life Sci. *7:* 107–112 (1968).

13 DALLMAN, M. F. and YATES, F. E.: Anatomical and functional mapping of central neural input and feedback pathways of the adrenocortical system. Mem. Soc. Endocrin. *17:* 39–71 (1968).

14 ENDROCZI, E.; SCHREIBERG, G., and LISSAK, K.: The role of central nervous activating and inhibitory structures in the control of pituitary-adrenocortical function. Effect of intracerebral cholinergic and adrenergic stimulation. Acta physiol. Acad. Sci. Hung. *24:* 211–221 (1963).

15 FRIEDMAN, A. H. and WALKER, C. A.: Circadian rhythms in rat mid-brain and caudate nucleus biogenic amine levels. J. Physiol., Lond. *197:* 77–85 (1968).

16 Fuxe, K. and Hökfelt, T.: The influence of central catecholamine neurons on the hormone secretion from the anterior and posterior pituitary; in Stutinsky Neurosecretion, pp. 165–177 (Springer, New York 1967).

17 Fuxe, K. and Hökfelt, T.: Catecholamines in the hypothalamus and the pituitary gland; in Ganong and Martini Frontiers in neuroendocrinology, 1969, pp. 47–96 (Oxford University Press, New York 1969).

18 Fuxe, K.; Hökfelt, T.; Jonsson, G., and Lidbrink, P.: Brain-endocrine interaction. Are some effects of ACTH and adrenocortical hormones on neuroendocrine regulation and behaviour mediated via central catecholamine neurons? Proc. 2nd Congr. of the Int. Soc. of Psychoneuroendocrinology (in press).

19 Fuxe, K.; Hökfelt, T.; Jonsson, G.; Levine, S.; Lidbrink, P., and Lofstrom, A.: Brain and pituitary-adrenal interactions: Studies on central monoamine neurons. Symp. on Brain-Pituitary-Adrenal Interrelationships, Cincinnati 1973.

20 Gann, D.; Schoeffler, J. D., and Ostrander, L.: A finite state model for the control of adrenal corticosteroid secretion; in Mesarovic Systems symposium, pp. 62–93 (Springer, New York 1968).

21 Ganong, W. F. and Lorenzen, L. C.: Brain neurohumors and endocrine function; in Martini and Ganong Neuroendocrinology, vol. 2, pp. 583–640 (Academic Press, New York 1967).

22 Ganong, W. F.: Evidence for a central noradrenergic system that inhibits ACTH secretion; in Knigge, Scott and Weindl Brain-endocrine interaction. Median eminence. Structure and function, pp. 254–266 (Karger, Basel 1972).

23 Ganong, W. F.: Pharmacological aspects of neuroendocrine integration. Progr. Brain Res. (in press).

24 Gaunt, R.; Renzi, A. A., and Chart, J. J.: Endocrine pharmacology of methyl reserpate derivatives. Endocrinology *71:* 527–535 (1962).

25 Gordon, R.; Spector, S.; Sjoerdsma, A., and Udenfriend, S.: Increased synthesis of norepinephrine and epinephrine in the intact rat during exercise and exposure to cold. J. Pharmacol. exp. Ther. *153:* 440–447 (1966).

26 Harwood, C. T. and Mason, J. W.: Acute effects of tranquilizing drugs on the anterior pituitary-ACTH mechanism. Endocrinology *60:* 239–246 (1957).

27 Hirsch, G. H. and Moore, K. E.: Brain catecholamines and the reserpine-induced stimulation of the pituitary-adrenal system. Neuroendocrinology *3:* 398–405 (1968).

28 Javoy, F.; Glowinski, J., and Kordon, C.: Effects of adrenalectomy on the turnover of norepinephrine in the rat brain. Europ. J. Pharmacol. *4:* 103–104 (1968).

29 Kaplanski, J.; Dorst, W., and Smelik, P. G.: Pituitary-adrenal activity and depletion of brain catecholamines after α-methyl-*p*-tyrosine administration. Europ. J. Pharmacol. *20:* 238–240 (1972).

30 Kaplanski, J. and Smelik, P. G.: Pituitary-adrenal activity and depletion of brain catecholamines after central administration of 6-hydroxy-dopamine. Res. Commun. chem. path. Pharmacol. *5:* 263–271 (1973).

31 KREIGER, D. T.: Lack of responsiveness to L-dopa in Cushing's disease. J. clin. Endocrin. *36:* 277 (1973).

32 LIPPA, A. S.; ANTELMAN, S. M.; FAHRINGER, E. E., and REDGATE, E. S.: Relationship between catecholamines and ACTH. Effects of 6-hydroxydopamine. Nature New Biol. *241:* 24–25 (1973).

33 MAICKEL, R. P.; WESTERMANN, E. O., and BRODIE, B. B.: Effects of reserpine and cold-exposure on pituitary-adrenocortical function in rats. J. Pharmacol. exp. Ther. *134:* 167–75 (1961).

34 MANGILI, G.; MOTTA, M., and MARTINI, L.: Control of adrenocorticotropic hormone secretion; in MARTINI and GANONG Neuroendocrinology, vol. 1, pp. 297–370 (Academic Press, New York 1966).

35 MARKS, B. H.; HALL, M. M., and BHATTACHARYA, A. N.: Psychopharmacological effects and pituitary-adrenal activity. Progr. Brain Res. *32:* 57–70 (1970).

36 MAYNERT, E. W. and LEVI, R.: Stress-induced release of brain norepinephrine and its inhibition by drugs. J. Pharmacol. exp. Ther. *143:* 90–95 (1964).

37 MCKINNEY, W. T., jr.; PRANGE, A. J., jr.; MAJCHOWICZ, E., and SCHLESINGER, K.: Plasma corticosterone changes following alterations in brain norepinephrine and serotonin. Dis. nerv. Syst. *32:* 308–313 (1971).

38 NAUMENKO, E. V.: Hypothalamic chemoreactive structures and the regulation of pituitary-adrenal function. Effects of local injections of norepinephrine, carbachol, and serotonin into the brain of guinea pigs with intact brains and after mesencephalic transection. Brain Res. *11:* 1–10 (1968).

39 NAUMENKO, E. V.; MASLOVA, L. N., and POPOVA, N. K.: Influence of amphetamine on the hypothalamic-pituitary-adrenal system of rats with deafferented hypothalamus. Biull. eksp. Biol. Med. U.S.S.R. (in press).

40 PETROVIC, V. M. and DAVIDOVIC, V.: Effects of hydrocortisone and ACTH feedback on catecholamines in the diencephalon in rats. J. Physiol., Paris *62:* suppl. 3, pp. 427–428 (1970).

41 PFEIFER, A. K.; VIZI, E.; SATORY, E., and GALAMBOS, E.: The effect of adrenalectomy on the norepinephrine and serotonin content of the brain and on reserpine action in rats. Experientia *19:* 482–483 (1963).

42 SCAPAGNINI, U.; VAN LOON, G. R.; MOBERG, G. P.; PREZIOSI, P., and GANONG, W. F.: Evidence for central norepinephrine-mediated inhibition of ACTH secretion in the rat. Neuroendocrinology *10:* 155–160 (1972).

43 SCAPAGNINI, U. and PREZIOSI, P.: Role of brain norepinephrine and serotonin in the tonic and phasic regulation of hypothalamic hypophyseal adrenal axis. Arch. int. Pharmacodyn. *196:* suppl., pp. 205–220 (1972).

44 SCHAEPDRYVER, A. F. DE et PREZIOSI, P.: Iproniazide et effets pharmacologiques sur la médullo-cortico-surrénale. Arch. int. Pharmacodyn. *119:* 506–510 (1959).

45 SCHAEPDRYVER, A. F. DE; PREZIOSI, P., and SCAPAGNINI, U.: Brain monoamines and adrenocortical activation. Brit. J. Pharmacol. *35:* 460–467 (1969).

46 SMELIK, P. G.: ACTH secretion after depletion of hypothalamic monoamines by reserpine implants. Neuroendocrinology *2:* 247–254 (1967).

47 STEINER, F. A.; RUF, K., and AKERT, K.: Steroid-sensitive neurones in rat brain. Anatomical localization and responses to neurohumors and ACTH. Brain Res. *12:* 74–85 (1969).

48 THIERRY, A. M.; FEKETE, M., and GLOWINSKI, J.: Effects of stress on the metabolism of noradrenaline, dopamine and serotonin in the central nervous system of the rat. I. Modifications of norepinephrine turnover. J. Pharmacol. exp. Ther. *163:* 163–171 (1968).

49 ULRICH, R. S. and YUWILER, A.: Failure of 6-hydroxydopamine to abolish the circadian rhythm of serum corticosterone. Endocrinology *92:* 611–614 (1973).

50 UTEVSKII, A. M.; GAISINSKAYA, M. Y.; RAISIN, M. S., and BRAUDE, I. Y.: Effect of adrenalectomy on the circulation of noradrenaline in rat brain and heart. Biull. eksp. Biol. Med. U.S.S.R. *70:* 1388–1389 (1971).

51 VAN LOON, G. R. and KRAGT, C. L.: Effect of dopamine on the biological activity and *in vitro* release of ACTH and FSH. Proc. Soc. exp. Biol. Med. *133:* 1137–1141 (1970).

52 VAN LOON, G. R. and KRAGT, C. L.: Neuroendocrine relationships. Progr. Neurol. Psychiat., vol. 26, pp. 261–310 (Grune & Stratton, New York 1971).

53 VAN LOON, G. R.; HILGER, L.; KING, A. B.; BORYCZKA, A. T., and GANONG, W. F.: Inhibitory effect of L-dihydroxyphenylalanine on the adrenal venous 17-hydroxycorticosteroid response to surgical stress. Endocrinology *88:* 1404–1414 (1971).

54 VAN LOON, G. R.; SCAPAGNINI, U.; COHEN, R., and GANONG, W. F.: Effect of intraventricular administration of adrenergic drugs on the adrenal venous 17-hydroxycorticosteroid response to surgical stress in the dog. Neuroendocrinology *8:* 257–272 (1971).

55 VAN LOON, G. R.; SCAPAGNINI, U.; MOBERG, G. P., and GANONG, W. F.: Evidence for central adrenergic neural inhibition of ACTH secretion in the rat. Endocrinology *89:* 1464–1469 (1971).

56 VAN LOON, G. R.; NICHOLSON, W., and BROWN, R.: Drug-induced hypersecretion of ACTH in rats and its treatment with L-dopa. Progr. 53rd Meet. of the Endocrine Soc., San Francisco 1971.

57 VAN LOON, G. R.: Brain catecholamines and ACTH secretion. In: Frontiers in neuroendocrinology, 1973, pp. 209–247 (Oxford University Press, New York 1973).

58 VERNIKOS-DANELLIS, J.: The parmacological approach to the study of the mechanisms regulating ACTH secretion; in BACK, MARTINI and PAOLETTI Pharmacology of hormonal polypeptides and proteins, pp. 175–189 (Plenum Press, New York 1968).

59 WERDER, K. VON; VAN LOON, G. R.; YATSU, F., and FORSHAM, P. H.: Corticosteroid and growth hormone secretion in patients treated with L-dopa. Klin. Wschr. *48:* 1454–1456 (1970).

60 WIED, D. DE: Chlorpromazine and endocrine function. Pharmacol. Rev. *19:* 251–288 (1967).

Author's address: Dr. G. R. VAN LOON, Room 3334, Medical Sciences Building, University of Toronto, *Toronto, Ontario M5S 1A8* (Canada)

Recent Studies of Hypothalamic Function
Int. Symp. Calgary 1973, pp. 114–133 (Karger, Basel 1974)

Human Growth Hormone and Somatomedin[1]

R. M. BALA
Division of Medicine, Faculty of Medicine, University of Calgary, Calgary

Control of Growth Hormone Secretion and Acromegaly

The concept of hypothalamic control of pituitary tropic hormone synthesis and secretion has been intensively investigated. Several major symposia have been published [25, 26, 44, 51, 52] and other authors in this symposium have summarized recent data. Growth hormone (GH) secretion and synthesis is presumed to be stimulated by GH-releasing factor (GRF). The ventromedial nucleus in the hypothalamus has been reported to be a site of production of GRF [40, 43]. A dual control of GH secretion, by GRF and by GH-inhibiting factor (GIF) has been postulated. The hypothalamic-hypophysiotropic area is presumed to be the final common pathway in control of pituitary GH secretion by GRF and GIF. It is also presumed that modulating influences on the hypothalamic-hypophysiotropic neurosecretory neurons occur from extra hypothalamic areas such as the limbic system and the 'primitive' brain. Neurotransmitters such as norepinephrine and dopamine are presumed to be relevant in synaptic mediation relevant to GRF secretion [55]. L-dopa has been shown to increase GH release [14, 16, 49, 53, 58, 75] while chlorpromazine decreases GH release [31, 39, 64]. The adrenergic hypothalamic receptors have been postulated to be important in GRF control, α-blockade will decrease whereas β-blockade will increase GH release in response to hypoglycemia [12, 15, 38].

1 Supported by the Medical Research Council of Canada Grant No. MA-4107.

Various tests have been devised for clinical assessment of human pituitary functional status by assessing the amount of HGH released after various provocative stimuli. These stimuli include a decrease in blood glucose by insulin [57] or after glucose ingestion [69], amino acids such as arginine [47], various proteins given orally [36, 70], glucagon [48, 54, 66], L-dopa [14, 16, 49, 53, 58, 75], exercise [63], sleep [27] and others [5]. The most widely accepted tests, until recently, included insulin hypoglycemia (ITT) and arginine monohydrochloride infusion intravenously (APT). Hypoglycemia presumably causes release of GRF [37] which has been identified in the human hypothalamus [62]. The arginine provocative tests for HGH release is less effective in adult males without prior estrogen treatment [47]. The mechanism of arginine stimulation of GH release is speculative, with some suggestions that the response to arginine is non specific and occurs due to the stress of venopuncture alone [11, 15, 65, 67]. Sleep-induced release of HGH cannot be suppressed by infusion of glucose suggesting that extra hypothalamic brain areas modulate HGH release [28, 43].

In acromegaly, the pituitary tumor has been assumed to be autonomous in its secretion and therefore serum HGH levels should not be suppressed by glucose ingestion or increased by various substances used to stimulate release of HGH in normal subjects. Many patients with acromegaly have normal basal levels of pituitary tropic hormones other than HGH, as well as normal reserves of ACTH. This suggests that a variable number of normal somatotropes is present.

We have studied the response of serum immunoreactive HGH (IR-HGH) in 15 acromegalic patients using glucose, ITT, arginine or glucagon. Nine patients were studied before treatment while 6 patients had been previously treated with either radiation or pituitary surgery. Some patients were studied at intervals after treatment as indicated in figure 1. Patients with levels of serum IR-HGH greater than 5 ng/ml which did not decrease after hyperglycemia were designated as *active* acromegaly. Serum IR-HGH levels were measured by radioimmunoassay [3] at –30, 0, 30, 60, 90 and 180 min (pituitary HGH kindly supplied by National Institute of Health, Bethesda). An indwelling venous catheter was inserted at approximatively 45 min prior to test substance administration and was kept open with physiological saline. Blood samples were obtained from the venous catheter and the serum was kept frozen at –20 °C until processed. Patients who were studied were kept fasting and in bed preceding and during the test. Glucose (75 g) was given orally at time 0

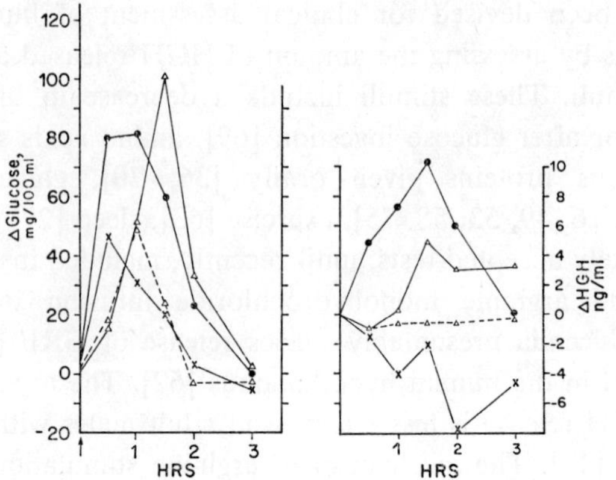

Fig. 1. The mean change in serum glucose and immunoreactive human growth hormone (IR-HGH) for each group of acromegalic patients is shown after 75 g of glucose were given orally to acromegalic patients (number of patients in each group in brackets). Three patients in the pretreatment group were restudied in the postradiation *active* group at least 1 year postradiation. Two postradiation *active* group patients were restudied in the postradiation *inactive* group at a later date.

GTT in acromegaly	Basal level	
	HGH	glucose
Pretreatment (9) (●—●)	29	92
Postradiation *'active'* (8) (×—×)	30	91
Postradiation *'inactive'* (3) (△---△)	2.7	73
Posthypophysectomy *'active'* (4) (△—△)	19	73

GTT = glucose tolerance test; HGH = human growth hormone.

over a 5-min period. Arginine monohydrochloride (R-Gene, Cutter), 0.5 g/kg (maximum 30 g) was given intravenously from time 0 to 30 min. Crystalline insulin was given as a bolus injection at time 0, in a dose of 0.15 U/kg of body weight to acromegalic patients judged *active* from clinical criteria, while 0.10 U/kg was given to other patients. One milligram of glucagon was given subcutaneously at time 0.

Five of the 12 acromegalic patients had abnormal glucose tolerance. The mean increase in plasma glucose for the various groups is shown in figure 1. The pretreatment and the *active* posthypophysectomy acromegalic patient groups showed a paradoxical increase in the mean level of IR-HGH after glucose ingestion. Two of the *active* posthypophysectomy, acromegalic patients presumably had a functional pituitary stalk since they showed normal ACTH reserve after testing with metopirone. One patient in this group, who had a transfrontal hypophysectomy showed no response in serum IR-HGH after glucose ingestion. The *active* postradiation acromegalic patients showed less glucose intolerance and showed partial suppression of serum IR-HGH after glucose ingestion, in spite of high basal serum IR-HGH levels comparable to the other *active* groups. The *inactive* acromegalic patients did not show a significant response of serum IR-HGH after glucose ingestion.

The mean decrease in serum glucose of the various groups in response to insulin is shown in figure 2. All patients showed glucose decrease greater than 25 mg/100 ml serum. The *active* acromegalic patient treated by transfrontal hypophysectomy did not show a significant IR-HGH response, whereas all other groups showed an increase in the mean serum IR-HGH after hypoglycemia. The pretreatment group showed a greater rise in serum IR-HGH than the postradiation group. However, all groups revealed a subnormal response.

The mean increase in serum IR-HGH in response to arginine infusion was greater in 3 untreated than in 2 postradiation *active* acromegalic patients. After glucagon, a substantial rise in serum IR-HGH occurred in 2 untreated acromegalic patients whereas no change occurred in one *active* acromegalic patient treated by transfrontal hypophysectomy.

If the secretion of GH by a pituitary tumor is autonomous in acromegaly, then various changes in response to glucose or stimulation tests are most likely due to an influence on the hypothalamic control of HGH secretion by nontumor somatotropes. This latter assumption is supported by lack of change in serum IR-HGH to stimulation or suppression tests in the patient with *active* acromegaly after a transfrontal hypophysectomy (*disconnected* pituitary). The paradoxical rise in serum IR-HGH in response to hyperglycemia, in acromegaly, has been described by others and is similar to that observed in neonatal and premature infants [17, 41, 50]. Increased levels of serum HGH have been shown to decrease normal pituitary secretion of HGH [1].

In untreated acromegaly it is possible that high levels of GRF secretion by the hypothalamus occurs and is not suppressed by the high levels of serum IR-HGH or glucose, unless a deficient GIF response is postulated [32]. The suppression of serum IR-HGH by hyperglycemia in the post-

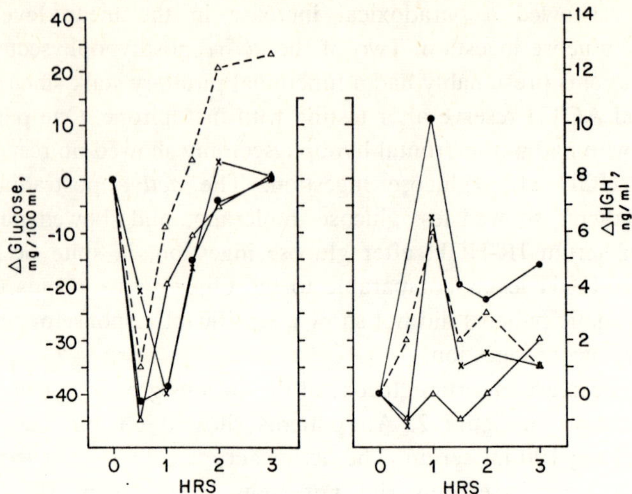

Fig. 2. The mean changes in blood glucose and IR-HGH for each group of acromegalic patients are shown (number of patients in brackets). Crystalline insulin was given as a bolus injection intravenously at time O; 0.15 U/kg to acromegalic patients designated as *active* on the basis of clinical criteria, while other patients received 0.10 U/kg.

ITT acromegaly	Basal level	
	HGH	glucose
Pretreatment (4) (●—●)	22	102
Postradiation *'active'* (4) (×—×)	17	88
Postradiation *'inactive'* (1) (△---△)	4	69
Posthypophysectomy *'active'* (1) (△—△)	11	103

ITT = insulin hypoglycemia; HGH = human growth hormone.

radiaction *active* group, associated with lesser glucose intolerance, might suggest more efficient response to glucose by the hypothalamic receptors, since altered IR-HGH responses to glucose also have been observed in diabetes mellitus and uremia [35, 61]. The high level of GRF secretion in acromegaly can be further increased by hypoglycemia, arginine and glucagon as reflected by increased secretion of HGH by nontumor somatotropes. A rise in IR-HGH in response to hypoglycemia and arginine was less in the postradiated *active* acromegalic group compared to the untreated group, suggesting either a diminished ability of the hypothalamus to increase GRF secretion because of radiation injury to the hypothalamus or alternatively a decreased nontumor somatotrope response after pituitary radiation. The similar serum IR-HGH response shown by the postradiation *active* and *inactive* groups, to hypoglycemia, in spite of markedly differing basal levels of IR-HGH suggests that the high level of serum IR-HGH in acromegaly may not suppress hypothalamic GRF secretion.

Summary

Changes of serum IR-HGH in response to suppression or stimulation tests before and after treatment in acromegaly support the hypothesis that acromegaly is initially a hypothalamic disease associated with excessive GRF secretion. Further studies of IR-HGH responses to sleep, L-dopa, and chlorpromazine before and after treatment of acromegaly are required.

Molecular Size of Plasma HGH

The molecular size of immunoreactive HGH (IR-HGH) in plasma has been subject to controversy. Earlier studies by BERSON and YALOW [9] and BOUCHER [3] suggested that IR-HGH and extracted pituitary HGH (epHGH) were similar in electrophoretic mobility and size, respectively. In 1969 [2, 4, 5] and 1970 [6] we first reported the presence of IR-HGH in plasma, in at least 3 different molecular size species.

Large serum samples, up to 350 ml, from 3 normal volunteers and from 15 acromegalic patients were fractionated on large columns of Sephadex G-75 as described [6]. The column effluent fractions were concentrated by Diaflo (Amicon) ultrafiltration. The IR-HGH in the concentrated fractions was determined by a solid phase radioimmunoassay [3]. More than 50 % of IR-HGH was eluted before distribution coefficient or Kd 0.25, in a molecular size range greater than epHGH. The fractions with the greatest potency of IR-HGH per milligram protein were eluted in the epHGH molecular size area corresponding to a Kd interval

$$(\mathrm{Kd} = \frac{V_e - V_o}{V_i})$$

of 0.25–0.35. A greater relative amount of large IR-HGH was eluted, before Kd 0.25, in the normal as compared to acromegalic patient plasmas. This may suggest a different form of IR-HGH in the acromegalic patient plasma or less aggregation as the levels of IR-HGH rise in plasma. The relatively larger amount of the large IR-HGH found after lyophilization suggested that lyophilization increases adsorption of IR-HGH to other plasma proteins or aggregation. The presence of IR-HGH in plasma of a much smaller molecular size than that of epHGH was not an artifact of various salt concentrations [7].

We further studied the molecular size of IR-HGH in plasma from normals as well as patients with Laron dwarfism (LD) and hypopituitarism associated with gigantism (HPG) [8]. Large volumes of serum were similarly fractionated on Sephadex G-200. IR-HGH was again found in at least 3 molecular size forms, that similar to epHGH as well as a larger and a smaller molecular size in all plasmas studied, as shown in figure 3. The maximal total amount of IR-HGH per 0.10 distribution coefficient (Kav) interval in normal plasma occurred in the 0.65 to 0.75 Kav interval, corresponding to a molecular size of 25,000–17,000. Again, a large proportion of IR-HGH was eluted in the molecular size range greater than epHGH. The distribution of IR-HGH in LD plasma was not different from normal other than a slightly lesser amount of IR-HGH occurring

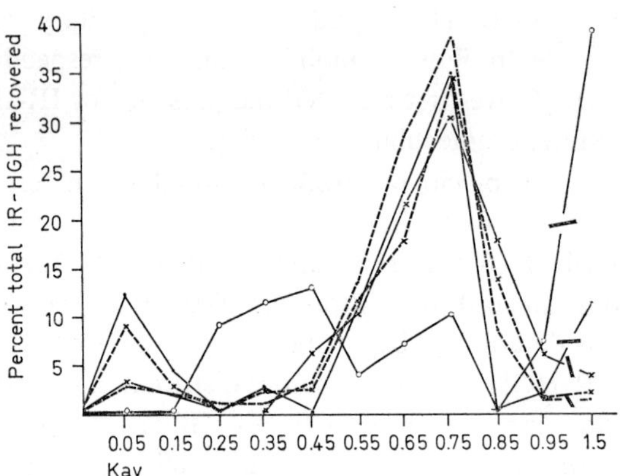

Fig. 3. Fractionation of various plasmas as designated by gel filtration on Sephadex G-200. The total IR-HGH recovered per distribution coefficient Kav interval is shown as percentage of the total IR-HGH recovered in each experiment with the different samples as noted, and plotted at the end of the Kav interval. After distribution coefficient Kd 0.95 the total fractions were pooled up to at least Kav 1.5. ●—● = normal pool; ●---● = extraced pituitary human growth hormone (epHGH) and plasma; ×—× = Laron, female; ×---× = Laron, male; o—o = fractional hypopituitarism.

with a molecular size smaller than epHGH. The HPG plasma IR-HGH differed from the normal and LD plasma in revealing a peak in the 50,000–120,000 (Kav 0.25–0.45) and in the very small molecular size area (Kav greater than 0.85). The difference in the relative amounts of very small IR-HGH in LD and HPG plasma may imply that the very small IR-HGH is important in the induction of HGH-dependent biological growth factors found in plasma. More than one molecular size form of HGH may be secreted by the pituitary. Extraction procedures used in preparation of epHGH could disrupt or alter a larger unit structure of endogenous pituitary HGH. Alternatively, the appearance of large IR-HGH after addition of rechromatographed epHGH to plasma (fig. 3), and the relatively greater amounts of larger IR-HGH in acromegalic plasma after lyophilization [6], would suggest the possibility of aggregation or adsorption of IR-HGH to other plasma proteins. Fractionation of plasmas obtained from hypopituitary patients, at varying time-periods after injection of epHGH, revealed increasing relative proportions of very small and large size IR-HGH with time. The very small molecular size IR-HGH in plasma may represent metabolic fragments of HGH which have an immunoreactive site intact. STUART *et al.* [68] have shown that a lesser proportion of the total IR-HGH in plasma occurs in the molecular size area smaller and larger than epHGH after stimulation of endogenous HGH secretion, than occurs in the basal state.

Recently, others [10, 29, 30] have also reported studies showing the presence of *big* IR-HGH in plasma in a larger molecular size form than epHGH. BERSON and YALOW [10] found up to 25 % of the total IR-HGH, present in normal and acromegalic plasma, in a molecular size form greater than albumin. GOODMAN *et al.* [29] found that approximately 35 % of plasma IR-HGH existed in a *big* form, with a molecular size twice as large as epHGH, in normal, acromegalic and LD plasma. They also found that freezing and storage resulted in conversion of the big IR-HGH to a form corresponding to the molecular size of the major component of epHGH which they have designated as *little* HGH. A corresponding increase in both the *big* and *little* forms in plasma occurred after stimulation of endogenous release of HGH. They also found *big* and *little* HGH in fractionation studies of pituitary extracts. GORDON *et al.* [30], in fractionation studies similar to GOODMAN *et al.,* have also shown two peaks of IR-HGH in plasma. They found up to 30 and 49 % of the total IR-HGH in the big form, in nonlyophilized and lyophilized normal plasma, respectively. Significant amounts of IR-HGH in the molecular size larger than the big IR-HGH occurred in some of their fractionations, but they have ignored this finding. These studies [10, 29, 30] agree with our findings of a greater amount of IR-HGH in a larger molecular size in normal plasma compared to acromegalic patient plasma. All these recent studies agree in the findings of IR-HGH in plasma in a molecular size similar to epHGH as well as in a larger molecular size form. However, the differences in regard to the molecular size of the large IR-HGH are still unresolved. GOODMAN *et al.* and GORDEN *et al.* used small (1–7 ml) serum samples. We chose to use larger volumes of plasma for fractionation, followed by concentration of the eluted proteins to allow detection of IR-HGH present in low concentration, in areas other than the molecular size area of epHGH, which otherwise might not be detected. However, the use of the large sample volumes may have resulted in some zone spreading on Sephadex, even though the sample volume was less than 5 % of the imbibed volume (Vi). It is possible that others have detected relatively less of the total IR-HGH in the large molecular size area due to the small starting samples, or have chosen to ignore it. Moreover, it is possible that some of the differences in these studies are due to the various types of radioimmunoassays used. We have avoided the use of the charcoal-dextran separation procedure in the radioimmunoassay for IR-HGH since increasing protein concentrations markedly decrease the concentration of IR-HGH detected. Additionally, some of the differences may be due to serum samples from persons after stimulation of release of endogenous IR-HGH as noted by STUART *et al.* [68]. Some of the differences in our own studies would tend to support this, in that in our initial studies [8] which revealed a lesser proportion of the large IR-HGH, we used nonfasting patient serum samples. Lypophilization may have resulted in insolubility of the larger IR-HGH and decreased the amount measured. Alternatively, the markedly different elution buffer conditions may be the main reason for the difference in the results obtained. The large IR-HGH may represent a separate pituitary GH or a *pro*-GH [29] whereas the very large IR-HGH may represent adsorption of IR-HGH to plasma proteins. Alternatively, and more likely the larger forms of IR-HGH represent polymers of HGH in plasma.

Our findings of IR-HGH of smaller molecular size than epHGH have only been noted by STUART *et al.* [68]. Small sample volumes might not have allowed detection in other studies. Concentration using dialysis tubing may have resulted in losses of this small molecular size IR-HGH. Lyophilization may have resulted in significant losses of the small molecules due to adsorption to glass or adsorption to other proteins. Alternatively, the very small IR-HGH may not have been fully eluted from the columns.

Summary

The existence of IR-HGH in plasma in a molecular size form similar to epHGH and a large molecular size form has been reasonably confirmed. The presence of IR-HGH in plasma in a molecular size form smaller than that of epHGH has been reported in several studies [6, 68]. The biological significance of the various molecular size forms of IR-HGH in plasma requires further study.

Serum Somatomedin

GH lacks biological activity *in vitro* in physiological doses [19]. Normal serum contains biological growth factors (BGF), which are active *in vitro,* as measured by stimulation of incorporation of radioactive sulfate [59], thymidine [20], uridine [60] and proline [18], into chondroitin sulfate, DNA, RNA and hydroxyproline, respectively, by cartilage from hypophysectomized rats. This activity in plasma is GH-dependent and in general reflects the levels of plasma IRGH. These BGF in normal plasma which stimulate sulfate and thymidine incorporation by cartilage cells *in vitro* have been designated as sulfation factor (SF) and thymidine factor (TF), respectively. Both SF and TF activity reflect synthesis of protein in protein-polysaccharide complexes in cartilage [60]. TF stimulates protein synthesis in other cells [24]. SF and TF activity in plasma fractions seems to occur in parallel and the term 'somatomedin' (SM) has been recently designated to describe both SF and TF [23]. In GH deficiency, the plasma SF is low, and can be restored to normal, following a time lag, after *in vitro* injection of GH [19]. The SF activity persists for a longer time in plasma than does IRGH. This suggests that GH induces the production of SF by some tissue or organ, followed by secretion of SF into plasma and subsequent transport to cartilage and other tissues, where SF acts as a BGF. SF production can be induced by substances other than GH, such as the worm factors produced by certain tape worms [22]. Discrepancies between plasma levels of IR-HGH and BGF in patients with LD, HPG and occasionally in craniopharyngioma patients postoperatively have been noted [21, 76]. Hall [34] has published an excellent review of somatomedin studies to date.

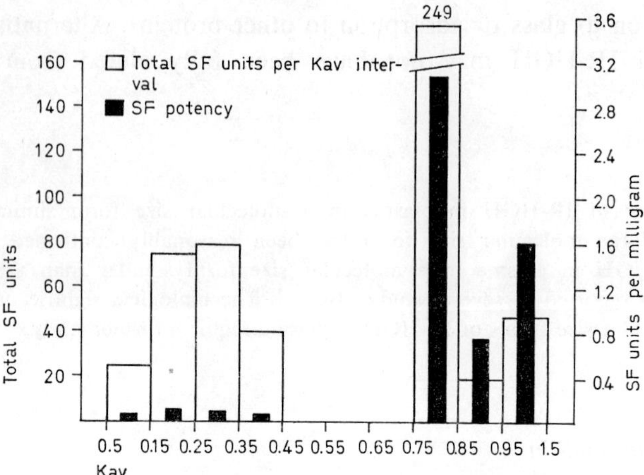

Fig. 4. Fractionation of normal serum by gel filtration on Sephadex G-200. The total sulfation factor (SF) units per Kav interval are shown by the open bars, the SF potency (SF units per milligram protein) is shown by the solid bars.

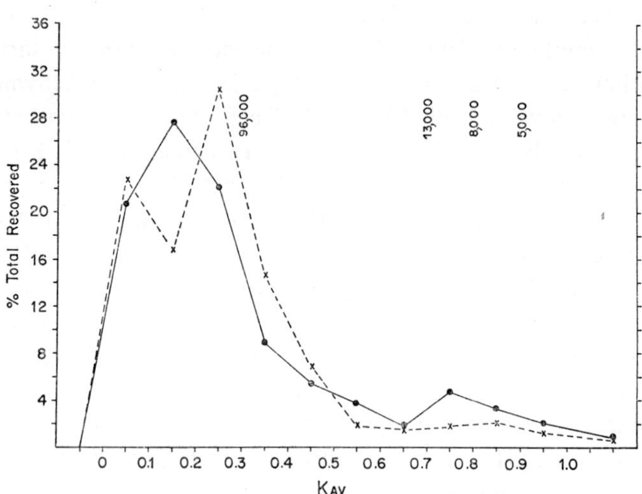

Fig. 5. Fractionation of serum from 12 normal and acromegalic patients on Sephadex G-150, using 0.05 M phosphate buffer, pH 7.4. The mean percentage of the total SF and thymidine factor (TF) recovered in each Kav interval is plotted. Approximate molecular size ranges are indicated. ●—● = SF; ×---× = TF.

In 1969 [2, 4] and 1970 [6] we first reported fractionation studies of normal and acromegalic patient plasmas with quantitation of SF in various protein fractions after Sephadex gel filtration. These studies were carried out, in parallel, on the same samples described above in the study of molecular size distribution of IR-HGH in plasma [6]. The greatest relative amount of SF appeared in the large molecular size range greater than 10,000 with a second peak, of greater potency, in the molecular size range less than 8,000 Daltons. Acromegalic patient plasma contained a greater relative total amount of more potent SF, compared to normal plasma, in the large molecular size area. SF and IR-HGH, total activity and potency, revealed some degree of parallelism. However, IR-HGH could not be identical to SF, since epHGH was shown to have no SF activity *in vitro,* nor could SF activity in plasma be inhibited by addition of HGH antisera. Insulin has been shown to stimulate sulfate incorporation into chondroitin sulfate. Measurement of immunoreactive insulin (IRI) and insulin-like activity (ILA) in the small molecular weight fractions revealed that only a part of the total SF activity could be accounted for by IRI or ILA activity. In these initial studies we were primarily interested in evaluating SF in the 5,000 to 50,000 molecular-size range. We subsequently fractionated large volumes of sera obtained from normals and patients with LD and HPG, using Sephadex G-200 gel filtration using slightly different conditions [8]. As shown in figure 4, two peaks of SF activity were found, one in molecular size range smaller than 17,000 and another greater than 55,000 Daltons. These studies revealed a greater amount and potency of SF in the smaller molecular size range when compared to our previous studies [6]. This difference may have occurred due to the use of lyophilization, in the later study [8], resulting in lesser solubility of the large molecular size proteins, as reflected by only 27 % recovery of SF after fractionation. In subsequent studies we have avoided lyophilization on the first fractionation of plasma. Also, we have used 0.05 M phosphate buffer eluant at pH 7.4, rather than the 0.15 M NH_4HCO_3 buffer at pH 8.2 as used in the previous studies. 12 further large samples of sera from normal and acromegalic patients have been fractionated using Sephadex G-150. SF and TF were assayed separately using an *in vitro* bioassay as previously described [6] with modifications including use of Medium 199, as described by ESANU *et al.* [24] and we have further increased serine, glutamine and sulfate concentrations. As shown in figure 5, at least two peaks of SF and TF activity are present, in two molecular size ranges,

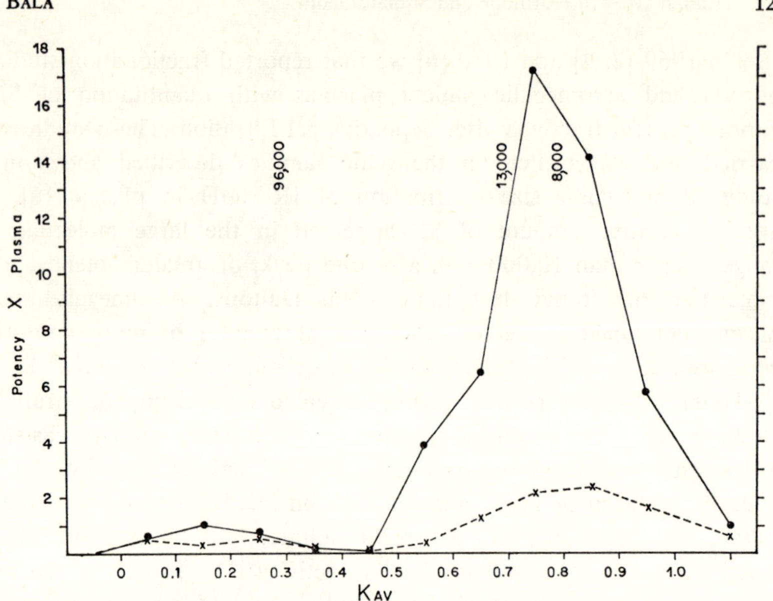

Fig. 6. The potency of serum fractions, as shown in figure 5, is shown in terms of SF (●—●) and TF (×--- ×) relative to a reference serum pool. Approximate molecular size ranges are indicated. Sephadex G-150.

one greater and one smaller than 17,000. The finding of SF and TF total activity mainly in the large molecular size range, again, may be due to the buffer conditions used and nonlyophilization with greater than 50 % recovery of SF. As shown in figure 6, the SF and TF potency of the eluted proteins was greatest in the small molecular size fractions. TF appeared to be reflected by only a slightly smaller molecular size as compared to SF. Recoveries and potencies of TF in the plasma fractions were lower as compared with SF. Freezing of serum samples prior to fractionation resulted in an increase of the small molecular size form of SF and TF. Refractionation of the most potent SF and TF containing fractions of molecular size smaller than 35,000 resulted in a further increase in potency by more than 500 times. Both the SF and TF with maximal potency occurring in the molecular size range near 5,000. These potent SF and TF substances were still highly nonhomogeneous on starch gel electrophoresis. Further purification to achieve homogeneity will be reported elsewhere.

Others [34, 72, 73] have also studied SF and TF in human plasma after various fractionation procedures. LIBERTI [42] reported that SF in bovine plasma was approximately 5,000 in molecular weight. VAN DEN BRANDE *et al.* [72] suggested a molecular size greater than 21,000 but subsequently have revised this to 8,000. HALL *et al.* [34] and VAN WYK *et al.* [73] have succeeded in isolating and purifying the small molecular size form of SM from pre-extracted plasma. In these studies recoveries of SM as low as 10 %, compared with the original plasma, suggest that a substantial amount of SM of larger molecular size may have been lost in the extraction procedures followed by gel filtration and lyophilization. These studies [34, 72, 73] have also shown that SM occurs in a larger molecular size form in unextracted plasma with predominantly a small form found in pre-extracted plasmas. It is likely that SM exists in various molecular size forms in plasma, either as different substances or adsorbed to other plasma proteins. Aggregation or polymerization of SM molecules is possible. Alternatively, the small SM in plasma may be a subunit component of the large SM molecule found in the large molecular size area. It is likely that the various pre-extraction procedures are either dissociating SM-protein complexes or fracturing the SM molecule into smaller unit structures. We are studying the large molecular size SM further in this regard. The imminent development of a radioimmunoassay for the small molecular size SM will greatly expedite further studies of SM. It is highly probable that multiple BGF exist in plasma in various molecular sizes other than the small molecular size form of SM. These BGF may have different relationships to various endogenous hormones that have been shown to stimulate biological growth in various tissues. Preliminary evidence suggests that SM may account for part of the nonsuppressible insulin-like activity (NSILA) in plasma and that SM may compete with insulin for binding with the insulin receptor [56, 74].

The site of production of SM in the body is a subject of increasing study. HALL *et al.* [33] demonstrated increased SF activity in media after muscle incubation, after addition of HGH *in vitro,* but this has not been further confirmed. MCCONAGHEY and SLEDGE [45] reported an increase in SF in medium perfused through rat liver when GH was added to the perfusate. This suggests that the liver is a site of production of SF and that the production of SF is stimulated by GH, as further supported by studies of UTHNE and UTHNE [71]. In preliminary studies we have found that SM activity in plasma from patients with severe renal insufficiency is decreased and is not restored to normal by hemodialysis, whereas the SF activity is increased after renal transplantation. This would suggest that the low SF was not due to retention of metabolites in plasma and may suggest that the kidney participates in production of SF. The kidney has been shown to contain SF activity [46].

Summary

SM is a BGF in plasma. It is likely that GH and perhaps other hormones stimulate the liver and possibly other tissues to produce and secrete SM. In plasma, SM appears in at least several molecular size forms. The most potent SM in plasma appears to have a molecular size similar to or smaller than insulin. SM in plasma then acts as a BGF in most body tissues, especially cartilages, where it primarily stimulates protein synthesis. The concentrations of SM in plasma are in the range similar for most other peptide hormones. It is unlikely that SM acts through a feedback mechanism to regulate HGH secretion directly or through GRF, since it does not fluctuate acutely in plasma. It is more likely that SM serves, as an intermediate longer-acting hormone, in effecting GH actions particularly as related to cartilage growth and protein synthesis. SM may, however, influence the overall GH secretion, by modulating the response of the hypothalamus in GRF release to various stimuli.

Acknowledgements

The studies on the molecular size on HGH in plasma were carried out in collaboration with Dr. K. A. FERGUSON and Dr. J. C. BECK, McGill University Clinic, Montreal. Dr. H. FRIESEN, McGill University Clinic, kindly carried out some of the radioimmunoassays. Dr. JOHN DUPRÉ, McGill University Clinic, kindly carried out the studies on IRI and ILA. The author is much indebted to D. ANDERSON, D. YUNG, D. BIRNIE and J. DE ROSENROLL for skilled technical assistance in the other studies reported here and in preparation of this manuscript.

References

1 ABRAMS, R. L.; GRUMBACH, M. M., and KAPLAN, S. L.: The effect of administration of human growth hormone on the plasma growth hormone, cortisol, glucose and free fatty acid response to insulin. Evidence for growth hormone autoregulation in man. J. clin. Invest. *50:* 940 (1971).

2 BALA, R. M.; FERGUSON, K. A., and BECK, J. C.: Human growth hormone in plasma. J. clin. Res. *18:* 641 (1969).

3 BALA, R. M.; FERGUSON, K. A., and BECK, J. C.: Modified solid phase (tube) radioimmunoassay of human growth hormone. Canad. J. Physiol. Pharmacol. *47:* 803 (1969).

4 BALA, R. M.; FERGUSON, K. A., and BECK, J. C.: Plasma biological HGH dependent and IR-HGH like activity. 51st meeting of the endocrine society, New York 1969.

5 BALA, R. M.; BURGUS, R.; FERGUSON, K. A.; GUILLEMIN, R.; KUDO, C. F.; OLIVER, G. C.; RODGER, N. W., and BECK, J. C.: Control of growth hormone secretion; in MARTINI, MOTTA and FRASCHINI The hypothalamus, pp. 401–448 (Academic Press, New York 1970).

6 BALA, R. M.; FERGUSON, K. A., and BECK, J. C.: Plasma biological and immunoreactive human growth hormone-like activity. Endocrinology *87:* 506 (1970).

7 BALA, R. M. and BECK, J. C.: Human growth hormone in urine. J. clin. Endocrin. *33:* 799 (1971).

8 BALA, R. M. and BECK, J. C.: Human growth hormone in plasma. Ann. Roy. Coll. Physicians Surg. Canada *6* (1): 15 (1973).

9 BERSON, S. A. and YALOW, R. S.: State of human growth hormone in plasma and changes in stored solutions of pituitary growth hormone. J. biol. Chem. *241:* 5745 (1966).

10 BERSON, S. A. and YALOW, R. S.: State of human growth hormone in plasma and changes in stored solutions of pituitary growth hormone. Proc. 11th Reunion of the French-Speaking Endocrinologists, pp. 105–135 (Masson, Paris 1971).

11 BEST, J.; CATT, K. J., and BURGER, H. G.: Non-specificity of arginine infusion as a test for growth hormone secretion. Lancet *ii:* 124 (1968).

12 BLACKARD, W. G. and HEIDINGSFELDER, S. A.: Adrenergic receptor control mechanism for growth hormone secretion. J. clin. Invest. *47:* 1407 (1968).

13 BOUCHER, R. J.: The molecular weight of radioimmunoassayable growth hormone in human serum. J. Endocrin. *42:* 153 (1968).

14 BOYD, A. D.; LEBOVITZ, H. E., and PFEIFFER, J. B.: Stimulation of human growth hormone secretion by L-dopa. New Engl. J. Med. *283:* 1425 (1970).

15 CAVAGNINI, F. and PERACCHI, M.: Effect of reserpine on growth hormone response to insulin hypoglycaemia and to arginine infusion in normal subjects and hyperthyroid patients. J. Endocrin. *51:* 651 (1971).

16 CAVAGNINI, F.; PERACCHI, M.; SCOTTI, G.; RAGGI, U.; PONTIROLI, A. E., and BANA, R.: Effect of L-dopa administration on growth hormone secretion in normal subjects and parkinsonian patients. J. Endocrin. *54:* 424–433 (1972).

17 CRYER, P. E. and DAUGHADAY, W. H.: Regulation of growth hormone secretion in acromegaly. J. clin. Endocrin. *31:* 502 (1970).

18 DAUGHADAY, W. H. and MARIZ, I. K.: Conversion of proline-U-C^{14} to labeled hydroxyproline by rat cartilage *in vitro*. Effects of hypophysectomy, growth hormone and cortisol. J. Lab. clin. Med. *59:* 741 (1962).

19 DAUGHADAY, W. H. and KIPNIS, D. M.: The growth promoting and anti-insulin actions of somatotropin. Recent Progr. Hormone Res. *22:* 49 (1966).

20 DAUGHADAY, W. H. and REEDER, C.: Synchronous activation of DNA synthesis in hypophysectomized rat cartilage by growth hormone. J. Lab. clin. Med. *68:* 357 (1966).

21 DAUGHADAY, W. H.; LARON, Z.; PERTZELAN, A., and HEINS, J. N.: Defective sulfation factor generation. A possible etiological link in dwarfism. Trans. Ass. amer. Physicians *82:* 129 (1969).

22 DAUGHADAY, W. H.: Sulfation factor regulation of skeletal growth. Amer. J. Med. *50:* 277 (1971).

23 DAUGHADAY, W. H.; HALL, K.; RABEN, M. S.; SALMON, W. D.; VANDENBRANDE, J. L., and VANWYK, J. J.: Somatomedin. Proposed designation for sulphation factor. Nature, Lond. *235:* 107 (1972).

24 ESANU, C.; MURAKAWA, S.; BRAY, G. A., and RABEN, M. S.: DNA synthesis in human adipose tissue *in vitro.* I. Effect of serum and hormones. J. clin. Endocrin. *29:* 1027 (1969).

25 GANONG, W. F. and MARTINI, L. (eds.): Frontiers in neuroendocrinology (Oxford University Press, New York 1969).

26 GANONG, W. F. and MARTINI, L. (eds.): Frontiers in neuroendocrinology (Oxford University Press, New York 1971).

27 GLICK, S. M.: Normal and abnormal secretion of growth hormone. Ann. N.Y. Acad. Sci. *148:* 471 (1968).

28 GLICK, S. M. and GOLDSMITH, S.: The physiology of growth hormone secretion; in PECILE and MULLER Growth hormone, pp. 84–88 (Excerpta Medica, Amsterdam 1968).

29 GOODMAN, A. D.; TANENBAUM, R., and RABINOWITZ, D.: Existence of two forms of immunoreactive growth hormone in human plasma. J. clin. Endocrin. *35:* 868 (1972).

30 GORDON, P.; HENDRICKS, C. M., and ROTH, J.: Evidence for 'big' and 'little' components of human plasma and pituitary growth hormone. J. clin. Endocrin. *36:* 178 (1973).

31 GRUENER, R. and NARAHASHI, T.: The mechanism of excitability blockade by chlorpromazine. J. Pharmacol. exp. Ther. *181:* 161 (1972).

32 HAGEN, T. C.; LAWRENCE, A. M., and KIRSTEINS, L.: *In vitro* release of monkey pituitary growth hormone by acromegalic plasma. J. clin. Endocrin. *33:* 448 (1971).

33 HALL, K. A.; HOLMGREN, H., and LINDAHL, U.: Purification of a sulphation factor from skeletal muscle of rat. Biochim. biophys. Acta *201:* 39 (1970).

34 HALL, K.: Partial purification of sulphation factor and thymidine factor from plasma. Acta endocrin., Kbh. *70:* suppl. 163, pp. 1–52 (1972).

35 HANSEN, A. P.: Serum growth hormone patterns in female juvenile diabetics. J. Clin. Endocrinol. Metab. *36:* 638 (1973).

36 JACKSON, D.; GRANT, D. B., and CLAYTON, B. E.: A simple oral test of growth hormone secretion in children. Lancet *ii:* 373 (1968).

37 KATZ, S. H.; DHARIWAL, A. P. S., and MCCANN, S. M.: Effect of hypoglycemia on the content of pituitary growth hormone (GH) and hypothalamic growth hormone-releasing factor (GHRF) in the rat. Endocrinology *81:* 333 (1967).

38 KEBABIAN, J. W. and GREENGARD, P.: Dopamine-sensitive adenyl cyclase. Role in synaptic transmission. Science. *174:* 1346 (1971).

39 Kolodny, H. D.; Sherman, L.; Singh, A.; Sooseng, K., and Benjamin, F.: Acromegaly treated with chlorpromazine. New Engl. J. Med. *284:* 819 (1971).

40 Krulich, L.; Illner, P.; Fawcett, C. P.; Quijada, M., and McCann, S. M.: Dual hypothalamic regulation of growth hormone secretion. Growth and growth hormone (Excerpta Medica, Amsterdam 1972).

41 Lawrence, A. M.; Goldfine, I. D., and Kirsteins, L.: Growth hormone dynamics in acromegaly. J. clin. Endocrin. *31:* 239 (1970).

42 Liberti, J. P.: Partial purification of bovine sulfation factor. Biochem. biophys. Res. Commun. *39:* 356 (1970).

43 Martin, J. B.; Kantor, J., and Mead, P.: Plasma GH responses to hypothalamic, hippocampal and amygdaloid electrical stimulation. Effects of variation in stimulus parameters and treatment with α-methyl-*p*-tyrosine (α-MT). Endocrinology *92:* 1354 (1973).

44 Martini, L.; Motta, M., and Fraschini, F. (eds.): The hypothalamus (Academic Press, New York 1970).

45 McConaghey, P. and Sledge, C. B.: Production of 'sulphation factor' by the perfused liver. Nature, Lond. *225:* 1249 (1970).

46 McConaghey, P. and Dehnel, J.: Preliminary studies of 'sulphation factor' production by rat kidney. J. Endocrin. *52:* 587 (1972).

47 Merimee, T. J.; Rabinowitz, D., and Fineberg, S. E.: Arginine-initiated release of human growth hormone. New Engl. J. med. *280:* 1434 (1969).

48 Mitchell, M. L.; Byrne, M. J.; Sanchez, Y., and Sawin, C. T.: Detection of growth hormone deficiency. New Engl. J. Med. *282:* 539 (1970).

49 Muller, E. E.; Pecile, A.; Felici, M., and Cocchi, D.: Norepinephrine and GH-releasing factor. Endocrinology *86:* 1376 (1970).

50 Nakagawa, K.; Horiuchi, Y., and Mashimo, K.: Effect of dexamethasome on plasma growth hormone levels in acromegaly. J. clin. Endocrin. *31:* 502 (1970).

51 Pecile, A. and Muller, E. E. (eds.): Growth hormone (Excerpta Medica, Amsterdam 1968).

52 Pecile, A. and Muller, E. E. (eds.): Growth and growth hormone (Excerpta Medica, Amsterdam 1972).

53 Perlow, M. J.; Sassin, J. F.; Boyar, R.; Hellman, L., and Weitzman, E. D.: Release of human growth hormone, follicle stimulating hormone, and luteinizing hormone in response to L-dihydroxyphenylalanine (L-dopa) in normal man. Dis. nerv. Syst. *33:* 804 (1972).

54 Podolsky, S. and Sivaprasad, R.: Assessment of growth hormone reserve. Comparison of intravenous arginine and subcutaneous glucagon stimulation tests. J. clin. Endocrin. *35:* 580 (1972).

55 Poivola, P. T. K. and Gale, C. C.: Central adrenergic regulation of growth hormone secretion in baboons. Int. J. Neurosci. *4:* 53 (1972).

56 Raben, M. S.; Murakawa, S., and Matute, M.: Some observations concerning serum 'thymidine factor'; in Pecile and Muller Growth and growth hormone, pp. 124–131 (Excerpta Medica, Amsterdam, 1972).

57 ROTH, J.; GLICK, S. M.; YALOW, R. S., and BERSON, S. A.: Hypoglycemia. A potent stimulus to secretion of growth hormone. Science *140:* 987 (1963).

58 ROOT, A. W. and RUSS, R. D.: Effect of L-dihydroxyphenylalanine upon serum growth hormone concentrations in children and adolescents. J. Paediat. *81* (4): 808 (1972).

59 SALMON, W. D. and DAUGHADAY, W. H.: A hormonally controlled serum factor which stimulates sulfate incorporation by cartilage *in vitro*. J. Lab. clin. Med. *49:* 825 (1957).

60 SALMON, W. D. and DUVALL, M. R.: A serum fraction with 'sulfation factor activity' stimulates *in vitro* incorporation of leucine and sulfate into protein-polysaccharide complexes, uridine into RNA, and thymidine into DNA of costal cartilage from hypophysectomized rats. Endocrinology *86:* 721 (1970).

61 SAMAAN, N. A. and FREEMAN, R. M.: Growth hormone levels in severe renal failure. Metabolism *19:* 102 (1970).

62 SCHALLY, A. V.; ARIMURA. A.; BOWERS, C. Y.; WAKABAYASHI, I.; KASTIN, A. J.; REDDING, T. W.; MITTLER, J. C.; NAIR, R. M. G.; PIZZOLATO, P., and SEGAL, A. J.: Purification of hypothalamic releasing hormones of human origin. J. clin. Endocrin. *31:* 291 (1970).

63 SCHALCH, D. S.: The influence of physical stress and exercise on growth hormone and insulin secretion in man. J. Lab. clin. Med. *69:* 256 (1967).

64 SHERMAN, L.; SOOSENG, K.; BENJAMIN, F., and KOLODNY, H. D.: Effect of chlorpromazine on serum growth-hormone concentration in man. New. Engl. J. Med. *284:* 72 (1971).

65 SHIZUME, K.; MATSUZAKI, F., and SAWANO, S.: Reproducibility of plasma growth hormone and insulin responses after double infusion of arginine solution. Endocrinology, Japan *18:* 253 (1971).

66 SNODGRASS, G. and STIMMLER, L.: The early rise of plasma growth hormone (HGH) and immunoreactive insulin (IRI) in children following intravenous glucagon. J. clin. Endocrin. *34:* 410 (1972).

67 SPITZ, I.; GONEN, B., and RABINOWITZ, D.: Growth hormone release in man revisited. Spontaneous vs. stimulus-initiated tides; in PECILE and MULLER Growth and growth hormone, pp. 371–381 (Excerpta Medica, Amsterdam 1972).

68 STUART, M. C.; LAZARUS, L., and FERGUSON, K. A.: The circulating forms of growth hormone. 2nd Int. Symp. on Growth Hormone, Milan, 1971.

69 STIMMLER. L.; MCARTHUR, R. G., and BROWN, G. A.: Plasma growth hormone in children of short stature following an oral glucose load. Canad. med. Ass. J. *97:* 1159 (1967).

70 SUKKAR, M. Y.; HUNTER, W. M., and PASSMORE, R.: Changes in plasma levels of insulin and growth-hormone levels after a protein meal. Lancet *ii:* 1020 (1967).

71 UTHNE, K. and UTHNE, T.: Influence of liver resection and regeneration on somatomedin (sulphation factor) activity in sera from normal and hypophysectomized rats. Acta endocrin., Kbh. *71:* 255 (1972).

72 VANDENBRANDE, J. L.; VANWYK, J. J.; WEAVER, R. P., and MAYBERRY, H. E.: Partial characterization of sulphation and thymidine factors in acromegalic plasma. Acta endocrin., Kbh. *66:* 65 (1971).

73 VanWyk, J. J.; Hall, K., and Weaver, R. P.: Partial purification of sulphation factor and thymidine factor from plasma. Biochem. biophys. Acta *192:* 560–562 (1969).

74 VanWyk, J. J.; Hintz, R. L.; Hall, K., and Uthne, K.: Somatomedin. The growth hormone dependent sulfation and thymidine factor. 4th Int. Congr. of Endocrinology, Washington 1972.

75 Weldon, V. V.; Gupta, S. K.; Haymond, M. V.; Pagliara, A. S.; Jacobs, L. S., and Daughaday, W. H.: The use of L-dopa in the diagnosis of hyposomatotropism in children. J. clin. Endocrin. *35:* 42 (1973).

76 Zimmerman, T. S.; White, M. G.; Daughaday, W. H., and Goetz, F. C.: Hypopituitarism with normal or increased height. Report of two cases, with measurement of plasma growth hormone levels. Amer. J. Med. *42:* 146 (1967).

Author's address: Dr. R. M. Bala, Division of Medicine, University of Calgary, *Calgary, Alberta T2N 1N4* (Canada)

Recent Studies of Hypothalamic Function
Int. Symp. Calgary 1973, pp. 134–146 (Karger, Basel 1974)

The Control of Prolactin Secretion[1]

G. Tolis and H. G. Friesen

Department of Experimental Medicine, McGill University Clinic,
Royal Victoria Hospital, Montreal, Quebec

The development of specific and sensitive radioimmunoassay for human prolactin (hPRL) [22] has made possible the measurement of serum hPRL concentrations in patients under various physiological and pathological circumstances [11]. As a result of these studies it has been possible to draw some tentative conclusions about the factors controlling hPRL secretion. These inferences in large measure are possible because the control of prolactin (PRL) secretion in man and other species seems very similar. Hence, the concepts of the regulation of PRL secretion which are based mainly on studies in rats have proved to be most helpful in designing experiments to unravel the mechanisms which influence hPRL secretion. However even in animal studies the data on which the ideas are based are not as direct as one would like. For example, the hypothalamus contains a prolactin inhibiting factor (PIF), that appears to be the prime factor which tonically suppresses prolactin secretion, yet very little information is available regarding its chemistry, its site of production or its secretory rate. Thus, concepts which refer to increased or decreased PIF secretion are really based on very indirect evidence and caution and judgement must be exercised before they are accepted as dogma. Similarly, it is alleged that PIF secretion is influenced by dopaminergic fibres in the hypothalamus but again the suggestion is based on experiments in which pharmacological agents which influence catecholamines were used and no one in fact has measured secretory rates either of the catecholamines in the hypothalamus, and certainly not of PIF. Despite these difficulties and

1 This review was aided by grants from the Medical Research Council of Canada and USPHS Department of Child Health and Human Development (HD-01727-08).

uncertainties it is possible to make some limited inferences regarding the factors controlling PRL secretion. This presentation will emphasize those studies which relate to the control of hPRL secretion, in order to develop a framework which will be helpful in understanding the factors influencing PRL secretion in health and disease in human subjects.

PRL secretion appears to be controlled both by inhibitory and stimulatory signals from the hypophysiotropic areas of the hypothalamus. In mammalian species, PRL synthesis and release is primarily under tonic hypothalamic inhibition exerted via the secretion of PIF [47]. PIF activity has been demonstrated in the hypothalamus of many mammalian species including man [39] and is detectable in the hypophyseal portal blood of intact [26] or in the peripheral blood of hypophysectomized animals [34]. Along with this inhibitory control, the existence of a stimulatory factor, PRL releasing factor (PRF), has been reported recently [29, 37].

There is evidence [35] that the secretion of these hypophysiotropic factors is controlled by the hypothalamic monoamines dopamine (DA), norepinephrine (NE) and 5-hydroxytryptamine (5HT) elaborated at dopaminergic [13] and serotoninergic neurons [19].

Any condition that interferes (1) with the synthesis, storage or release of the neurotransmitters; (2) with their action at the synaptic level; (3) with the integrity of the hypophysiotropic area; (4) with the transport of the hypophysiotropic factors to the pituitary and (5) finally, with the functional status of the pituitary prolactin cells, will result in an alteration in serum PRL concentration provided that the metabolic clearance rate of PRL is constant.

Experimental evidence based on morphological and histochemical studies in rats suggests that DA is the major amine present in the tubero-infundibular neurons, the cell bodies of which are mainly found in hypothalamic nuclei [14, 15]. The terminal axons of these cells end on the portal capillary plexus of the median eminence. Because both DA and NE cause an inhibition of PRL release *in vitro* [3], the possibility exists that DA itself may directly control prolactin secretion. However in one study, DA could not be detected in the pituitary portal circulation of rats [55]; hence, it is unlikely that DA is the direct physiologic regulator of PRL secretion. Indeed more direct evidence that DA acts at the hypothalamic level is provided by the elegant experiments in which DA was infused into the third ventricle of rats and portal capillary blood was collected. PIF appeared in the portal circulation promptly after the infusion of DA

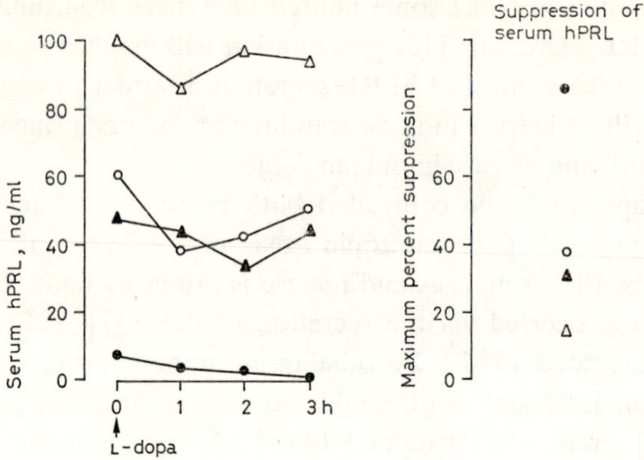

Fig. 1. Effect of L-dopa (500 mg *per os*) on serum hPRL concentrations in normal individuals (●, n = 4) and patients on phenothiazine (Trifluorperazine 10 mg). ○ = α-Methyldopa medications (△ = 5 g; ▲ = 1.5 g).

into the third ventricle whereas when equivalent amounts of DA were infused into the portal vessels no effect on PRL secretion was noted. The administration of L-dopa – an agent which is converted to DA in the hypothalamus, evokes PIF release by suppressing pituitary prolactin secretion [31]. In man, an acute suppression of serum hPRL concentrations follows the administration of L-dopa [10, 33] and such a response may suggest that the hypothalamic-pituitary prolactin axis is intact. A diminished response to L-dopa may be indicative of hypothalamic diseases [49] such as sarcoidosis trauma, tumors, etc. Inhibition of the decarboxylation of L-dopa to DA by α-methyldopa (Aldomet) would be expected to decrease endogenous DA resulting in an increase of basal hPRL secretion, whereas the presence of false neurotransmitters (Aldomet) or DA competitors such as phenothiazines [45] will prevent DA action at the receptor site resulting in limited effectiveness of L-dopa in suppressing serum hPRL concentrations; figure 1 shows that the results which are predicted from the theoretical considerations outlined are in fact observed experimentally.

As was recently shown in rats [7], the degree of PRL suppression by L-dopa under these circumstances presumably depends upon the extent to which the receptor sites are occupied, and upon the ability of the competitors to reverse the process. Elevated serum PRL concentrations, which occur in animals bearing hypophyseal transplants or tumors, or

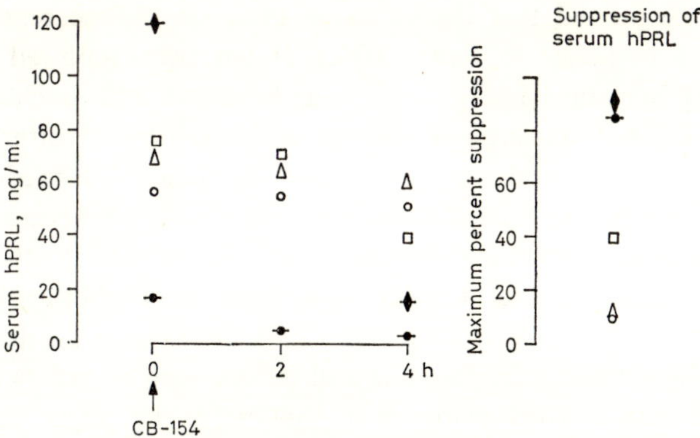

Fig. 2. Serum hPRL concentration following the administration of 2.5 mg 2-Br-α-ergocryptine (CB-154) to normal individuals (-●- mean, n = 8), healthy postpartum females (-◆- mean, n = 4), and to patients on psychotropic medications (□ = phenothiazine 100 mg; △ = thioridazine 25 mg; ○ = trifluoperazine 10 mg).

which are found after the intravenous injection of PRL, markedly increase DA turnover in the tuberoinfundibular neurons [17]. Whether the increased turnover of DA is also the reason why PRL secreting tumors in man are resistant to L-dopa treatment [9, 33] is at present speculative. The observation [48] that such tumors do respond to chronic administration of 2-Br-α-ergocryptine (CB-154) may support such a concept since this agent has been reported to reduce DA turnover [21]. CB-154 has also been shown to have a direct effect on pituitary prolactin cells [40] decreasing PRL release initially and subsequently also inhibiting PRL synthesis [18]. CB-154 suppresses PRL secretion in normal individuals under basal conditions [8], blocks the stimulatory effect of thyrotropin releasing hormone TRH [12], and suppresses PRL secretion in hyperprolactinemic states [12, 32]. However, in patients with elevated serum hPRL concentrations secondary to psychotropic drug intake, CB-154 was found to be significantly less effective in suppressing hPRL secretion when compared to normals (fig. 2). These results suggest that CB-154, in addition to a direct action on the pituitary, also acts on the hypothalamus. Similar findings have been reported in experimental animals [56].

When PRL is implanted into the median eminence, a decrease in pituitary PRL secretion follows [5]; hence, it is assumed that a short loop feedback exists by which PRL controls its own secretion. Since PRL affects DA turnover in the hypothalamus and PIF secretion is DA-dependent it has been suggested that the effect of PRL on PIF secretion is partly indirect via a change in monoamines. It has been reported that follicle stimulating hormone (FSH) - luteinizing hormone (LH) - releasing hormone (RH) as well as PIF, is present in portal blood after the infusion of DA into the third ventricle suggesting that a DA-mediated reciprocal relationship may exist between PRL and gonadotropins. Therefore, in hyperprolactinemic states, such as patients with the Chiari-Frommel syndrome, del Castillo syndrome or Forbes-Albright syndrome, one might expect disrupted menstrual function in the female or altered libido and potency in the male and indeed such disorders have been reported [2, 12]. Further evidence in support of this suggestion is provided by studies in patients in which there is a cessation of galactorrhea and resumption of normal menses in patients with prolactin secreting tumors after the suppression of PRL secretion, either by the selective removal of microadenomas of the pituitary or after treatment with CB-154 [12] or L-dopa [50]. Similarly, galactorrhea and amenorrhea and elevated hPRL concentrations are frequently encountered in patients who are receiving Aldomet, reserpine or phenothiazines, agents which are known to deplete the hypothalamus of catecholamines. Cessation of galactorrhea and a return of hPRL to normal levels occurs after the discontinuation of such medicaments. Although the exact mode of action of psychotropic drugs, i.e. haloperidal, chlorpromazine and reserpine, is not entirely clear, it has been suggested on the basis of animal studies that all these agents share a similar chemical configuration which results in the activation of a hypothalamic receptor which suppresses PIF release [46].

We have speculated previously that the longer duration of *post partum* amenorrhea in lactating, as compared to non-nursing mothers may be related to the intermittent surges of PRL secretion which follow breast-feeding [52]. As the result of a direct or indirect action of PRL on the hypothalamus, cyclic discharge of FSH and LH is prevented, resulting in amenorrhea or anovulatory cycles. The validity of this hypothesis has been tested more recently in baboons. When these primates were given TRH orally daily during the menstrual cycle, no midcycle LH and FSH peak was noted and no increase in progesterone in the luteal phase was detected [53]. Presumably, the increases in PRL which occur after TRH is given account for the change in gonadotropin secretion. Since not all patients with hyperprolactinemia are amenorrheic, other factors undoubtedly influence the release of LH and FSH, but at the very least we suggest that PRL may be one of them.

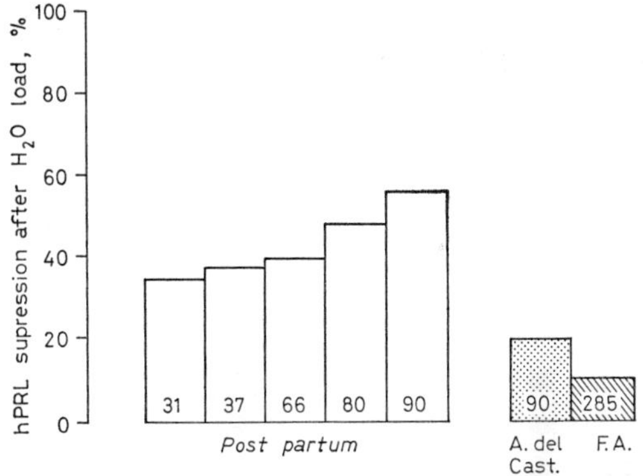

Fig. 3. Suppression of serum hPRL following a water load (1,000 ml *per os*) in 5 postpartum females and 2 patients with the Argonz del Castillo (A. del Cast.) and Forbes-Albright (F.A.) syndromes.

Nonpsychotropic agents which enhance PRL secretion in animals and man are the gonadal steroids. These substances, which have certain chemical similarities with the psychoactive drugs, can effectively raise PRL secretion by both a hypothalamic [16, 42] as well as a direct pituitary effect [36]. At the hypothalamic level they suppress PIF secretion either by a direct effect or by influencing monoamine concentrations in the hypothalamus as well [16]. The demonstration that estrogen administration to postmenopausal women can raise serum hPRL concentrations and suppress gonadotropin secretion may be explained by both postulates. It is of interest that when estrogens were given to individuals in other categories [30] somewhat contradictory results were obtained, suggesting that many factors determine the responsiveness of pituitary PRL cells to a given stimulus. In this respect we have observed that a water load given to *post partum* females with hyperprolactinemia suppresses serum hPRL whereas in a patient with Argonz del Castillo syndrome and the Forbes-Albright syndrome the same stimulus had little effect (fig. 3).

Whereas CPZ stimulates hPRL secretion by inhibiting PIF release, TRH stimulates the prolactin cells of the pituitary by a direct action [6, 20, 28]. The response to TRH is greater in females than in males and a variable response is found in patients with PRL secreting tumors (fig. 4). In some, but not all, subjects the TRH-induced increase of hPRL can be blocked by CB-154. In patients with primary hyperthyroidism or in patients who receive excessive amounts of thyroid hormones ($T_4 + T_3$),

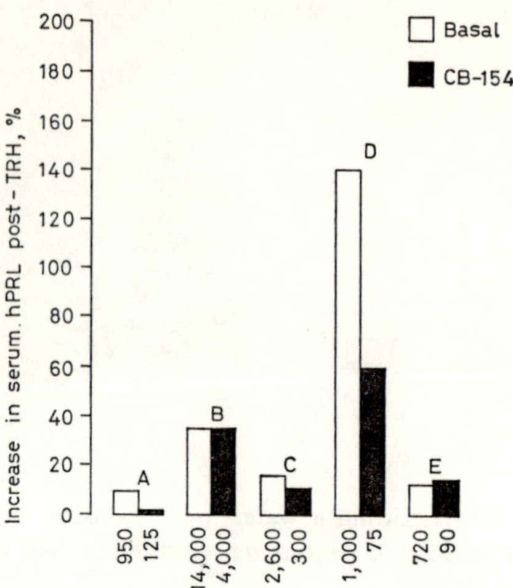

Fig. 4. Percent increase of serum hPRL following TRH-administration (400 μg i.v.) in patients with prolactin secreting tumors (A, B, C: ♂, D, E: ♀) before (□) and after (■) a week's administration of CB-154 (2.5 mg t.i.d.). Basal hPRL concentrations are indicated at the base of the bars.

the TRH effect on both TSH and hPRL secretion is inhibited. In patients with primary hypothyroidism an exaggerated release of both hormones occurs after the administration of TRH [4]. Since TRH, but not CPZ, can stimulate hPRL secretion even in patients with pituitary stalk secretion (fig. 5)[2] the failure of TRH to raise serum hPRL concentrations above basal levels should be taken as evidence of a deficiency of pituitary lactotropes, provided the patient is not hyperthyroid or on drugs which suppress PRL secretion. It is also worth mentioning that in hyperprolactinemic states TRH may be ineffective in further stimulating hPRL after TRH indicates the presence of pituitary PRL cells, whereas the CPZ-induced hPRL release requires an intact hypothalamus as well as a functioning pituitary. On the basis of the above information, it has been shown that in disorders of the hypothalamic pituitary system, CPZ and TRH tests are useful in detecting abnormalities of the hypothalamic-pituitary-prolactin axis and in localizing the disease process to the pituitary or hypothalamus [49].

2 Serum samples from this patient were kindly supplied by Dr. J. J. VAN WYK and Dr. R. C. LISTER of the University of North Carolina, Chapel Hill.

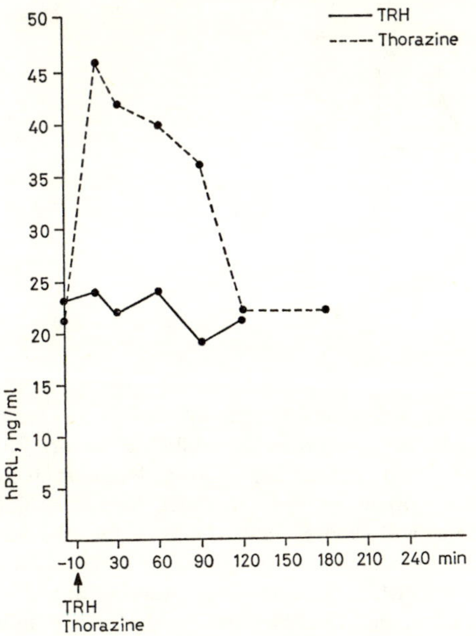

Fig. 5. Effect of TRH and thorazine in serum hPRL concentrations in a patient with a pituitary stalk section.

In contrast to the pharmacological agents which stimulate PRL, a major physiological factor for PRL release is breast stimulation. Elevations of serum hPRL concentrations are seen when the breast is manually or mechanically pumped [38] and galactorrhea has been reported following excessive breast manipulation (pumping and suckling). Following suckling, PRL release is seen in the rat [44] and also in women [52]. Although the mechanism by which suckling evokes hPRL release is not known, the fact that hPRL elevations are not accompanied by concomitant TSH increases [4] suggests that TRH is not the factor responsible for hPRL release. The acute elevation in hPRL which is seen immediately after the onset of nursing is more likely due to the release of a prolactin releasing factor (PRF) than of inhibition of PIF secretion although animal data on this point are contradictory. If one postulates that suckling induces the release of both PRF and the suppression of PIF then one is also obliged to postulate the existence of a neurotransmitter which is capable of a dual control mechanism. The observation that 5-HT (serotonin) has no direct action on the pituitary lactotropes whereas it elicits hPRL release when injected into the third ventricle [25, 26]

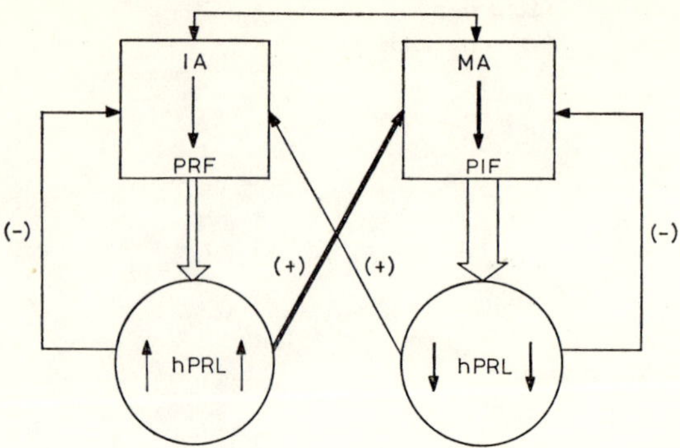

Fig. 6. Model of control of hPRL secretion. IA = Indoloamines, i.e. serotonin. MA = Monoamines, i.e. dopamine. PIF = Prolactin inhibiting factor. PRF = Prolactin releasing factor. (+) = Positive feedback. (–) = Negative feedback. Elevated serum hPRL concentrations (a) by negative feedback to the indoloaminergic neurons and/or the PRF center will suppress the release of the PRF and (b) by a positive feedback will enhance the elaboration of MA and/or PIF. As a result of the action of (a) + (b) pituitary hPRL secretion will be lowered and prolactin levels will return towards a normal range. The thickness of the arrows indicates the relative influence of each of the mechanisms controlling hPRL secretion.

suggests that serotonin stimulates PRL secretion either by triggering the release of PRF or by inhibiting PIF secretion. That both mechanisms may be operative is suggested by the recent demonstration of complete inhibition of PRL release following suckling in rats pretreated with an inhibitor of serotonin synthesis (*p*-chlorophenylalanine) [27] as well as the observation that 5-hydroxytryptophan can displace NE and DA from central nervous system neurons and *vice versa* [1]. Recent human data have shown the involvement of serotonin in the secretion of other anterior pituitary hormones [23] and in at least one study tryptophan and 5-hydroxytryptophan infusions stimulated PRL secretion [51]. Indirect evidence also has been provided by the finding of a diurnal variation of serum hPRL concentrations [30, 43], showing elevated levels at night during sleep which coincide with the rise in concentrations of serum tyrosine phenylalanine and tryptophan, all of which are precursors of serotonin [54], the significance of which in the sleep process has been well documented [24].

It should be apparent that our understanding of the multiple factors controlling hPRL secretion from the hypothalamus is still far from complete. The concept which is emerging from studies in experimental animals and more recently in man is that stimulation or inhibition of PRL secretion is regulated from moment to moment by two major regulatory systems: monoaminergic and serotoninergic. Activation of either system can alter PRL secretion probably via the release of the corresponding hypothalamic factor and concomitant suppression of the release of the opposing factor (fig. 6). The principal significance of the model is that it provides a useful framework for planning additional experiments and also sets forth a rational basis for the functional evaluation of the hypothalamic-pituitary PRL axis in clinical disorders.

Acknowledgements

We would like to thank Drs. J. J. VAN WYK and R. C. LISTER who allowed us to study their patient and Ms. JEAN PARODO and MONICA BACON for technical help.

References

1 AGHAJANIAN, G. K. and BLOOM, F. E.: Localization of tritiated serotonin in rat brain by electron-microscopic autoradiography. J. Pharmacol. exp. Ther. *156:* 23–30 (1967).

2 BESSER, G. M.; PARKE, L.; EDWARDS, C. R. V.; FORSYTH, A., and MCNEILY, A. S.: Galactorrhea. Successful treatment with reduction of plasma prolactin levels by 2-Br-α-ergocryptine. Brit. med. J. *iii:* 669–672 (1972).

3 BIRGE, C. A.; JACOBS, L. S.; HAMMER, C. T., and DAUGHADAY, W. H.: Catecholamine inhibition of prolactin secretion by isolated rat adenohypophyses. Endocrinology *86:* 120–130 (1970).

4 BOWERS, C. Y.; FRIESEN, H. G., and FOLKERS, K.: Further evidence that TRH is also a physiological regulator of prolactin (PRL) secretion in man. Biochem. biophys. Res. Commun. *51:* 512–521 (1973).

5 CLEMENTS, J. A. and MEITES, J.: Inhibition of hypothalamic prolactin implants on prolactin secretion, mammary growth and luteal function. Endocrinology *82:* 878–881 (1968).

6 DANNIES, P. S. and TASHJIAN, A. H., jr.: Stimulation of prolactin (PRL) synthesis by thyrotropin releasing hormone (TRH). Fed. Proc. *32:* 308 (1973).

7 DANON, A. and BEN-DAVID, M.: Effect of dopamine on perphenazine-induced release of prolactin in rats. 4th Int. Congr. of Endocrinology, abstract No. 124, Excerpta med. *256:* 50 (1972).

8 POZO, E. DEL; RE, R. BRUN DEL; VARGA, L., and FRIESEN, H. G.: The inhibition of prolactin secretion in man by CB-154 (2-Br-α-ergocryptine). J. clin. Endocrin. *35:* 768–771 (1972).

9 EDMONDS, M.; FRIESEN, H., and VOLPE, R.: The effect of levodopa on galactorrhea in the Forbes-Albright syndrome. Canad. med. Ass. J. *107:* 534–536 (1972).

10 FRIESEN, H. G.; GUYDA, H.; HWANG, P.; TYSON, J. E., and BARBEAU, A.: Functional evaluation of prolactin secretion. A guide to therapy. J. clin. Invest. *51:* 706–709 (1971).

11 FRIESEN, H. G.; HWANG, P.; GUYDA, H.; TOLIS, G.; TYSON, J., and MYERS, R.: In GRIFFITHS and BOYNS Prolactin and carcinogenesis. Proc. 4th Tenovus Workshop, Cardiff 1972 (Alpha-Omega-Alpha, Cardiff 1972).

12 FRIESEN, H. G.; TOLIS, G.; SHIU, R., and HWANG, P.: Studies on human prolactin. Chemistry, radioreceptor assay and clinical investigation. Excerpta med. 11–23 (1973).

13 FUXE, K.: Cellular localization of monoamines in the median eminence and infundibular stem of some mammals. Z. Zellforsch. Mikr.-Anat. *61:* 710–724 (1964).

14 FUXE, K.; HOKFELT, T., and NILLSON, O.: A fluorescent and electron microscopic study of certain brain regions rich in monoamine terminals. Amer. J. Anat. *117:* 33–45 (1965).

15 FUXE, K.; HOKFELT, T., and NILLSON, O.: Activity changes in the tubero-infundibular DA neurons of the rat during various states of the reproductive cycle. Life Sci. *6:* 2057–2061 (1967).

16 FUXE, K.; HOKFELT, T., and NILLSON, O.: Castration, sex hormones and tubero-infundibular dopamine neurons. Neuroendocrinology *5:* 257–270 (1969).

17 FUXE, K. and HOKFELT, T.: Aspects of neuroendocrinology, pp. 192–205 (Springer, Berlin 1970).

18 GAUTVIK, K. M.; HOYT, K. F., jr., and TASHJIAN, A. H.: Effects of colchicine and 2-Br-α-ergocryptine-methane-sulfonate (CB-154) on the release of prolactin and GH by functional pituitary tumor cells in culture. J. Cell Physiol. (in press).

19 HILLARP, N. A.; FUXE, K., and DAHLSTRÖM, A.: Mechanisms of release of biogenicamines, pp. 31–37 (Pergamon Press, London 1966).

20 HINKLE, P. M.; WOROCH, E. L., and TASHJIAN, A. H., jr.: Activity of analogs of thyrotropin releasing hormone (TRH) on prolactin (PRL) synthesis and binding to TRH receptors in pituitary cells in culture. Endocrine Society Meeting, Chicago 1973.

21 HOKFELT, T. and FUXE, K.: Brain endocrine interaction, pp. 181–223 (Karger, Basel 1972).

22 HWANG, P.; GUYDA, H., and FRIESEN, H.: A radioimmunoassay for human prolactin. Proc. Nat. Acad. Sci., Wash. *68:* 1902–1906 (1971).

23 IMURA, H.; NAKAI, Y., and YOSHIMI, T.: Effect of 5-hydroxy-tryptophan (5-HTP) on growth hormone and ACTH release in man. J. clin. Endocrin. *36:* 204–206 (1973).

24 JOUVET, M.: Biogenic amines and the states of sleep. Science *163:* 32–41 (1969).

25 KAMBERI, I. A.; MICAL, R. S., and PORTER, J. C.: Prolactin inhibiting activity in hypophysial stalk blood and elevation by dopamine. Experientia *26:* 1150–1151 (1970).

26 KAMBERI, I. A.; MICAL, R. S., and PORTER, J. C.: Effects of melatonin and serotonin on the release of FSH and prolactin. Endocrinology *88:* 1288–1293 (1971).

27 KORDON, C. A.; BLAKE, C. A., and SAWYER, C. H.: Participation of serotonin-containing neurons in the suckling-induced rise in plasma prolactin levels in lactating rats. 4th Int. Congr. Endocrinology, abstract No. 126. Excerpta *256:* 51 (1972).

28 LABELLA, F. S. and VIVIAN, S. R.: Effect of synthetic TRF on hormone release from bovine anterior pituitary *in vitro*. Endocrinology *88:* 787–789 (1971).

29 LABELLA, F. S.; DULAR, R., and VIVIAN, S. R.: Purification of bovine hypothalamic factors which inhibit (PIF) or enhance (PRF) the release of prolactin (PL) from bovine anterior pituitary *in vitro*. 4th Int. Congr. of Endocrinology, abstract No. 354. Excerpta med. *256:* 141–142 (1972).

30 L'HERMITE, M.; DELROYE, P.; NOKIN, J.; VEKEMANS, M., and ROBYN, C.: Human prolactin secretion as studied by radioimmunoassay. Some aspects of its regulation; in GRIFFITHS and BOYNS Prolactin and carcinogenesis, pp. 81–97 (Alpha-Omega-Alpha, Cardiff 1972).

31 LU, K. H. and MEITES, J.: Inhibition of L-dopa and monoamine oxidase inhibitors of pituitary prolactin release. Stimulation by methyldopa and D-amphetamine. Proc. Soc. exp. Biol. Med. *137:* 480–483 (1971).

32 LUTTERBECK, P. M.; PRYOR, J. S.; VARGA, L., and WENNER, R.: Treatment of non-puerperal galactorrhea with an ergot alkaloid. Brit. med. J. *iii:* 228–229 (1971).

33 MALARKEY, W. B.; JACOBS, L. S., and DAUGHADAY, W. H.: Levodopa suppression of prolactin in non-puerperal galactorrhea. New Engl. J. Med. *284:* 1160–1163 (1971).

34 MEITES, J.; NICOLL, C. S., and TALWALKER, P. K.: Advances in Endocrinology, pp. 238–277 (University of Chicago Press, Urbana 1963).

35 MEITES, J.; LU, K. H.; WUTTKE, W.; WELSCH, C. W.; NAGASAWA, H., and QUADRI, S. K.: Hypothalamic regulation of prolactin secretion. Recent Progr. Hormone Res. *28:* 499–526 (1972).

36 NICOLL, C. S. and MEITES, J.: Estrogen stimulation of prolactin production by rat adenohypophysis *in vitro*. Endocrinology *70:* 272–277 (1962).

37 NICOLL, C. S.; FIORINDO, R. P.; MCKENNEE, C. T., and PARSONS, J. A.: Hypophysiotropic hormones of the hypothalamus, pp. 115–150 (Williams & Wilkins, Baltimore 1970).

38 NOEL, G. L.; SUH, H. K., and FRANZ, A. G.: Induction of prolactin release by breast stimulation in humans. 4th Int. Congr. Endocrinology, abstract No. 256. Excerpta med. *256:* 103 (1972).

39 PASTEELS, J. L.: Recherches morphologiques et expérimentales sur la sécrétion de prolactine. Arch. Biol. *74:* 439–553 (1963).

40 PASTEELS, J. L.; DANGUY, A.; FREROTTE, M., et ECTORS, F.: Inhibition de la sécrétion de prolactine par l'ergocornine et la 2-Br-α-ergocryptine. Action directe sur l'hypophyse en culture. Ann. Endocrin., Paris *32:* 188–192 (1971).

41 RAMIREZ, V. D. and MCCANN, S. M.: Induction of prolactin secretion by implants of estrogen into the hypothalamo-hypophyseal region of female rats. Endocrinology *75:* 206–214 (1964).

42 RATNER, A. and MEITES, J.: Depletion of prolactin inhibiting activity of rat hypothalamus by estradiol or suckling stimulus. Endocrinology *75:* 377–382 (1964).

43 SASSIN, J. F.; FRANTZ, A. G.; WEITZMAN, E. D., and KAPEN, S.: Human prolactin. 24-hr pattern with increased release during sleep. Science *177:* 1205–1207 (1972).

44 SHIHINO, M.; WILLIAMS, G., and RENNELS, E. G.: Ultrastructural observation of pituitary release of prolactin in the rat by the suckling stimulus. Endocrinology *90:* 176–197 (1972).

45 SOURKES, T. L.: Actions of levodopa and dopamine in the central nervous system. J. amer. med. Ass. *218:* 1909–1911 (1971).

46 SULMAN, F. G.: Hypothalamic control of lactation, pp. 133–141 (Springer, Berlin 1971).

47 TALWALKER, P. K.; RATNER, A., and MEITES, J.: *In vivo* inhibition of pituitary synthesis and release by hypothalamic extract. Amer. J. Physiol. *205:* 213–218 (1963).

48 TOLIS, G.; DEL POZO, E., and GOLDSTEIN, M.: Dopamine and ergocryptine effects on hyperprolactinaemic states. Endocrine Society Meeting, Chicago 1973.

49 TOLIS, G.; GOLDSTEIN, M., and FRIESEN, H.: Functional evaluation of prolactin secretion in patients with hypothalamic-pituitary disorders. J. clin. Invest. *52:* 783–788 (1973).

50 TURKINGTON, R. W.: Inhibition of prolactin secretion and successful therapy of the Forbes-Albright with L-dopa. J. clin. Endocrin. *34:* 306–311 (1972).

51 TURKINGTON, R. W. and MACINDOE, J. H.: Stimulation of human prolactin secretion by intravenous infusion of L-tryptophan. J. clin. Invest. *51:* 98a (1972).

52 TYSON, J. E.; FRIESEN, H. G., and ANDERSON, M. S.: Human lactational and ovarian response to endogenous prolactin release. Science *177:* 897–899 (1972).

53 WHITE, W. F.: Hypothalamic releasing factors. Proc. 4th Ross Conf. Endocrine Milieu of Pregnancy, Puerperium and Childhood, St. Maarten 1972.

54 WURTMAN, R. J.; ROSE, C. M.; CHOU, C., and LARIN, F.: Daily rhythms in the concentrations of various amino acids in human plasma. New. Engl. J. Med. *279:* 171–175 (1968).

55 WURTMAN, R. J.: Hypophysiotropic hormones of the hypothalamus, pp. 184–194 (Williams & Wilkins, Baltimore 1970).

56 WUTTKE, W.; CASSELL, E., and MEITES, J.: Effects of ergocornine on serum prolactin and LH and on hypothalamic contents of PIF and LRF. Endocrinology *88:* 737–741 (1971).

Authors' addresses: Dr. G. TOLIS, Department of Experimental Medicine, McGill University Clinic, Royal Victoria Hospital, *Montreal 112, Quebec;* Dr. H. G. FRIESEN, Department of Physiology, Faculty of Medicine, University of Manitoba, *Winnipeg, Manitoba* (Canada)

Recent Studies of Hypothalamic Function
Int. Symp. Calgary 1973, pp. 147–165 (Karger, Basel 1974)

An Analysis of the Positive Feedback Action of Estrogen on LH Secretion in the Rat[1]

SANDRA J. LEGAN and F. J. KARSCH

Reproductive Endocrinology Program, Departments of Physiology and Pathology, University of Michigan, Ann Arbor, Mich.

I. Introduction

Numerous observations made in several species indicate that estrogen is an important steroidal component of two feedback systems which govern the secretion of luteinizing hormone (LH), the inhibitory, or negative, feedback loop, and the stimulatory, or positive, feedback system [8, 17, 21, 27, 33, 35, 41, 49, 50]. One hypothesis which accounts for the phenomenon that a single steroid hormone can exert two completely diametric actions on LH secretion is that there are two separate mechanisms, the tonic and phasic, which function independently and concurrently [4, 25, 44]. Evidence obtained in rhesus monkeys [26, 27] indicates that the tonic mechanism, which regulates basal LH concentrations throughout the cycle, is responsive to relatively low sustained levels of estrogen, whereas the phasic mechanism, which controls the preovulatory LH discharge, is activated by an increase in circulating estrogen to a level maintained for approximately 36 h. Furthermore, in the rat, each mode of LH secretion appears to be controlled by an anatomically distinct region of the hypothalamus, the ventromedial-arcuate area, controlling tonic LH secretion, and the anterior hypothalamic-preoptic area, which regulates phasic LH release [4, 22, 23]. Thus, the bimodal action of estrogen on LH release in the rat may be explained by its differential feedback action on these two centers, an inhibition of the tonic center and stimulation of the phasic center. The identification of two anatomic centers controlling LH release has not yet been established in nonrodent species.

1 Supported by a Program Project in Reproductive Endocrinology, NIH-HD-05318.

Although some evidence suggests a direct action of estrogen on the hypothalamus [11, 35], the possibility that estrogen may also act on the pituitary cannot be eliminated. In addition, little is known concerning the mechanism of action of estrogen at either or both of these sites, or the nature of the estrogen stimulus. In the studies to be described, the role played by estrogen in the induction of LH release in the rat was investigated.

II. General Methods

In all experiments, adult female Sprague-Dawley derived rats were housed with lights on from 0500 h to 1900 h and given food and water *ad libitum*. Blood samples were withdrawn either through a carotid cannula inserted as previously described [20], or by cardiac puncture. Results obtained employing both methods of blood collection within the same group of animals were identical. Plasma LH concentrations were determined by radioimmunoassay [36] and are expressed in terms of nanogram equivalents per milliliter of the rat pituitary LH standard supplied by the National Institute of Arthritis and Metabolic Diseases (NIAMD-LH-RP-1). In all experiments, the day of initiation of the estrogen stimulus for LH release, whether endogenous or exogenous, is designated as day 0.

III. Characteristics of Positive Feedback Action of Estrogen on LH Secretion in Ovariectomized Rats

A. 24-Hour Periodicity

In the first experiment, rats obtained from Spartan Farms (Haslett, Michigan) were treated with estradiol benzoate (EB, 50 μg/rat s. c. in 0.1 ml sesame oil) at 0700 h on the 15th day following ovariectomy (day 0). Within 1 h, plasma LH concentrations decreased approximately 50 % to levels 10- to 20-fold greater than those seen in diestrous rats, a result of negative feedback inhibition of LH release. LH concentrations in samples obtained at 1230 h on days 0–3 remained low despite presumably decreasing circulating estrogen levels. On each of the 3 days following estrogen injection, a large increase in plasma LH levels was ob-

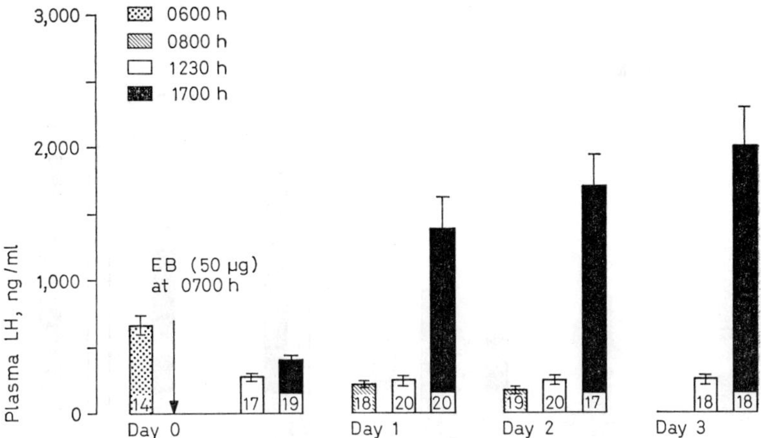

Fig. 1. Plasma luteinizing hormone (LH) concentrations (mean ± SE) in Spartan rats treated with 50 μg estradiol benzoate (EB, s.c. in 0.1 ml sesame oil) 2 weeks after ovariectomy. In all figures, the numbers at the base of the bars indicate the number of animals in each group.

served at 1700 h (fig. 1). This time coincides with the initiation of the preovulatory LH surge in proestrous rats (1600–1800 h) [19], which will subsequently be referred to as the 'release period'. These results indicate that the estrogen-induced LH release mechanism in ovariectomized rats has a 24-hour periodicity and functions only during the release period. Similar results were obtained, with minor differences, when the experiment was repeated using Holtzman and Charles River rats [29].

B. Estrogen Dependence

Repetitive surges of LH in estrogen-treated ovariectomized rats have also been observed by CALIGARIS *et al.* [8] and NEILL [35]. However, whether estrogen induces these surges by a positive feedback action on the phasic center, or whether it merely unmasks them by a negative feedback inhibition of the high rate of tonic LH secretion normally seen in ovariectomized rats has not yet been determined. The latter possibility is suggested by the recent observations of BEATTIE *et al.* [5] that circulating LH concentrations increased 3-fold during the release period in ovariectomized rats which did not receive estrogen, and that administration of

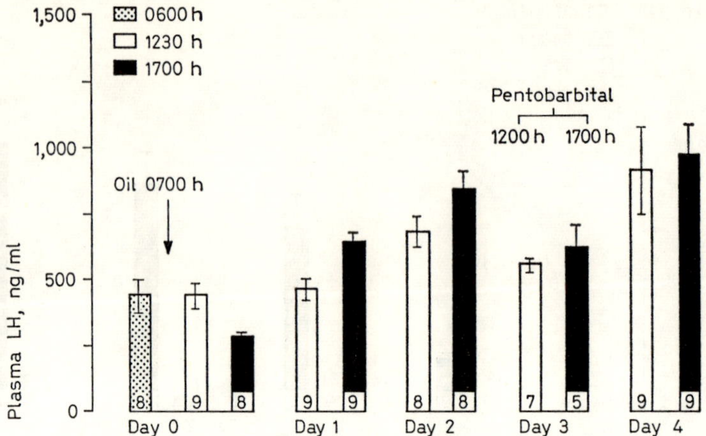

Fig. 2. Plasma LH concentrations (mean ± SE) in Spartan rats treated with 0.1 ml sesame oil 2 weeks after ovariectomy.

pentobarbital abolished these increases. To investigate this problem in our experimental model, rats were treated with 0.1 ml sesame oil on the 15th day following ovariectomy (day 0). As shown in figure 2, no consistent increase in plasma LH levels was observed between 1230 and 1700 h for 5 days. Even the highest levels observed in oil-treated ovariectomized rats were well below those seen at 1700 h in estrogen-treated rats. These results clearly demonstrate that the repetitive surges of LH we observe at 1700 h require a positive feedback action of estrogen, although our experimental design did not allow us to determine whether or not there is a daily spontaneous increment in circulating LH at another time during the release period in ovariectomized rats. The slight suppression of LH levels observed following pentobarbital administration on day 3 confirms previous observations [1, 5] that treatment with pentobarbital can inhibit tonic LH secretion. The gradual change in plasma LH concentrations from days 0 to 4 reflects the increase in LH secretion normally seen during the first 4 weeks after ovariectomy [6, 18, 48].

C. Neural Control

During the estrous cycle of the rat the preovulatory LH discharge can be inhibited by administration of neural blocking agents such as

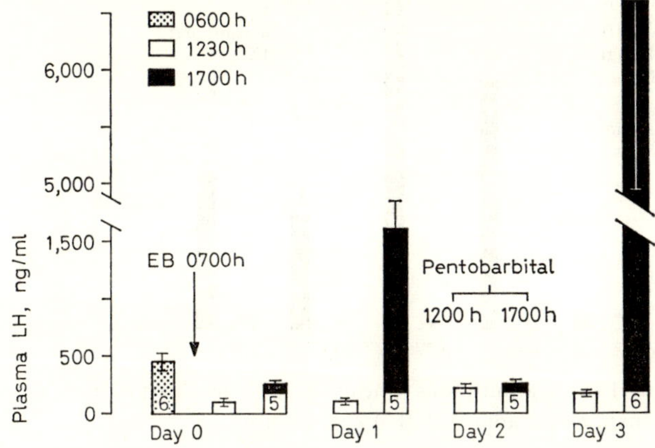

Fig. 3. Effect of pentobarbital on plasma LH concentrations (mean ± SE) in Spartan rats treated with 50 μg EB 2 weeks after ovariectomy.

pentobarbital [10], an effect which can be overcome by administration of luteinizing hormone releasing factor (LRF) [30; LEGAN, unpublished observations]. These and other observations [4, 11, 15, 22, 23, 31] provide compelling evidence that the proestrous LH surge is controlled by a neural pathway. To determine whether a similar mechanism is operative in rats treated with estrogen 2 weeks after ovariectomy, a dose of pentobarbital required to block a proestrous LH surge was administered at 1200 h on the 2nd day after injection of EB. As expected, an LH surge was observed at 1700 h on day 1 but there was no surge the next day (fig. 3), suggesting that estrogen-induced LH release in ovariectomized rats is controlled, at least in part, by a neural component. It is interesting to note that the surge recurred at 1700 h on day 3 and plasma LH levels were much greater than those usually seen at this time.

IV. Stimulus Duration Required to Induce LH Release

In the ovarian cycle of the rat and other species, the preovulatory LH surge is dependent upon an antecedent rise in circulating estrogen concentrations [17, 27, 28, 32, 46]. In the rhesus monkey, the increment in circulating estrogen levels has to be maintained for approximately 36 h. When the estrogen stimulus is withdrawn prior to initiation of the LH

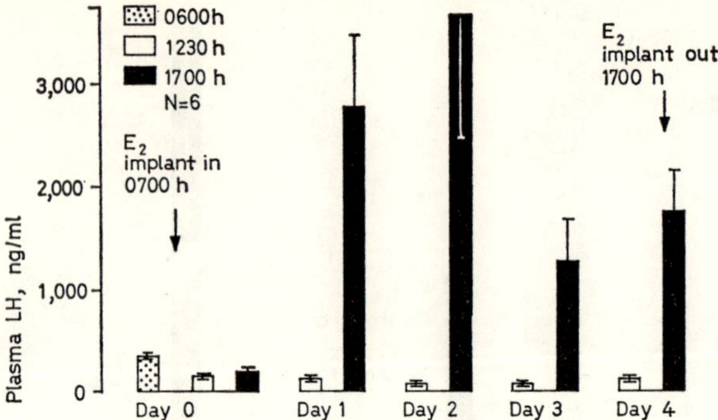

Fig. 4. Plasma LH concentrations (mean ± SE) in Spartan rats receiving Silastic implants containing estradiol-17β (E_2) 2 weeks after ovariectomy.

discharge, the surge is abolished [26]. In contrast, however, in the rat, withdrawal of the estrogen stimulus by means of ovariectomy as early as 1000 h on proestrus has no effect on the LH surge [43]. This striking difference prompted us to examine the nature of the estrogen stimulus required for induction of daily LH surges in ovariectomized rats.

In subsequent experiments, precise control of stimulus duration was achieved by administering estradiol-17β in Silastic capsules implanted s.c. Thus, as described previously [26], it was possible to induce a rise in serum estrogen concentrations to a level which was maintained, and to terminate the stimulus abruptly. To estimate the dosage of estradiol achieved by this treatment the implants were incubated *in vitro* for 2 days in 30 ml of a medium containing 0.1 % gelatin in phosphate-buffered saline with constant agitation at 37 °C. The medium was changed every 4 h. Under such conditions the capsules released 1.5 μg estradiol/day as measured by radioimmunoassay [14]. While this release rate provides an estimate of the daily dosage, it should be recognized that it may not replicate that occurring *in vivo*. When implants were inserted on the 15th day following ovariectomy and left in place for 4 days, daily LH surges were observed which were similar to those seen after s.c. injection of 50 μg EB in oil (fig. 4). Therefore, it appears that a large dose of estrogen is not required to elicit this response and that the relatively moderate increase in plasma estrogen concentrations presumably achieved by our implant is sufficient for induction of LH release in ovariectomized rats.

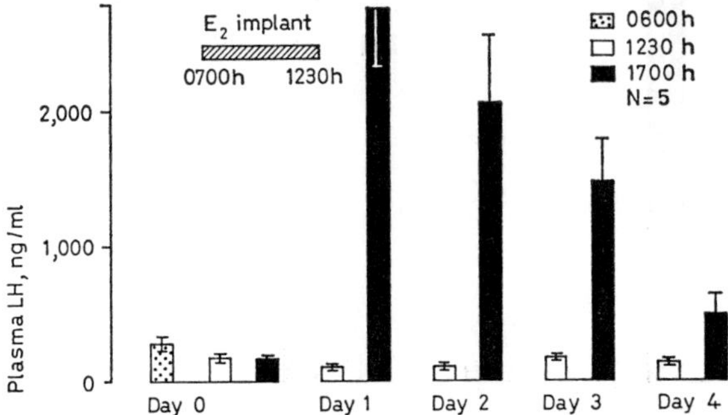

Fig. 5. Plasma LH concentrations (mean $\pm$ SE) in Spartan rats receiving Silastic implants containing estradiol-17β 2 weeks after ovariectomy.

To determine whether the repetitive LH surges in ovariectomized rats require continuously elevated plasma estrogen concentrations, Silastic capsules containing estradiol-17β were implanted s.c. at 0700 h and removed at 1230 h the next day, a stimulus duration similar to that occurring during the estrous cycle. An LH surge was observed during the release period on days 1–4 (fig. 5), demonstrating that a rise and sustained elevation of estrogen for only 29 $^1/_2$ h triggers an LH surge on at least 4 consecutive days in ovariectomized rats. Not only do these observations strengthen earlier findings that the LH surge in the rat can be initiated following withdrawal of the estrogen stimulus, they introduce the concept of memory in an endocrine feedback system. It is important to note, moreover, that both the negative and positive feedbacks of estrogen are maintained, or 'remembered' in ovariectomized rats.

V. Positive Feedback of Estrogen in Rats Ovariectomized on Diestrus or Proestrus

A. Endogenous *versus* Exogenous Stimuli

These results suggest the intriguing possibility that the preovulatory increase in circulating estrogen levels which occurs during the estrous cycle [7, 24, 45] might also trigger repetitive LH surges if the processes

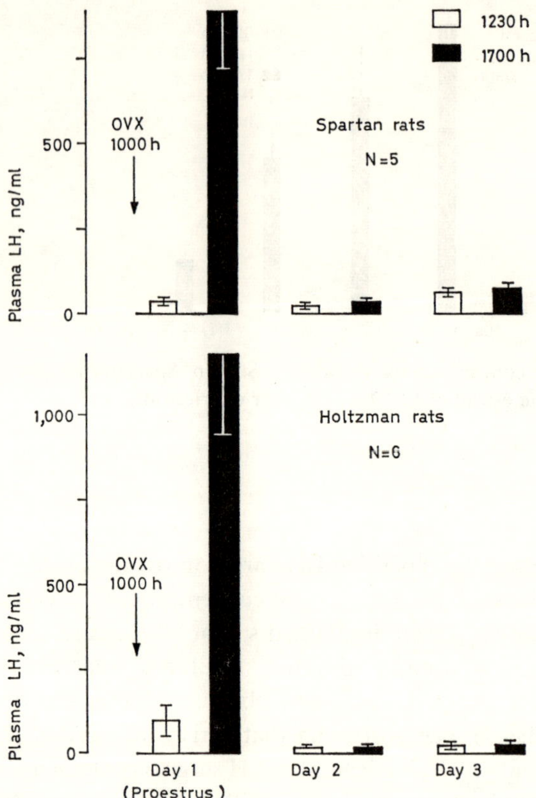

Fig. 6. Plasma LH concentrations (mean ± SE) in rats ovariectomized (OVX) on proestrus.

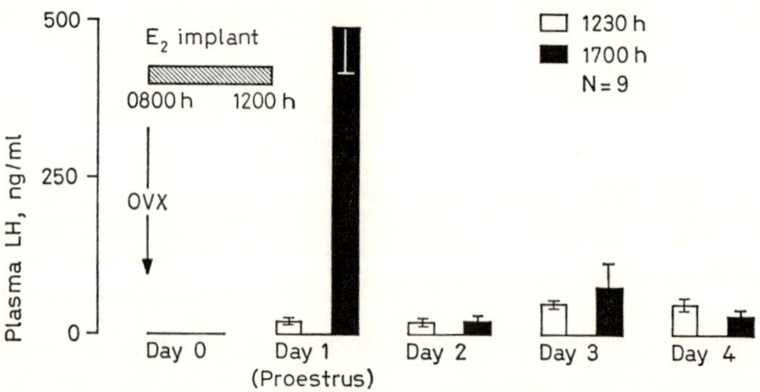

Fig. 7. Plasma LH concentrations (mean ± SE) in pseudo-intact rats.

of ovulation and corpus luteum formation were prevented. To investigate this possibility, intact female rats (Spartan and Holtzman) were ovariectomized at 1000 h on proestrus (day 1) and blood samples were withdrawn at 1230 and 1700 h for the next 2 days. On day 1, plasma LH concentrations at 1230 h were characteristic of those observed prior to initiation of the proestrous LH surge (fig. 6). The prominent increase in LH levels observed at 1700 h indicates that ovariectomy at 1000 h did not interfere with the preovulatory LH surge, a finding which agrees with previous observations [43]. On the next 2 days, however, no discharge of LH occurred (fig. 6).

This absence of daily LH surges in rats ovariectomized on proestrus may be attributed to a difference between either the competency of the LH release mechanisms in long-term ovariectomized (2 weeks) and recently ovariectomized rats, or the nature of the endogenous estrogen stimulus and that imposed by Silastic implants. The latter hypothesis was investigated by inserting Silastic capsules containing estradiol-17β s.c. immediately following ovariectomy at 0800 h on the day before proestrus. This experimental model will subsequently be referred to as the 'pseudo-intact' rat. On the following day (expected proestrus), the implants were removed at 1200 h; thus, the duration of the estrogen stimulus (28 h) was similar to that which had induced daily surges in long-term ovariectomized rats. Although an LH surge occurred during the release period on the day of expected proestrus (day 1), no subsequent LH surge was observed (fig. 7). This observation led to the conclusion that the failure to elicit daily LH surges in rats ovariectomized on proestrus or in pseudo-intact rats is a function of the competency of the LH release mechanism and not the type of estrogen stimulus.

B. Effect of Pentobarbital Administration

The foregoing observations in ovariectomized rats, in conjunction with the earlier findings of Everett and Sawyer in intact rats [16], indicate that a daily neural signal for LH release is given on at least 4 consecutive days following initiation of the estrogen stimulus. The failure to observe a memory in the estrogen-induced LH release mechanism of pseudo-intact rats or rats ovariectomized on proestrus does not necessarily mean that these daily signals are not given. Alternatively, it is possible that the occurrence of one LH surge renders the LH release mecha-

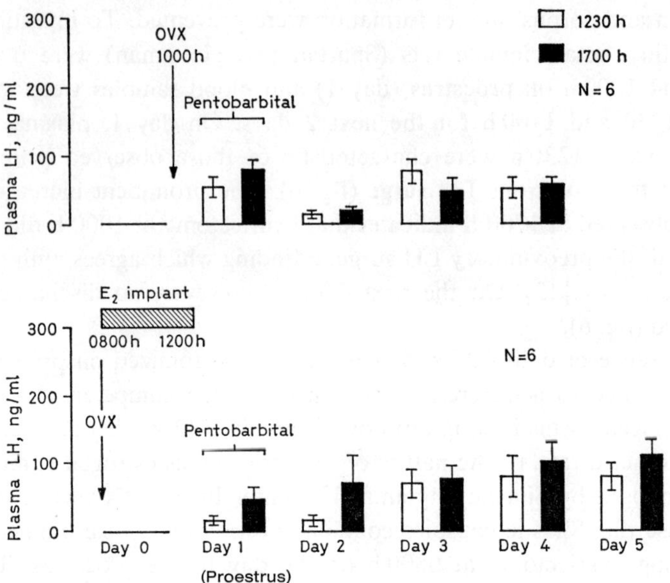

Fig. 8. Effect of administration of pentobarbital on plasma LH concentrations (mean ± SE) in rats ovariectomized on proestrus and in pseudo-intact rats.

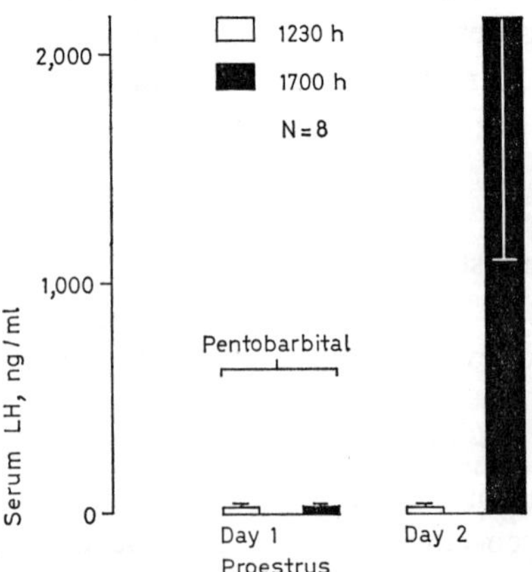

Fig. 9. Serum LH concentrations (mean ± SE) in intact rats given pentobarbital on proestrus.

nism in such rats unresponsive to these signals. To test this possibility, pentobarbital was administered during the first release period following withdrawal of the estrogen stimulus to rats ovariectomized at 1000 h on proestrus, and to pseudo-intact rats receiving estrogen implants from 0800 h on the day before proestrus until 1200 h the next day. No LH surge was observed during the release period on the day of pentobarbital blockade or on any of the next 3 days (fig. 8). The absence of an LH surge on the day following administration of pentobarbital was somewhat surprising in the light of EVERETT and SAWYER's observation [16] that ovulation, and presumably the preovulatory LH discharge, was delayed 1 day in intact rats treated with pentobarbital on proestrus. Therefore, we repeated EVERETT and SAWYER's experiment, and verified that an LH surge occurred during the first release period after pentobarbital administration (fig. 9), a result which strengthens the argument for the existence of a neural signal for LH release on at least 2 consecutive days in intact rats.

C. Daily LH Surges in Pseudo-Intact Rats

What accounts for the difference between intact rats and those ovariectomized on proestrus which determines whether or not an LH surge will occur on the day following pentobarbital blockade? The continuous presence of estrogen prior to the release period may be the distinguishing factor. As indicated by the biologic evidence of EVERETT and SAWYER [16] and direct radioimmunoassay measurements made by NAFTOLIN *et al.* [34], circulating estrogen concentrations in intact rats remained high for at least 1 day following administration of pentobarbital on proestrus. If this hypothesis is tenable, daily LH surges should be elicited in pseudo-intact rats merely by leaving the Silastic implants in place, rather than removing them after 28 h. That this was indeed the case may be seen in figure 10. This observation leads to the natural conclusion that the memory in the mechanism which mediates the positive feedback action of estrogen in long-term ovariectomized rats does not exist in pseudo-intact rats. Furthermore, these results show for the first time that in the pseudo-intact rat, an experimental model for the intact rat, LH surges and their antecedent neural signals can occur on at least 2 *consecutive* days.

The foregoing findings in pseudo-intact rats have important implications concerning the activation of the transducer which converts the

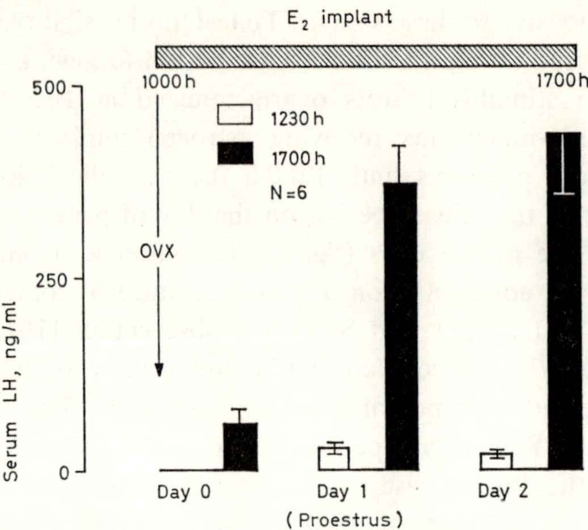

Fig. 10. Serum LH concentrations (mean ± SE) in pseudo-intact rats.

estrogen signal into an LH surge. Three general modes of input which are not necessarily mutually exclusive have been postulated: (1) an increase in circulating estrogen levels, which activates a rate sensor [26, 27, 33, 44, 50]; (2) attainment of a threshold concentration, which activates a level sensor [26, 44], and (3) a decrease in estrogen levels, which releases the hypothalamo-hypophyseal unit from negative feedback inhibition [33, 44, 50]. The latter possibility can be eliminated by the present observations in rats, and previous findings in sheep [21] and monkeys [26], that LH surges can be initiated prior to a decline in circulating estrogen levels. As shown above, in pseudo-intact rats, maintenance of a threshold estrogen concentration is indispensable for the surge(s) subsequent to the initial discharge of LH. Regarding a rate sensor, we postulate that either there is not a rate-sensitive element or, if there is, the increase in circulating estradiol concentrations is 'remembered' for at least 2 days in pseudo-intact rats.

VI. Modulation of Pituitary Responsiveness to LRF by Estrogen

At this point, we investigated the site of the positive feedback action of estrogen on LH secretion in the rat. Estrogen may induce LH release by increasing hypothalamic LRF secretion [9, 13, 38, 47], by changing

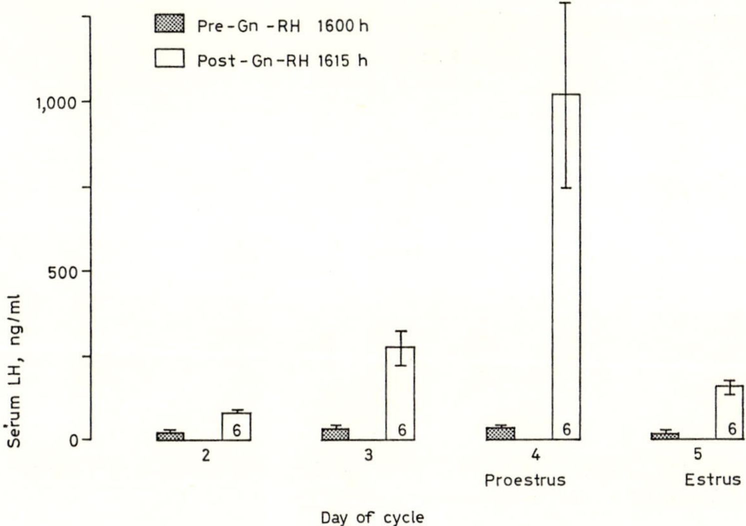

Fig. 11. Serum LH concentrations (mean ± SE) during the 5-day estrous cycle in response to Gn-RH (200 ng) in pentothal-treated Holtzman rats.

pituitary sensitivity to LRF [9, 13, 38, 47], or both. To investigate whether estrogen modulates pituitary responsiveness to LRF, synthetic LRF (Gn-RH)[2] was administered to Holtzman rats at 1600 h on 4 days of the 5-day estrous cycle. To eliminate possible variations in endogenous LH and LRF which occur at 1600 h, pentothal was administered at 1200 h. As shown in figure 11, pituitary responsiveness to Gn-RH remained low with minor variations, except on proestrus, when there was a dramatic increase in sensitivity. This increased responsiveness was abolished when the preovulatory increase in estrogen levels was prevented by ovariectomy at 0800 h on the day before proestrus (fig. 12). A similar potentiation of pituitary sensitivity to exogenous Gn-RH during the preovulatory phase of the estrous cycle, when circulating estrogen levels are high, has been demonstrated in sheep [39] and hamsters [2]. These observations, in addition to earlier findings that estrogen acts directly on the pituitary to stimulate LH release *in vitro* [37, 42] and that it causes

2 The term 'Gn-RH' is used in this discussion to designate the synthetic decapeptide generously supplied by W. F. WHITE, Abbott Laboratories. The term 'LRF' designates the natural LH releasing hormone(s). The relationship between the activities of these substances remains to be established.

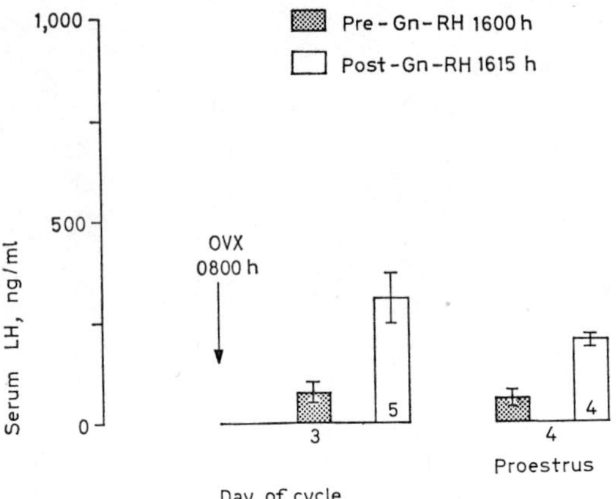

Fig. 12. Response to Gn-RH (200 ng) in pentothal-treated Holtzman rats ovariectomized 1 day before proestrus.

increased sensitivity to LRF [3, 12, 40], support the hypothesis that the preovulatory increment in estrogen secretion causes an increase in pituitary responsiveness to LRF on proestrus, although the possibility that other ovarian steroids modulate this response has not yet been eliminated.

In the light of these and the above observations, it may be postulated that the presence of a memory in the estrogen-induced LH release mechanism in long-term ovariectomized rats, and its absence in pseudo-intact rats or rats ovariectomized on proestrus, may be attributed to the duration of the estrogen-induced increase in pituitary responsiveness to LRF. Experiments to test this hypothesis are now in progress.

VII. The Pseudo-Intact Rat – A Versatile Experimental Model

In the above discussion, we have described a new model for examining the control of LH secretion, the pseudo-intact rat. The characteristics of precise timing of both stimulus and response, and the repetitiveness of an easily measured event, the LH surge, render the pseudo-intact rat eminently useful for studying biological clock mechanisms, as well as for investigating the sites and mechanisms of feedback action of steroids on gonadotropin release in the whole animal, and at the cellular or molecular level. The periodicity inherent in the model permits individual animals to serve as their own controls. In addition, nonspecific short-term effects of surgical stress and/or anesthesia on the LH release mechanism can readily be distinguished from the specific effects of experimental manipulations. This model also provides an opportunity to investigate a unique phenomenon, that of memory in an endocrine feedback system. Finally, it should provide a wealth of information about the physiology of the estrous cycle, especially if physiologic concentrations of circulating steroids are replicated.

Summary

The phasic mode of LH secretion in the rat is triggered prior to the release period by an increase in circulating estrogen concentrations to a level which is maintained for a finite interval. This LH release mechanism functions with a 24-hour periodicity, is expressed only during the release period, and has a neural component. In long-term ovariectomized rats, the estrogen stimulus can be 'remembered' for at least 3 days after its withdrawal, whereas in pseudo-intact rats, a continuous estrogen stimulus is required for the appearance of daily LH surges. The data presented here suggest that the positive feedback action of estrogen on LH secretion can be accounted for, at least in part, by an increase in pituitary sensitivity to LRF. Finally, the foregoing observations are consonant with the hypothesis that daily neural signals for an LH surge are given throughout the estrous cycle of the rat and that the LH discharge occurs only when the appropriate estrogen stimulus has sensitized the pituitary to LRF, thus obviating the need for an action of estrogen on the hypothalamic surge center and localizing the positive feedback action of estrogen at the level of the pituitary.

Acknowledgements

The authors would like to express their appreciation and gratitude to Dr. A. Rees Midgley, jr., for his genuine interest, enthusiastic encouragement and perceptive suggestions throughout our effort, and for unselfishly providing the opportunity to present these studies. We would also like to thank Gary Coon and Cathie Stepien for their excellent technical assistance and for those invaluable contributions which only students can give and teachers can recognize.

References

1 Ajika, K.; Kalra, S. P.; Fawcett, C. P.; Krulich, L., and McCann, S. M.: The effect of stress and Nembutal on plasma levels of gonadotropins and prolactin in ovariectomized rats. Endocrinology *90:* 707–715 (1972).

2 Arimura, A.; Debeljuk, L., and Schally, A. V.: LH release by LH-releasing hormone in golden hamsters at various stages of estrous cycle. Proc. Soc. exp. Biol. Med. *140:* 609–612 (1972).

3 Arimura, A. and Schally, A. V.: Augmentation of pituitary responsiveness to LH-releasing hormone (LH-RH) by estrogen. Proc. Soc. exp. Biol. Med. *136:* 290–293 (1971).

4 Barraclough, C. A.: Modification in the CNS regulation of reproduction after exposure of prepubertal rats to steroid hormones. Recent Progr. Hormone Res. *22:* 503–539 (1966).

5 Beattie, C. W.; Campbell, C. S.; Nequin, L. G.; Soyka, L. F., and Schwartz, N. B.: Barbiturate blockade of tonic LH secretion in the male and female rat. Endocrinology *90:* 1634–1638 (1973).

6 Blackwell, R. E. and Amoss, M. S., jr.: A sex difference in the rate of rise of plasma LH in rats following gonadectomy. Proc. Soc. exp. Biol. Med. *136:* 11–14 (1971).

7 Brown-Grant, K.; Exley, D., and Naftolin, F.: Peripheral plasma oestradiol and luteinizing hormone concentrations during the oestrous cycle of the rat. J. Endocrin. *48:* 295–296 (1970).

8 Caligaris, L.; Astrada, J. J., and Taleisnik, S.: Release of luteinizing hormone induced by estrogen injection into ovariectomized rats. Endocrinology *88:* 810–815 (1971).

9 Chowers, I. and McCann, S. M.: Content of luteinizing hormone-releasing factor and luteinizing hormone during the estrous cycle and after changes in gonadal steroid titers. Endocrinology *76:* 700–708 (1965).

10 Daane, T. A. and Parlow, A. F.: Periovulatory patterns of rat serum follicle stimulating hormone and luteinizing hormone during the normal estrous cycle: effects of pentobarbital. Endocrinology *88:* 653–663 (1971).

11 Davidson, J. M.: Feedback control of gonadotropin secretion; in Ganong and Martini Frontiers in neuroendocrinology, 1969, pp. 343–388 (Oxford University Press, New York 1969).

12 Debeljuk, L.; Arimura, A., and Schally, A. V.: Effect of estradiol and progesterone on the LH release induced by LH-releasing hormone (LH-RH) in intact diestrous rats and anestrous ewes. Proc. Soc. exp. Biol. Med. *139:* 774–777 (1972).

13 Döcke, F. and Dörner, G.: The mechanism of the induction of ovulation by estrogens. J. Endocrin. *33:* 491–499 (1965).

14 England, B. G.; Niswender, G. D., and Midgley, A. R., jr.: Radioimmunoassay of estradiol-17β without chromatography (in press).

15 Everett, J. W.: Central neural control of reproductive functions of the adenohypophysis. Physiol. Rev. *44:* 373–431 (1964).

16 Everett, J. W. and Sawyer, C. H.: A 24-hour periodicity in the 'LH-release apparatus' of female rats, disclosed by barbiturate sedation. Endocrinology *47:* 198–218 (1950).

17 Ferin, M.; Tempone, A.; Zimmerung, P. E., and Vande Wiele, R. L.: Effect of antibodies to 17β-estradiol and progesterone on the estrous cycle of the rat. Endocrinology *85:* 1070–1078 (1969).

18 Gay, V. L. and Midgley, A. R., jr.: Response of the adult rat to orchidectomy and ovariectomy as determined by LH radioimmunoassay. Endocrinology *84:* 1359–1364 (1969).

19 Gay, V. L.; Midgley, A. R., jr., and Niswender, G. D.: Patterns of gonadotropin secretion associated with ovulation. Fed. Proc. *29:* 1880–1887 (1970).

20 Gay, V. L.; Rebar, R. W., and Midgley, A. R., jr.: Constant monitoring of plasma luteinizing hormone by radioimmunoassay in individual rats following injection of hypothalamic extract. Proc. Soc. exp. Biol. Med. *130:* 1344–1347 (1969).

21 Goding, J. R.; Catt, K. J.; Brown, J. M.; Kaltenbach, C. C.; Cumming, I. A., and Mole, B. J.: Radioimmunoassay for ovine luteinizing hormone. Secretion of luteinizing hormone during estrus and following estrogen administration in the sheep. Endocrinology *85:* 133–142 (1969).

22 Gorski, R. A.: Gonadal hormones and the perinatal development of neuroendocrine function; in Martini and Ganong Frontiers in neuroendocrinology, 1971, pp. 237–290 (Oxford University Press, New York 1971).

23 Halasz, B.: The endocrine effects of isolation of the hypothalamus from the rest of the brain; in Ganong and Martini Frontiers in neuroendocrinology, 1969, pp. 307–342 (Oxford University Press, New York 1969).

24 Hori, T.; Ide, M., and Miyake, T.: Ovarian estrogen secretion during the estrous cycle and under the influence of exogenous gonadotropins in rats. Endocrin. Jap. *15:* 215–222 (1968).

25 Karsch, F. J.; Dierschke, D. J.; Weick, R. F.; Yamaji, T.; Hotchkiss, J., and Knobil, E.: Positive and negative feedback control by estrogen of luteinizing hormone secretion in the rhesus monkey. Endocrinology *92:* 799–804 (1973).

26 Karsch, F. J.; Weick, R. F.; Butler, W. R.; Dierschke, D. J.; Krey, L. C.; Weiss, G.; Hotchkiss, J.; Yamaji, T., and Knobil, E.: Induced LH surges in the rhesus monkey: strength-duration characteristics of the estrogen stimulus. Endocrinology *92:* 1740–1747 (1973).

27 Knobil, E.; Dierschke, D. J.; Yamaji, T.; Karsch, F. J.; Hotchkiss, J., and Weick, R. F.: Role of estrogen in the positive and negative feedback control of LH secretion during the menstrual cycle of the rhesus monkey; in Saxena, Beling and Gandy Gonadotropins, pp. 72–86 (Wiley, New York 1972).

28 Labhsetwar, A. P.: Role of estrogens in ovulation: a study using the estrogen-antagonist, ICI 46,474. Endocrinology *87:* 542–551 (1970).

29 Legan, S. J.; Coon, G. A., and Midgley, A. R., jr.: Evidence for a 24-hour periodicity of LH and FSH release in estrogen-treated ovariectomized rats. Fed. Proc. *32:* 240 (1973).

30 Martin, J. E.; Tyrey, L.; Everett, J. W., and Fellows, R. E.: Variation in pituitary response to synthetic luteotropin releasing factor (LRF) in the cycling rat. Fed. Proc. *32:* 240 (1973).

31 McCann, S. M.; Dhariwal, A. P. S., and Porter, J. C.: Regulation of the adenohypophysis. Annu. Rev. Physiol. *30:* 589–640 (1968).

32 MIYAKE, T.: Causal relationship between ovarian steroid secretion and pituitary luteinizing hormone release in the rat estrous cycle. Endocrin. Jap. Suppl. 1, pp. 83–92 (1969).

33 MONROE, S. E.; JAFFE, R. B., and MIDGLEY, A. R., jr.: Regulation of human gonadotropins. XII. Increase in serum gonadotropins in response to estradiol. J. clin. Endocrin. *34:* 342–347 (1972).

34 NAFTOLIN, F.; BROWN-GRANT, K., and CORKER, C. S.: Plasma and pituitary luteinizing hormone and peripheral plasma oestradiol concentrations in the normal oestrous cycle of the rat and after experimental manipulation of the cycle. J. Endocrin. *53:* 17–30 (1972).

35 NEILL, J. D.: Sexual differences in the hypothalamic regulation of prolactin secretion. Endocrinology *90:* 1154–1159 (1972).

36 NISWENDER, G. D.; MIDGLEY, A. R., jr.; MONROE, S. E., and REICHERT, L. E., jr.: Radioimmunoassay for rat luteinizing hormone with antiovine LH serum and ovine LH-^{131}I. Proc. Soc. exp. Biol. Med. *128:* 807–811 (1968).

37 PIACSEK, B. E. and MEITES, J.: Effects of castration and gonadal hormones on hypothalamic content of luteinizing hormone releasing factor (LRF). Endocrinology *79:* 432–439 (1966).

38 RAMIREZ, V. D. and SAWYER, C. H.: Fluctuations in hypothalamic LH-RF (luteinizing hormone-releasing factor) during the rat estrous cycle. Endocrinology *76:* 282–289 (1965).

39 REEVES, J. J.; ARIMURA, A., and SCHALLY, A. V.: Pituitary responsiveness to purified luteinizing hormone-releasing hormone (LH-RH) at various stages of the estrous cycle in sheep. J. Anim. Sci. *32:* 123–126 (1971).

40 REEVES, J. J.; ARIMURA, A., and SCHALLY, A. V.: Changes in pituitary responsiveness to luteinizing hormone-releasing hormone (LH-RH) in anestrous ewes pretreated with estradiol benzoate. Biol. Reprod. *4:* 88–92 (1971).

41 SCARAMUZZI, R. J.; TILLSON, S. A.; THORNEYCROFT, I. H., and CALDWELL, B. V.: Action of exogenous progesterone and estrogen on behavioral estrus and luteinizing hormone levels in the ovariectomized ewe. Endocrinology *88:* 1184–1189 (1971).

42 SCHNEIDER, H. P. G. and MCCANN, S. M.: Estradiol and the neuroendocrine control of LH release *in vitro*. Endocrinology *87:* 330–338 (1970).

43 SCHWARTZ, N. B.: Acute effects of ovariectomy on pituitary LH, uterine weight, and vaginal cornification. Amer. J. Physiol. *207:* 1251–1259 (1964).

44 SCHWARTZ, N. B.: A model for the regulation of ovulation in the rat. Recent Progr. Hormone Res. *25:* 1–55 (1969).

45 SHAIKH, A. A.: Estrone and estradiol levels in the ovarian venous blood from rats during the estrous cycle and pregnancy. Biol. Reprod. *5:* 297–307 (1971).

46 SPIES, H. G. and NISWENDER, G. D.: Effect of progesterone and estradiol on LH release and ovulation in rhesus monkeys. Endocrinology *90:* 257–261 (1972).

47 WEICK, R. F. and DAVIDSON, J. M.: Localization of the stimulatory feedback effect of estrogen on ovulation in the rat. Endocrinology *87:* 693–700 (1970).

48 YAMAMOTO, M.; DIEBEL, N. B., and BOGDANOVE, E. M.: Analysis of initial and delayed effects of orchidectomy and ovariectomy on pituitary and serum LH levels in adult and immature rats. Endocrinology *86:* 1102–1111 (1970).

49 YEN, S. S. C. and TSAI, C. C.: The biphasic pattern in the feedback action of ethinyl estradiol on the release of pituitary FSH and LH. J. clin. Endocrin. *33:* 882–887 (1971).
50 YEN, S. S. C. and TSAI, C. C.: Acute gonadotropin release induced by exogenous estradiol during the mid-follicular phase of the menstrual cycle. J. clin. Endocrin. *34:* 298–305 (1972).

Authors' address: Dr. SANDRA J. LEGAN and Dr. F. J. KARSCH, Reproductive Endocrinology Program, Department of Pathology, University of Michigan, *Ann Arbor, MI 48104* (USA)

Recent Studies of Hypothalamic Function
Int. Symp. Calgary 1973, pp. 166–179 (Karger, Basel 1974)

Neurohumoral Regulation of Neuroendocrine Cells in the Hypothalamus

J. N. HAYWARD

Departments of Neurology and Anatomy, Reed Neurological Research Center and Brain Research Institute, University of California, Los Angeles, Calif.

Spontaneous Activity of Magnocellular Neuroendocrine Cells

The systems in the hypothalamus which control pituitary secretions are the 'parvicellular' neuroendocrine cells which synthesize, transport and secrete hypothalamic releasing hormones, and the 'magnocellular' neuroendocrine cells which synthesize, transport and release neurohypophysial hormones and neurophysins. Although the exact location of the cell bodies and axonal processes of the 'parvicellular' neurons in the basal hypothalamus is not known, some information about the 'periodic' nature of their firing patterns can be deduced from he 'pulsatile' type of release of adenohypophysial hormones [5]. Otherwise we must rely on information from 'magnocellular' neurons for information about the elusive 'parvicellular' neuroendocrine cell.

Magnocellular neuroendocrine cells of reptiles, birds and mammals evolve from the preoptic nucleus of cyclostomes and fishes [3, 12] which divides into the supraoptic (NSO) and paraventricular (NPV) nuclei and the internuclear zone of Greving (INZ) [13, 14]. In mammals magnocellular neuroendocrine cells synthesize and package into neurosecretory vesicles the three neurohypophysial hormones, vasopressin, oxytocin and coherin [10], along with the three neurophysins, I, II and III [29, 34]. These membrane-bound vesicles are transported via axoplasmic flow to the posterior pituitary gland for release by an action potential, secretion-coupled process of exocytosis [23]. Anatomical and chemical data [34] suggest the presence of two or possibly three chemically distinct magnocellular neuroendocrine cells in the hypothalamus, each responsible for a set of secretory products: neurophysin I-oxytocin; neurophysin II-vasopressin and neurophysin III-coherin [13, 14, 29, 34]. Under such a hypothesis the input-output requirements for synthesis and release of each particular neurohypophysial hormone would differ for each of these cell types, resulting in several functional classes of magnocellular neuroendocrine cells.

Functional cell types. In our initial study of 'nonidentified' osmosensitive supraoptic neurons in the hypothalamus of the unanesthetized monkey, we found but one type of spontaneous cell-firing pattern, the 'continuously active' neuron [17, 32]. Later, studying antidromically 'identified' magnocellular neuroendocrine cells in the NSO and the INZ in the unanesthetized monkey, we found three spontaneous firing patterns: 'silent' cells (3 %) discovered only by pituitary gland stimulation (fig. 1), 'continuously active' cells (63 %) with regularly irregular activity (fig. 1) and 'burster' neurons (21 %) with alternate periods of discharge and

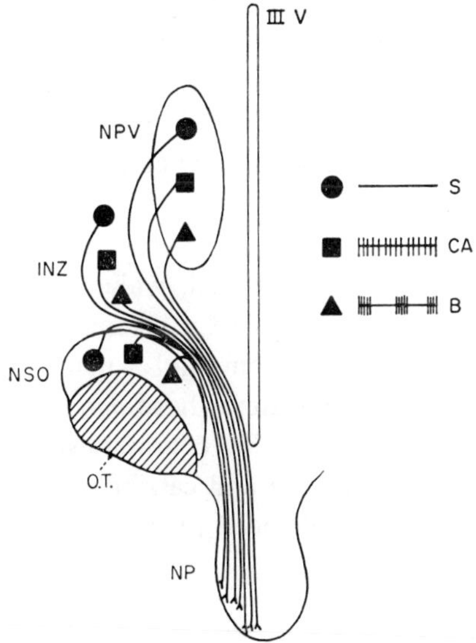

Fig. 1. Schematic drawing of the distribution of the three functional types of magnocellular neuroendocrine cells in the hypothalamus of the mammal. Spontaneous firing patterns, 'silent' (S), 'continuously active' (CA) and 'burster' (B) neurons, distributed throughout the supraoptic (NSO) and paraventricular nuclei (NPV) and the internuclear zone of Greving (INZ) in the hypothalamus with axons descending into the posterior pituitary gland (NP). III V = third ventricle. O.T. = optic tract.

silence (fig. 1–3) [13, 18]. These cells are distributed in an apparent random manner without 'nuclear', laminar or columnar grouping throughout the NSO and INZ (fig. 1). Stimulation of the posterior pituitary gland evoked antidromic potentials in neurons in NSO and INZ at stable latencies ranging from 5.2 to 12.0 m/sec, at conduction velocities ranging from 0.58 to 1.35 m/sec [13, 18].

'Burster' (B) neurons. The B neurons showed regular, repetitive periods of firing with a mean burst duration of 5 sec, at mean rates of 5 spikes/sec, alternating with periods of silence of mean 12 sec or a mean cycle length of 17 sec (fig. 1–3) [33]. B neuroendocrine cells maintained a regular, rhythmic pattern of repetitive discharge in time, uninterrupted by mildly arousing sensory stimuli or sleep-waking behavior. Serial interspike interval histograms showed no particular pattern of burst discharge, such as the 'parabolic burster' of Aplysia [28], beyond an alternation of long and short intervals. The physiological basis for this phasic discharge of these B neurons may be a recurrent collateral inhibitory pathway (fig. 4), an 'endogenous' pacemaker mechanism [28] or an exogenous 'phasic' input. We suggest that these B magnocellular neuroendocrine cells have 'specific' input connections and are responsible for the 'pulsatile' release of one of the three neurohypophysial hormones (vasopressin, oxytocin, coherin) and of one of the neurophysins (I–III) [13, 14].

Dynamic secretory state. In the lightly restrained, waking monkey, each of these three cell types should be responsible for a separate set of secretory products, i.e. a neurohypophysial hormone and a neurophysin. We speculate that the 'silent' cell represents a state of hormone synthesis, packaging in vesicles and axoplasmic transport, the 'continuously active' cell reflects a stage of 'tonic' release of hormone and carrier protein, the B neuron, a cell involved in the 'pulsatile' enhanced release of secretory products.

Osmotic Input to Magnocellular Neuroendocrine Cells

Blood-brain barrier. In his classical Croonian lecture on the antidiuretic hormone, VERNEY [30] described three main physiological stimuli that determine the release of vasopressin from the neurohypophysis: (1) osmotic pressure of the carotid arterial blood; (2) volume of blood in the vascular system, and (3) behavioral state of the mammal. At the present time we know more about his 'osmometric' hypothesis [11]. Yet we still do not know whether the 'osmoreceptors' of VERNEY are supraoptic neurons [4, 19, 30], separate cells in the perinuclear zone synaptically connected to the supraoptic neurons [1, 17, 32] or elements lying outside or inside the blood-brain barrier such as a sodium detector in the vicinity of the third ventricle [7]. On the basis of his *short-term* and *long-term* intracarotid infusions of hypertonic solutions in the conscious dog VERNEY [30] concluded that substances such as sodium salts and sucrose, unable to penetrate through the hypothetical hypothalamic 'osmoreceptors', caused dehydration and volume reduction of these receptors with a consequent antidiuretic hormone release. When he found that intracarotid hypertonic urea did not produce an antidiuretic response, he did not consider the blood-brain barrier [24], but concluded that this substance was freely diffusible through the 'osmoreceptor' membrane. Recent studies indicate that while the blood-brain barrier is impermeable to low concentrations of urea [24, 25], thereby possibly negating VERNEY's hypothesis, infusions of intracarotid urea in high concentrations (2 M) may damage the blood-brain barrier and make it permeable to urea and other substances [25].

Osmoreceptors of VERNEY [30] and SAWYER [27]. To resolve some of these conflicts regarding VERNEY'S 'osmoreceptors' [30] and the blood-brain barrier we chose to use our stylized technique of single cell recording in the hypothalamus of the chronically prepared mammal. Initially we found that 60 % of the hypothalamic neurons changed firing rates with a shift in sleep-waking behavior [9]. When we injected hypertonic sodium chloride intracarotid in the unanesthetized monkey we discovered a variety of 'continuously active' osmosensitive neurons [17]. On the basis of the anatomical location of the cells, the pattern of discharge to intracarotid osmotic stimuli and the response to arousing sensory stimuli, we divided the osmosensitive cells into two major groups, 'specific' and 'nonspecific' osmosensitive cells [17]. We labeled 50 % of the osmosensitive cells as 'specific' because they responded to an intracarotid injection of hypertonic sodium chloride, generally did not respond to non-noxious arousing sensory stimuli and were located in or near the NSO. Of these we found two subtypes, 20 % of supraoptic cells with 'biphasic' osmosensitive responses, that is, acceleration followed by inhibition, and 30 % of cells in the perinuclear zone (PNZ) of the NSO with 'monophasic' osmosensitive responses, acceleration or inhibition. 50 % of our osmosensitive cells we labeled as 'nonspecific' because they responded both to intracarotid injections of hypertonic sodium chloride and also to mildly arousing sensory stimuli with 'monophasic' excitatory or inhibitory responses [17].

We concluded from this study of 'continuously active' 'nonidentified' osmosensitive cells (fig. 4) that the 'specific' 'biphasic' osmosensitive neurons in NSO were the magnocellular neuroendocrine cells; the 'specific' 'monophasic' osmosensitive neurons in the PNZ were the 'osmoreceptors' of VERNEY [30]; the 'nonspecific' 'monophasic' osmosensitive neurons in the PNZ were the 'osmoreceptors' of SAWYER [27]. We speculated that the 'osmoreceptors' of VERNEY were separate neurons lying adjacent to the NSO [17, 32] with osmotic input across the blood-brain barrier of the capillaries and connected to the magnocellular neuroendocrine cells by excitatory axosomatic synapses [32] (fig. 4). We further suggested that the 'osmoreceptors' of SAWYER [27] might be involved in the nonendocrine and behavioral effects of intracarotid osmotic stimulation, such as drinking and arousal [11] (fig. 4).

Antidromically 'identified' neurons. In our antidromically 'identified' magnocellular neuroendocrine cells in the NSO and the INZ we find 'specific' 'biphasic' osmosensitive responses to intracarotid hypertonic NaCl in each of the three spontaneous functional types: 'silent' (S), 'continuously active' (CA) and 'burster' (B) [14, 16] (fig. 2). With appropriate sodium chloride loading we could shift some of the S cells transiently to CA firing, cause an occasional CA cell to 'burst' briefly, and make a few B cells develop a 'hyperburster' state (fig. 2) [14, 16]. Intracarotid infusions of hypertonic solutions of D-glucose (1.2–2.5 M) produce a 'specific' 'biphasic' osmosensitive response in these cells (fig. 3) [16]. In the same magnocellular neuroendocrine cell it required a 2- to 3-times higher level of osmolality of the intracarotid infusion of D-glucose to produce the same degree of 'biphasic' osmosensitive response as found with intracarotid hypertonic NaCl [16]. Others also report intracarotid sodium chloride 2- to 3-times more effective, osmol for osmol, than intracarotid D-glucose in driving 'nonidentified' supraoptic neurons [4, 19, 21] and producing shifts in hypothalamic steady potentials [8]. The behavioral sequence of EEG 'arousal', irregular sniffing respiration, lip and tongue smacking, chewing and associated mildly increased face, eye and body movements occurred with both intracarotid hypertonic D-glucose [16] and hypertonic NaCl [14, 16–18].

Osmoreceptors of Verney vs. cerebrospinal fluid (CSF) sodium ion detector. On the basis of our studies in the unanesthetized monkey where *5-sec* intracarotid injections of both hypertonic sodium chloride [14, 16–18] (fig. 2) and hypertonic D-glucose (fig. 3) produce an *immediate* (1- to 5-sec delay) increased firing rate in 'identified' and 'nonidentified' magnocellular neuroendocrine cells, it seems likely that such a stimulation is too rapid to be detected as a *slow* change in sodium ion in the CSF [7]. It is more likely to have been a *rapid* 'osmotic' effect across the endothelial barrier of the rich capillary bed of these hypothalamic neurons. While these results with *short-term* intracarotid infusions suggest an 'osmoreceptor' mechanism in VERNEY's sense [30], the results of *long-term* infusions by VERNEY [30] and ERIKSSON *et al.* [7] may favor a basically different mechanism such as a sodium ion detector in the vicinity of the third ventricle. It is not known whether other potentially osmotically effective agents, such as urea and the hexoses [7, 24], can be injected intracarotid in such a way as not to disturb the blood-brain barrier [25, 30] and still activate the osmoreceptor-supraoptic nuclear complex in the unanesthetized monkey.

Behavior and the Magnocellular Neuroendocrine Cells

Slow sleep. The behavioral state of a mammal is a major determinant for the release of the antidiuretic hormone, vasopressin, from the neurohypophysis. For example, the paradoxical antidiuresis of sleep [15] occurs in the face of a falling brain temperature and increasing intrathoracic blood volume [11], physiological changes which are associated with a water diuresis in waking mammals [11]. The 'continuously active' magnocellular neuroendocrine cells fired during waking in a regularly irregularly pattern without response to mildly arousing sensory stimuli [13]. During slow sleep these same cells developed a distinctly rhythmic, periodic discharge [15]. In addition, during brief periods of higher voltage EEG these cells showed accelerated discharge at 10 spikes/sec. With lower voltage EEG these cells slowed to 2 spikes/sec with a burst frequency of 1 burst every 10 sec [15]. The relationship of these 'sleep' bursts of magnocellular neuroendocrine cells to vasopressin release is not known.

Nociceptor stimuli. Pain, a powerful determinant of the behavioral state, causes the release of vasopressin from the posterior pituitary gland.

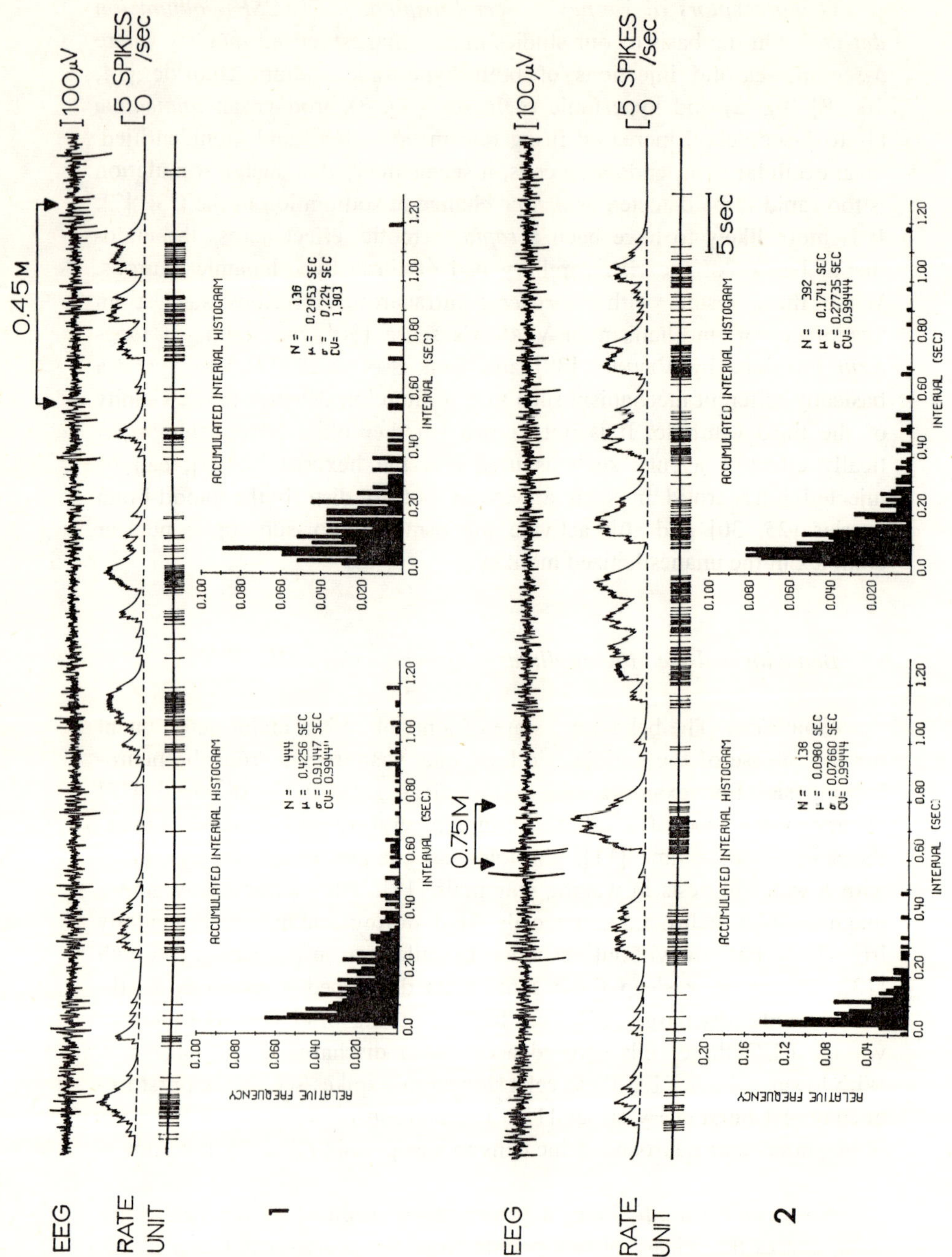
EEG
RATE
UNIT
1
0.45M
100μV
5 SPIKES /sec
0
ACCUMULATED INTERVAL HISTOGRAM
N = 444
μ = 0.4256 SEC
σ = 0.91447 SEC
CV= 0.9944
ACCUMULATED INTERVAL HISTOGRAM
N = 136
μ = 0.2053 SEC
σ = 0.224 SEC
CV= 1.903
RELATIVE FREQUENCY
INTERVAL (SEC)
EEG
RATE
UNIT
2
0.75M
100μV
5 SPIKES /sec
0
5 sec
ACCUMULATED INTERVAL HISTOGRAM
N = 138
μ = 0.0980 SEC
σ = 0.07660 SEC
CV= 0.99444
ACCUMULATED INTERVAL HISTOGRAM
N = 392
μ = 0.1741 SEC
σ = 0.27735 SEC
CV= 0.99444
RELATIVE FREQUENCY
INTERVAL (SEC)

VERNEY [30] applied electrical stimuli to subcutaneous areas of conscious dogs to the point of 'resentment', attributing the prolonged antidiuretic response to 'emotional stress'. Electrical stimulation of central pain pathways [26] and sites in the midbrain and limbic system suspected of motivational and affective linkage [11, 26] caused the release of antidiuretic hormone from the neurohypophysis. When we applied natural, non-noxious, mildly arousing sensory stimuli, such as light, sound and touch, to our unanesthetized monkeys (fig. 3) we found no change in the discharge rates of our 'nonidentified' 'specific' 'biphasic' osmosensitive supraoptic neurons [17, 32] or in our 'identified' magnocellular neuroendocrine cells [13, 15, 18]. These results supported studies of CROSS and GREEN [4], JOYNT [19], and KOIZUMI and YAMASHITA [22] and contradicted the results of KOIZUMI *et al.* [21]. When we applied noxious stimuli to the skin we found accelerated cell discharge in some of these B neurons (fig. 3) [15]. Since these responses appeared to be 'conditioned' [2] and showed 'habituation', we concluded, as did VERNEY [30] and CORSON [2], that an important element in the nociceptor activation of supraoptic neurons is the degree of 'emotional stress' associated with the cutaneous stimulus [15].

Fig. 2. Osmosensitivity of an antidromically 'identified' 'burster' magnocellular neuroendocrine cell in the caudal-lateral part of the supraoptic nucleus in the conscious monkey. *Osmotic stimulation.* Upper 1: intracarotid infusion of 5 ml of 0.45 M NaCl solution in 30 sec (0.17 ml/sec) induces a period of enhanced 'bursting' with an overall acceleration of mean firing rate (two injection periods) from a control of 3.5 spikes/sec to the 'hyperburster' state of 4.9 spikes/sec. Note similar control (left) and osmotic (right) accumulated interspike interval histograms. Lower 2: 'triphasic' osmosensitive response to intracarotid infusion of 1 ml of 0.75 M NaCl in 5 sec (0.20 ml/sec) with initial 'biphasic', excitatory-inhibitory discharge followed by an enhanced 'hyperburster' state, all lasting for a total of 50 sec. Accumulated statistics and histograms for four such intracarotid injections and 'triphasic' sequences show mean firing rate of 10.2 spikes/sec during the initial 'biphasic' osmosensitive response (left) and a mean firing rate of 5.8 spikes/sec during the subsequent 'hyperburster' state (right) with similar 'assymetric' histograms. Control bursting of this supraoptic neuron was 4.8-sec duration, 4.8 spikes/sec, 22 spikes per burst with a new burst every 12.8 sec. Electrical stimuli applied to the posterior pituitary gland evoked antidromic potentials of 7.0 msec latency with collision of orthodromic and antidromic spikes. EEG = biparietal electrocorticogram; rate = analog output proportional to the rate of unit discharge; unit = pulse output from pulse height discriminator triggered by action potentials of the spike in the window; N = number of intervals; μ = mean interspike intervals; σ = standard deviation; CV = coefficient of variation [from HAYWARD and JENNINGS, (upper 1: unpublished; lower 2: modified from ref. No 14)].

2.5 M Dex
1.2 M Dex
EEG
Move
1
Rate
Unit
100 µV
1 mV
22
0
Spikes/sec
Sound
Pain
0.15 M NaCl
EEG
Move
2
Rate
Unit
100 µV
1 mV
22
0
Spikes/sec
5 sec

3

Aff. Non. Sp.
Osm. S.
D. Ne.
Ct
OSMORECEPTORS OF
SAWYER, VERNEY
BBB
H2O
Osm. V.
Hy. Ec.
III
CSF
Osm.
Hy. Ec.
Aff. V.
Ef
Aff. B
NEUROENDOCRINE
CELLS
Mag. Ne.
('S', 'CA', 'B')
R
INTERNEURONS
Col.
A
POSTPITUITARY
Cf

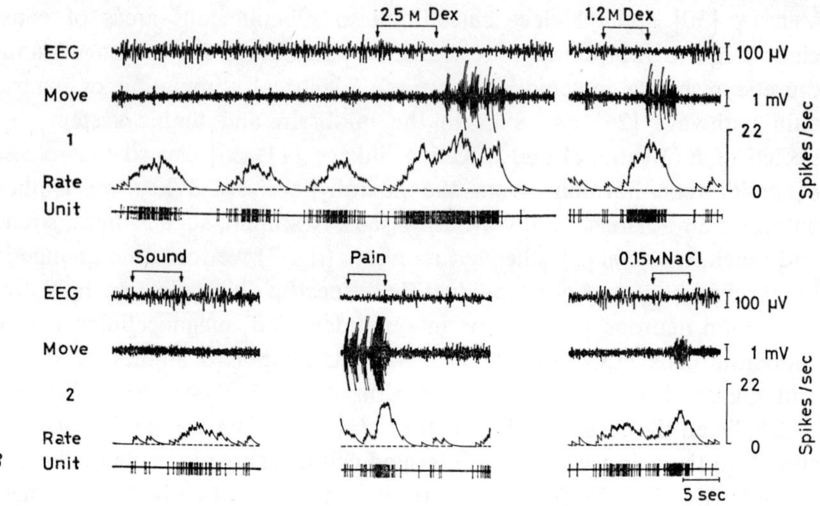

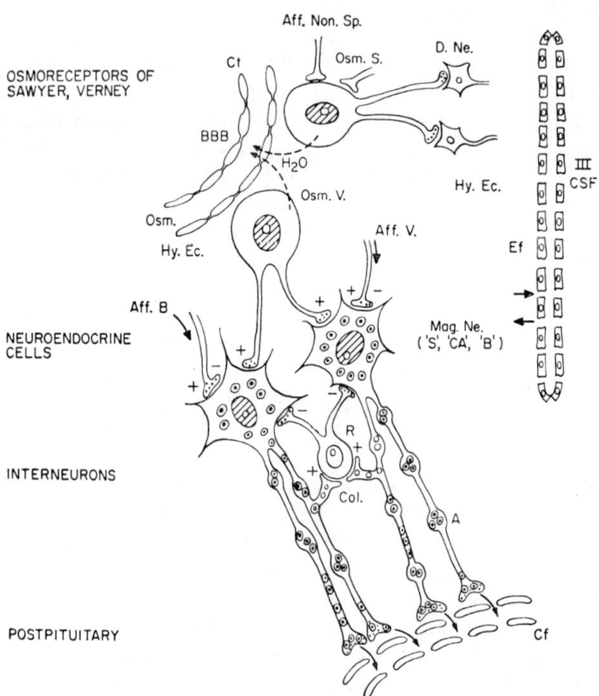

4

Fig. 3. Osmotic and behavioral responses of a 'nonidentified' 'burster' (B) magnocellular neuroendocrine cell in the rostral part of the supraoptic nucleus (NSO) of the waking monkey. *Hypertonic-D-glucose stimulation.* Upper 1: intracarotid infusion of D-glucose in 2.5 M (left) and 1.2 M (right) concentrations over 7.5 and 6 sec, respectively, interrupts control 'bursting' with increased discharge rates from control of 2.4 spikes/sec to 12.2 spikes/sec during osmotic stimulation. Note the multiphasic and prolonged response to 2.5 M D-glucose in comparison to the 'brief biphasic' response to 1.2 M D-glucose. EEG 'arousal', increased eye and body movement accompanies osmotic stimulation. *Behavioral stimuli.* Lower 2: rapid, repeated, light sticks with a pin on the leg over 5 sec (center) causes increase in mean firing rate from control of 2.4 spikes/sec to 8.5 spikes/sec during noxious stimulation. Non-noxious, mildly arousing sensory stimuli such as tapping on the chamber (sound, left) and intracarotid infusion of isotonic NaCl (0.15 M NaCl, right) produce EEG 'arousal', increased eye and body movement without change in cell firing. Spontaneous firing rate of this B supraoptic neuron demonstrated burst duration of 7.0 sec, mean firing rate during bursts of 8.0 spikes/sec with a cyclicity of one burst every 12.6 sec. Dex = D-glucose. Other abbreviations as in figure 2. [HAYWARD and JENNINGS, unpublished observations].

Fig. 4. A schematic interpretation of the possible cellular elements and connections in the osmoreceptor-magnocellular nuclear complex in the hypothalamus of the monkey. We propose that the three functional types of magnocellular neuroendocrine cells (Mag. Ne., 'S'. 'CA', 'B') are under excitatory and inhibitory synaptic influences from: (1) 'osmoreceptors of Verney (Osm. V.); (2) 'specific' afferents related to behavioral state (Aff. B.); (3) 'specific' afferents related to blood volume (Aff. V.), and (4) recurrent collateral (Col.) inhibitory pathway with neuroendocrine Renshaw cell (R). Intravascular osmotic pressure (Osm.) causes dehydration of 'osmoreceptors' of Verney (Osm. V.) and Sawyer (Osm. S.) across 'tight junctions' of hypothalamic capillaries (Ct), the blood-brain barrier (BBB), with activation, respectively, of magnocellular neuroendocrine cells (Mag. Ne.) and neurons involved in drinking (D. Ne.). Note the presence of a barrier to free diffusion of molecules in the hypothalamic capillary 'tight junctions' (Ct) in contrast to the absence of such a barrier to diffusion between cerebrospinal fluid of the third ventricle (III-CSF) and the hypothalamic extracellular space (Hy. Ec.) across the fenestrated ependyma (Ef), and between the magnocellular axonal terminals and the fenestrated neurohypophysial capillaries (Cf). Magnocellular axons are bifurcated, have axon collaterals (Col.) and axon dilatations and transport neurosecretory vesicles. A. = bifurcated magnocellular neuroendocrine cell axons with dilatations; Aff. B. = 'specific' afferent input related to behavioral state; Aff. Non. Sp. = 'non-specific' afferent input related to behavioral state; Aff. V. = 'specific' afferent input related to blood volume; BBB = blood-brain barrier; Cf = fenestrated capillaries; Col. = recurrent collaterals; Ct = hypothalamic capillaries with 'tight junctions'; D. Ne. = neurons related to drinking; Ef = fenestrated ependyma; Hy. Ec. = hypothalamic extracellular space; Mag. Ne. = magnocellular neuroendocrine cells ('S' = silent, 'CA' = continuously active, 'B' = burster); Osm. = osmotic input; Osm. S. = 'osmoreceptors' of Sawyer; Osm. V. = 'osmoreceptors' of Verney; R = neuroendocrine Renshaw cell; III-CSF = cerebrospinal fluid in the third ventricle; H_2O = water; + = excitatory synaptic input; – = inhibitory synaptic input; → = direction of flow.

Recurrent Collaterals and Neuroendocrine Renshaw Cells

Following stimulation of the pituitary gland in the goldfish, KANDEL [20] found inhibitory postsynaptic potentials (IPSP) which, according to the latencies involved, seemed to him to represent recurrent collateral inhibition of preoptic magnocellular neuroendocrine cells without the involvement of an inhibitory interneuron, i.e. a neuroendocrine Renshaw cell [32]. Other workers have described inhibitory effects of pituitary gland stimulation on supraoptic neurons in the anesthetized rat [6], cat and dog [22] and in the waking monkey [13]. In the behaving monkey we find 'high-frequency' B neurons at the junction between the NSO and optic tract. These neurons cannot be driven by pituitary stalk stimulation, are inhibited by osmotic stimulation and excited by auditory stimuli [13, 14]. We do not know how these units are related to the magnocellular neuroendocrine cells and osmoregulation. Our finding of spontaneous 'bursting' activity [13, 14], 'biphasic' osmosensitive response [14, 16–18, 32] and 'bursting' during slow sleep [15] in magnocellular neuroendocrine cells in the unanesthetized monkey favors a recurrent collateral inhibitory pathway for these excitatory-inhibitory sequences [32]. KOIZUMI and YAMASHITA [22] described recordings in the vicinity of the NSO from rapidly firing small cells trans-synaptically driven by pituitary gland stimulation. These are prime candidates for the role of inhibitory interneurons, neuroendocrine Renshaw cells [32], in this recurrent collateral pathway (fig. 4).

Summary

Figure 4 summarizes some of the elements and their connections theoretically involved in the neurohumoral regulation of magnocellular neuroendocrine cells in the hypothalamus. From our studies on procion yellow (ICI) filled preoptic (NPO) magnocellular neuroendocrine cells in the goldfish, we suggest that these neurons have bifurcated and varicose axons [12] with recurrent collaterals [12, 20] and an associated neuroendocrine Renshaw cell [22]. In our monkey studies we find three functional types of magnocellular neuroendocrine cells with spontaneous firing patterns, 'silent', 'continuously active' and 'burster', which we believe indicate different dynamic states of secretion, respectively synthesis and transport (S), 'tonic' secretion (CA), and 'pulsatile' release (B) [13, 18]. The effectiveness of osmotic input to magnocellular neuroendocrine cells depends upon an intact blood-brain barrier with activation of 'specific' osmoreceptors of Verney for endocrine responses [11, 14, 16–18, 30] and activation of 'nonspecific' osmoreceptors of Sawyer [11, 27] for behavioral responses. We find that *short-term* infusions of both hypertonic sodium chloride and D-glucose can drive our magnocellular neuroendocrine cells [16].

In addition to osmometric and volumetric input [11] the 'specific' state of behavior also determines the release rate of vasopressin and the activity of supraoptic neurons [2, 15, 30, 31]. In certain magnocellular neuroendocrine neurons slow sleep [15], nociceptor-induced behavior [15] and drinking [31] result in 'activation' of supraoptic neurons in the unanesthetized monkey. Physiological evidence suggests that recurrent collaterals and an inhibitory pathway play a role in the excitatory-inhibitory sequence of these neurosecretory neurons [6, 13, 15–20, 22, 32]. Future studies should reveal that many functional characteristics of 'magnocellular' neuroendocrine cells are shared by the 'parvicellular' neuroendocrine cells in the hypothalamus of mammals.

Acknowledgements

This review was aided by grants from the US Public Health Service (NS-05638), National Institute of Neurological Disease and Stroke, and from the Ford Foundation.

The author thanks Mrs. E. CORNELL and Mrs. RUBYE S. LAWRENCE for valuable technical assistance, and his friends and colleagues who have contributed in so many ways to the ideas and content of this chapter, especially A. L. R. FINDLAY, D. P. JENNINGS, K. MURGAS and J. D. VINCENT.

References

1 BRIDGES, T. E. and THORN, N. A.: The effect of autonomic blocking agents on vasopressin release *in vivo* induced by osmoreceptor stimulation. J. Endocrin. *48:* 265–276 (1970).

2 CORSON, S. A.: Conditioning of water and electrolyte excretion. Publ. Ass. Res. nerv. ment. Dis. *43:* 140–198 (1966).

3 CROSBY, E. C. and SHOWERS, M. J. C.: Comparative anatomy of the preoptic and hypothalamic areas; in HAYMAKER, ANDERSON and NAUTA The hypothalamus, pp. 61–135 (Thomas, Springfield 1969).

4 CROSS, B. A. and GREEN, J. D.: Activity of single neurones in the hypothalamus. Effect of osmotic and other stimuli. J. Physiol., Lond. *148:* 554–569 (1959).

5 DIERSCHKE, D. J.; BHATTACHARYA, A. N.; ATKINSON, L. E., and KNOBIL, E.: Circhoral oscillations of plasma LH levels in the ovariectomized rhesus monkey. Endocrinology *87:* 850–853 (1970).

6 DREIFUSS, J. J. and KELLY, J. S.: Recurrent inhibition of antidromically identified rat supraoptic neurones. J. Physiol., Lond. *220:* 87–103 (1972).

7 ERIKSSON, L.; FERNANDEZ, O., and OLSSON, K.: Differences in the antidiuretic response to intracarotid infusions of various hypertonic solutions in the conscious goat. Acta physiol. scand. *83:* 554–562 (1971).

8 EULER, C. VON: A preliminary note on slow hypothalamic osmo-potentials. Acta physiol. scand. *29:* 133–136 (1953).

9 FINDLAY, A. L. R. and HAYWARD, J. N.: Spontaneous activity of single neurones in the hypothalamus of rabbits during sleep and waking. J. Physiol., Lond. *201:* 237–258 (1969).

10 GOODMAN, I. and HIATT, R. B.: Coherin. A new peptide of the bovine neurohypophysis with activity on gastrointestinal motility. Science *178:* 419–421 (1972).

11 HAYWARD, J. N.: The amygdaloid nuclear complex and mechanisms of release of vasopressin from the neurohypophysis; in ELEFTHERIOU The neurobiology of the amygdala, pp. 685–739 (Plenum Press, New York 1972).

12 HAYWARD, J. N.: Morphological identification of physiological defined neuroendocrine cells in the hypothalamus of the goldfish *(Carassius auratus).* Proc. 4th Int. Congr. of Endocrinology, Wash., D.C. *256:* 136 (1972).

13 HAYWARD, J. N. and JENNINGS, D. P.: Activity of magnocellular neuroendocrine cells in the hypothalamus of unanesthetized monkeys. I. Functional cell types and their anatomical distribution in the supraoptic nucleus and the internuclear zone. J. Physiol., Lond. *232:* 515–543 (1973).

14 HAYWARD, J. N. and JENNINGS, D. P.: Activity of magnocellular neuroendocrine cells in the hypothalamus of unanesthetized monkeys. II. Osmosensitivity of functional cell types in the supraoptic nucleus and the internuclear zone. J. Physiol., Lond. *232:* 545–572 (1973).

15 HAYWARD, J. N. and JENNINGS, D. P.: Influence of sleep-waking and nociceptor-induced behavior on the activity of supraoptic neurons in the hypothalamus of the monkey. Brain Res. *57:* 461–466 (1973).

16 HAYWARD, J. N. and JENNINGS, D. P.: Osmosensitivity of hypothalamic magnocellular neuroendocrine cells to intracarotid hypertonic D-glucose in the waking monkey. Brain Res. *57:* 467–472 (1973).

17 HAYWARD, J. N. and VINCENT, J. D.: Osmosensitive single neurones in the hypothalamus of unanesthetized monkeys. J. Physiol., Lond. *210:* 947–972 (1970).

18 JENNINGS, D. P. and HAYWARD, J. N.: Activity of identified neuroendocrine cells in the hypothalamus of the waking rhesus monkey. Physiologist *15:* 182 (1972).

19 JOYNT, R. J.: Functional significance of osmosensitive units in the anterior hypothalamus. Neurology, Minneap. *14:* 584–590 (1964).

20 KANDEL, E. R.: Electrical properties of hypothalamic neuroendocrine cells. J. gen. Physiol. *47:* 691–717 (1964).

21 KOIZUMI, K.; ISHIKAWA, T., and BROOKS, C. MC.: Control of activity of neurons in the supraoptic nucleus. J. Neurophysiol. *27:* 878–892 (1964).

22 KOIZUMI, K. and YAMASHITA, H.: Studies of antidromically identified neurosecretory cells of the hypothalamus by intracellular and extracellular recordings. J. Physiol., Lond. *221:* 683–705 (1972).

23 MATTHEWS, E. K.; LEGROS, J. J.; GRAU, J. D.; NORDMANN, J. J., and DREIFUSS, J. J.: Release of neurohypophysial hormones by exocytosis. Nature New Biol. *241:* 86–88 (1973).
24 OLDENDORF, W. H.: Brain uptake of radiolabeled amino acid, amines and hexoses after arterial injection. Amer. J. Physiol. *221:* 1629–1639 (1971).
25 RAPOPORT, S. I.; BACHMAN, D. S., and THOMPSON, H. K.: Chronic effects of osmotic opening of the blood-brain barrier in the monkey. Science, N. Y. *176:* 1243–1245 (1972).
26 ROTHBALLER, A. B.: Pathways of secretion and regulation of posterior pituitary factors. Publ. Ass. Res. nerv. ment. Dis. *43:* 86–131 (1966).
27 SAWYER, C. H. and GERNANDT, B. E.: Effects of intracarotid and intraventricular injections of hypertonic solutions on electrical activity of the rabbit brain. Amer. J. Physiol. *185:* 209–216 (1956).
28 STRUMWASSER, F.: Membrane and intracellular mechanisms governing endogenous activity in neurons; in CARLSON Physiological and biochemical aspects of nervous integration, pp. 329–341 (Prentice-Hall, Englewood Cliffs 1968).
29 UTTENTHAL, L. O. and HOPE, D. B.: The isolation of three neurophysins from procine posterior pituitary lobes. Biochem. J. *116:* 899–909 (1970).
30 VERNEY, E. B.: The antidiuretic hormone and the factors which determine its release. Proc. roy. Soc. B *135:* 25–106 (1947).
31 VINCENT, J. D.; ARNAULD, E., and BIOULAC, B.: Activity of osmosensitive single cells in the hypothalamus of the behaving monkey during drinking. Brain Res. *44:* 371–384 (1972).
32 VINCENT, J. D. and HAYWARD, J. N.: Activity of single cells in the osmoreceptor-supraoptic nuclear complex in the hypothalamus of the waking rhesus monkey. Brain Res. *23:* 105–108 (1970).
33 WAKERLEY, J. B. and LINCOLN, D. W.: Phasic discharge of antidromically identified units in the paraventricular nucleus of the hypothalamus. Brain Res. *25:* 192–194 (1971).
34 ZIMMERMAN, E. A.; HAU, K. C.; ROBINSON, A. G.; CARMEL, P. W.; FRANTZ, A. G., and TANNENBAUM, M.: Studies of neurophysin secreting neurons with immunoperoxidase techniques employing antibody to bovine neurophysin I. Light microscopic findings in monkey and bovine tissues. Endocrinology *92:* 931–940 (1973).

Author's address: Dr. J. N. HAYWARD, Department of Neurology, Reed Building, UCLA School of Medicine, *Los Angeles, CA 90024* (USA)

Recent Studies of Hypothalamic Function
Int. Symp. Calgary 1973, pp. 180–195 (Karger, Basel 1974)

Control of Anterior Pituitary Hormone Secretion

Hypothalamic Peptides and their Interactions[1]

G. Grant, W. Vale, P. Brazeau, J. Rivier, M. Monahan, C. Gilon, M. Amoss, C. Rivier, N. Ling, R. Burgus and R. Guillemin

The Salk Institute, San Diego, Calif.

I. Introduction

The topic of this discussion will concern 3 hypothalamic peptides which act in the control of secretion of 5 pituitary hormones and which have been extensively studied in our laboratory: (1) thyrotropin releasing factor (TRF), a tripeptide which stimulates the pituitary secretion of thyroid stimulating hormone (TSH) in all species examined [5, 26] and which also stimulates prolactin secretion in primates, cattle and marginally in experimentally treated rats [2, 8, 18, 30, 36]; (2) luteinizing hormone releasing factor (LRF), a decapeptide which increases the rate of secretion of both luteinizing hormone (LH) and follicle stimulating hormone (FSH) [6, 22], and (3) somatostatin or somatotropin release inhibiting factor (SRIF), a tetradecapeptide which blocks both growth hormone (GH) secretion and induced TSH secretion of the pituitary gland [3, 7, 21, 33, 37].

We would like to review the parameters of interaction of these peptides with the pituitary cells and to discuss the structure-activity relationships which have been found since their isolation and elucidation of their sequence.

1 This research is supported by AID Contract No. AID/csd 2785, Ford Foundation, Rockefeller Foundation and Edna McConnell Clark Foundation Grants to Dr. Roger Guillemin.

II. TRF

The primary sequence of TRF isolated from sheep and swine hypothalami is (pGlu-His-Pro-NH_2) [5, 26].

The *in vitro* systems used to study pituitary hormone secretion rates show that the minimal active dose of TRF is 10^{-10} M, a maximal secretion rate of TSH is obtained at 10^{-8} M (ca. 8–10 times the control rate) [16–34].

Binding studies of ^{3}H-TRF to thyrotropic cells show that TRF has an affinity constant of ca. 2×10^{-8} M [13] a somewhat higher concentration than the half-maximal biologically active dose of ca. 2×10^{-9} M [34].

Table I. Comparison of biological potency and affinity constants (as determined by competition binding experiments) of thyrotropin releasing factor (TRF) analog [14]

Analog	Biological potency, %[1]	Relative affinity constant, M[2]
TRF	100	2.0×10^{-8}
N^{τ}-Me-His^2-TRF	800	3.0×10^{-9}
TRF-Gly-NH_2	35	6.0×10^{-8}
TRF-NH-EtOH	16	2.0×10^{-7}
TRF-NH-Et	14	6.0×10^{-7}
TRF-methyl-ester	10	8.0×10^{-7}
N-Me-<Glu^1-TRF	1.7	3.0×10^{-6}
<Glu-His-prolinol	1.2	3.2×10^{-6}
<Glu-Met-Pro-NH_2	1.0	5.0×10^{-6}
<Glu-His-pyrrolidide	0.8	1.2×10^{-6}
TRF-ethyl-ester	0.4	1.0×10^{-6}
<Glu-His-hexamethylenimine	0.1	4.0×10^{-5}
<Glu-His-morpholide	0.3	1.2×10^{-5}

All analogs examined give identical maximal (plateau) biological responses when given at sufficiently high dose-levels and also completely displace bound [^{3}H]TRF from pituitary receptors.

1 Biological potency – reciprocal percentile of relative analog concentration with respect to TRF in the release of thyroid stimulating hormone (TSH) *in vivo* [1]. Calculated from 3- or 4-point bioassays, indices of precision ranging from 0.15 to 0.25.

2 Relative affinity constant – concentration of analog required to 50% compete for saturating amounts of bound [^{3}H]TRF [27, 32]. Standard error of mean of replicates is $< 10\%$.

The biological potency of a large number of TRF analogs has been examined both *in vivo* in mice [31], and *in vitro* by the direct measurement of TSH secreted by pituitary cells in culture by radioimmunoassay and/or bioassay.

Equilibrium competition of bound ^{3}H-TRF by TRF analogs gives apparent affinity constants proportional to the biological potencies obtained for each TRF derivative. The proportional relationship between membrane binding affinity and biological potency is shown in table I [14]. The observations, together with the fact that both the abilities of the cells to bind ^{3}H-TRF and to respond biologically by TSH secretion are sensitive to mild enzymatic treatment (as during cell dispersion) [34], suggest a membrane trigger for the action of TRF. The 10-fold excess of TRF required to saturate the binding sites may indicate a relatively large supply of reserve receptors, a not uncommon phenomenon in pharmacological systems [28].

In considering the interaction of TRF with its membrane receptor, it is necessary to examine not only the parameters of what comprises the receptor but also the 3-dimensional characteristics of the tripeptide.

It has been proposed recently that TRF exists in a 'hairpin turn' conformation stabilized by two intramolecular hydrogen bonds [1]; this model (fig. 1) was initially arrived at by semiempirical energy calculations based on the contributions to the potential energy of Van der Waal and Coulombic interactions as well as interatomic distances and imidazole potentiometric titration [11]. The model is supported by proton magne-

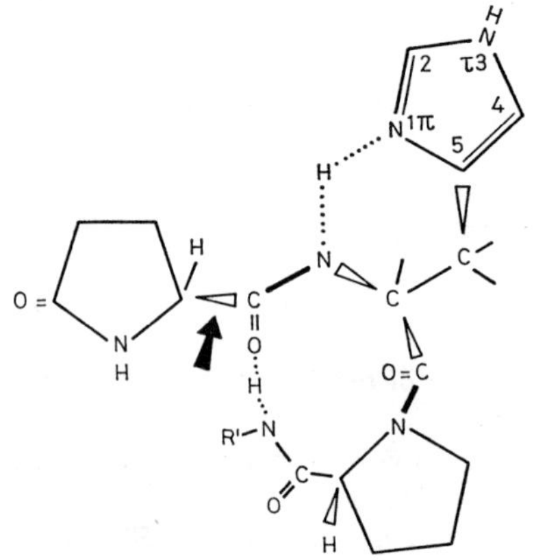

Fig. 1. Proposed conformation of thyrotropin releasing factor (TRF) showing two intramolecular hydrogen bonds [1]. R′ = H in TRF.

Table II. Biological potency of TRF analogs

Pyroglutamic modifications (R–His–Pro–NH_2)

TRF(a) potency, %	Analog	Structure
< 0.01	cyclopentane-carboxyl[1]–TRF	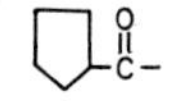
0.1	prolyl[1]–TRF	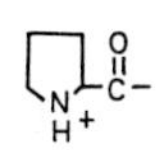
<0.10	N–acetyl Ala[1]–TRF	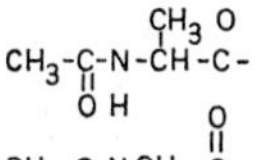 CH_3-C(=O)-N(H)-CH(CH_3)-C(=O)-
0.25	N–Acetyl Gly[1]–TRF	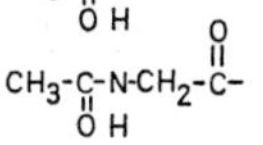 CH_3-C(=O)-N(H)-CH_2-C(=O)-
0.3	pyrrolidine–4–carboxyl[1]–TRF (Iso <Glu)	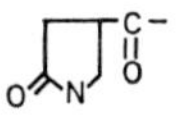
1.7	N–methyl <Glu[1]–TRF	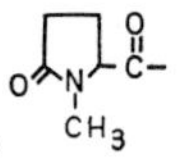
5.0	γ–hydroxy glutoryl lactone[1]–TRF	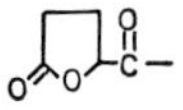
10.0	N–formyl–Pro[1]–TRF	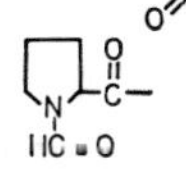

(b) Histidyl Modifications (<Glu–R–Pro NH_2)

TRF(a) potency, %	Analog	Structure
800	N^{τ}–methyl His[2]–TRF	Ala –
0.04	N^{π}–methyl His[2]–TRF	Ala –
1.0	phenylalanine[2]–TRF	Ala –
5.0	pyrazole[2]–TRF	Ala –
10.0	β–thienyl Ala[2]–TRF	Ala –
1.0	Met[2]–TRF	CH_3–S–CH_2–Ala –

(a) < Glu His Pro–NH_2 potency is 100% (half maximal activity at 2 x 10^{-9} M).

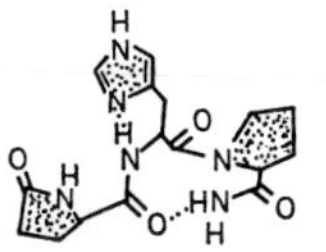

(b) Substitution of Tyr, Arg, Orn, Lys or Gly for His[2] of TRF yields analogues with <0.1% potency

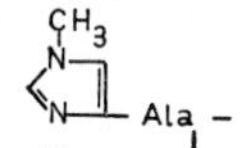

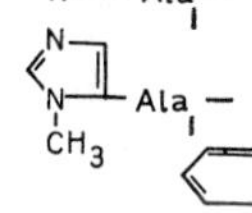

Proline modifications (< Glu–His–R)

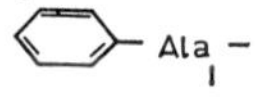

TRF potency, %	Analog	Structure
< 0.02	Gly[3]–TRF	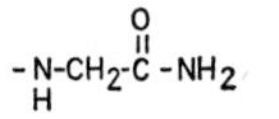 -N(H)-CH_2-C(=O)-NH_2
0.3	Sar[3]–TRF	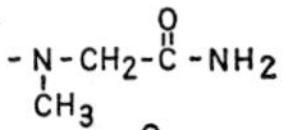 -N(CH_3)-CH_2-C(=O)-NH_2
0.04	Leu[3]–TRF	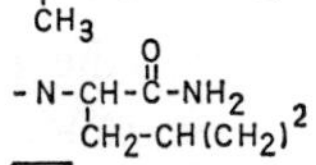 -N-CH(CH_2-CH$(CH_2)_2$)-C(=O)-NH_2
0.8	pyrrolidine[3]–TRF	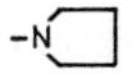
0.1	hexamethylene[3]–TRF	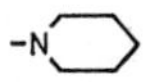
0.3	morpholine[3]–TRF	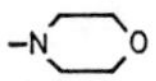
0.02	Pro[3]–TRF (TRF acid)	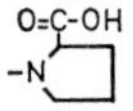 O=C-OH
1.2	prolinol[3]–TRF	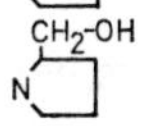 CH_2-OH

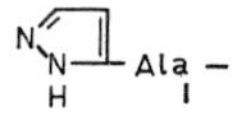

amide modifications (< Glu–His–Pro–R)

TRF potency, %	Analog	Structure
0.02	TRF–OH (free acid)	
10.0	TRF–O–CH_3	
14.0	TRF–N(H)–CH_2 CH_3	
16.0	TRF–N–CH_2 CH_2 OH	
16.0	TRF–N(H)–anilide	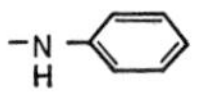
0.5	TRF–N–$(CH_3)_2$	
0.04	TRF–N–$(CH_2\ CH_3)_2$	
0.6	TRF–N–piperidine	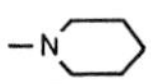
35.0	TRF–N–Gly NH_2	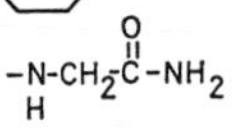 -N(H)-CH_2-C(=O)-NH_2
0.05	TRF–N–Ala NH_2 (D and L)	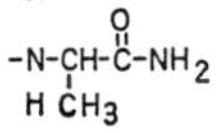 -N(H)-CH(CH_3)-C(=O)-NH_2

tic resonance data [1] although interpretations of the observations differ considerably [9].

Table II shows the biological potencies of many TRF derivatives. Lists of the biological activities and properties of TRF analogs have been published (table II) [17, 23, 29, 38]. We will attempt to define these biological data in terms of the conformational model of the TRF molecule. We will project the 3-dimensional model of the TRF molecule and define the boundaries of the apparently lipophilic envelope which recognizes this molecule. Many of the component parts of the TRF molecule stabilize the preferred structure and, when modified, the molecule can feasibly alter its shape and orientation and, hence, affect the ability of the membrane receptor to recognize it.

A. Pyroglutamyl Residue (<Glu)

The proposed model allows rotation of the pyroglutamyl residue on its carboxyl-α carbon bond with a suggested preferred position of the lactam function proximal to the carbonyl function of the <Glu His peptide bond. The biological data indicate a requisite for the lactam function (cis-orientation) as indicated by the relative inactivity (percent biological activity shown in parentheses) of the cyclopentane-carboxyl (<0.01 %), prolyl$^{(+)}$ (0.01 %), N-methyl <Glu (1.7 %), N-acetyl-Gly (0.25 %), and N-acetyl-Ala (0.025 %) analogs. Such substitutions may lack not only a sufficient dipole moment in this moiety but also the correct functional group orientation which might explain the lack of TRF activity of analogs with these substitutions.

The low activity of the TRF analog with a 4-carboxy-pyrrolidone

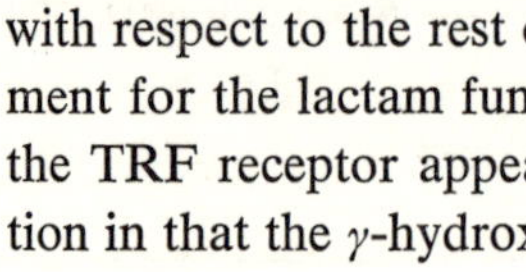

(0.3 %)
substituted for <Glu

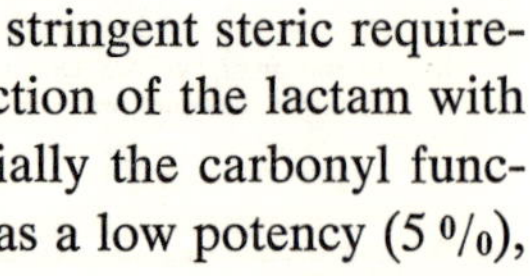

a change which affects both the position and plane of the pyrrolidone ring with respect to the rest of the molecule, suggests a stringent steric requirement for the lactam functional groups. The interaction of the lactam with the TRF receptor appears to involve at least partially the carbonyl function in that the γ-hydroxy-glutaryl-lactone[1]-TRF has a low potency (5 %), but is one of the most active <Glu substituted analogs. The other highly active <Glu substitute of interest is the N-formyl Pro[1] analog (10 %) in which the carbonyl is moved and the amide nitrogen blocked with a formyl group, yielding a 3° amide. A series of <Glu substituted analogs not shown in table II indicates a possible need for a nucleophilic function in this position: tetrahydrothiophene[1] (0.2 %), tetrahydrofuran[1] (0.01 %), cyclopentane[1] (<0.01 %) and proline (0.01 %). Pro[1] -TRF at physiological pH is, of course, protonated.

B. Histidyl Substitutions and Modifications

The most striking feature of the hairpin turn model of TRF is the close proximity of the imidazole ring to the peptide backbone which might be anticipated to stimulate repulsive forces. However, an imidazole nitrogen-amide hydrogen bond has been detected potentiometrically [11]. This hydrogen bond not only overcomes the repulsion forces between the imidazole ring and the backbone, but also aids in the stabilization of the molecule [1]. This hydrogen-bonded interaction may be further stabilized by methylation of the non-H bonded imidazole nitrogen (N^{τ}), apparently favoring the equilibrium of this peptide towards a hairpin turn and thereby producing a hyperactive TRF molecule (N^{τ}Me-His^{2}-TRF, 800 %) [32]. In contrast, methylation of the N^{π} position prevents the formation of an H-bond both in absolute and steric terms, yielding a virtually inactive compound (N^{π}-Me-His^{2}-TRF, 0.04 %). Only a few other histidyl-substituted analogs show appreciable activity; these include derivatives containing phenyl (1 %), pyrazole (5 %) [29] or thiophene (10 %) rings which are sterically similar to the imidazole ring but either are repelled from close proximity to the backbone, (possibly do not favor H-bonded interactions due to different ring resonance) or are too bulky and, hence, show less strictly defined steric parameters of TRF required for high affinity binding to the receptor. The activity of the thiophene ring which, although bulky, can feasibly H-bond to the peptide backbone, supports the model.

One other consideration particularly important, but not readily obvious, is the stabilization of one of the tautomers of the <Glu-histidyl amide bond by the hydrogen interaction with an electronegative group favoring the amide configuration and, hence, complementing the formation of a second intramolecular hydrogen bond between the <Glu-His

amide carbonyl and the C-terminal carboxyamide. Similar complementary hydrogen bonds are found in the α helix of many proteins.

C. Prolylamide Substitution

The affinity of TRF for its receptors appears to require the prolyl ring and the carboxamide group. Removal of either part of this amino acid drastically reduces the activity of the peptide. The glycyl (<0.02 %), sarcosyl (0.3 %), leucyl (0.04 %), pyrrolidine (0.3 %), hexamethyleneimine (0.04 %), morpholine (0.3 %), proline (0.02 %) and prolinol (1.2 %) analogs all suggest the specific recognition of the hydrophobic ring and carbonyl functions of the prolylamide in its preferred position, hydrogen bonded via the amide hydrogen to the <Glu-His-amide carbonyl oxygen.

D. Amide Function Modification

The requirement for both the carbonyl function and an H-bonded carboxamide has been reasoned on the basis that prolinol (1.2 %), able

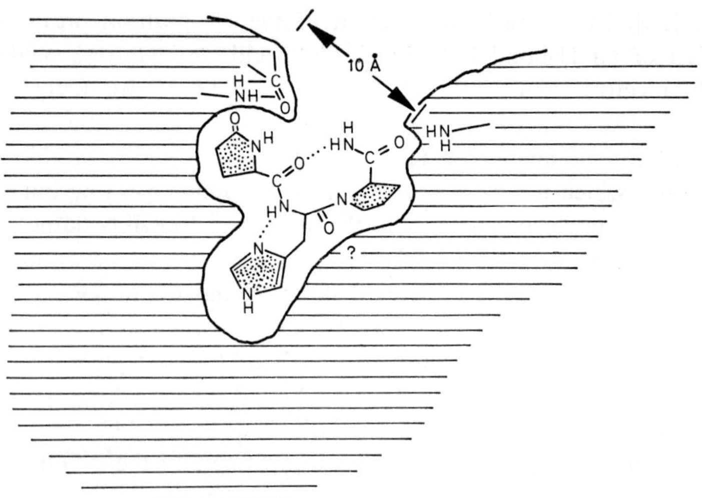

Fig. 2. Diagrammed summary of TRF-receptor interactions as deduced from structure-activity studies [1, 38].

to H-bond but not possessing a carbonyl function, has low biological activity and that the 2° TRF-NH-R′ amides, still able to form this H-bond, are among the most active analogs while the C-terminal tertiary amides are significantly less active. TRF-OCH_3 cannot form this bond and has high (10 %) activity so presumably must prefer a similar but nonbonded orientation. The size of the R′ group in the TRF C-terminal-2° amides appears important in the case of amino acids, as indicated by the activity of glycine *versus* L- or D-alanine, the first of which is 35 % active and both the latter, surprisingly, have low activity. Appraisal of atomic models (CPK) indicate that the D-L-alanine enantiomeric methyl groups sterically interfer with H-bonding of the carboxamide-hydrogen to the <Glu-His carbonyl oxygen; the methyl group, of course, does not exist in the glycine analog.

E. Summary of TRF Structure-Activity Studies

The conclusions from the examination of the data on biological potency of TRF analog suggest that (1) the plane and position of the <Glu lactam function (cis-) is required, but the role of the ring structure in recognition is unknown; (2) the size, position and a restricted plane of the imidazole ring is recognized by the receptor; (3) the orientation and hydrophobic ring structure of the proline moiety is required, and (4) there appears also to be a requirement for the carbonyl function of the carboxamide terminus by the receptor. These conclusions (summarized in fig. 2) help to define what we understand about the recognition of TRF by its cellular receptor. The 3-dimensional model, the interpretation of the conformation of TRF analogs, the interaction of analogs with the receptor and the deductions about the topography of the TRF receptor all provide and suggest the basis of many experiments, i.e. they are testable.

III. LRF

The control of pituitary gonadotropin secretion by the hypothalamus has been found to be due, at least in part, to a decapeptide with the primary sequence: pGlu-His-Trp-Ser-Tyr-Gly-Leu-Arg-Pro-Gly-NH_2 [6, 22]. Although the isolation from both sheep and pigs was based on the

Table III. Biological activity of luteinizing hormone releasing factor (LRF) analogues

Chain-shortened analogs	LRF potency, %
pGlu-His-Trp-Ser-Tyr-Gly-Leu-Arg-Pro-Gly-NH_2	100 (LRF)
pGlu-His-Trp-Ser-Tyr-Gly-Leu-Arg-Pro-Gly-OH_2	0.2
pGlu-His-Trp-Ser-Tyr-Gly-Leu-Arg-Pro-NH_2	11.0
pGlu-His-Trp-Ser-Tyr-Gly-Leu-Arg-NH_2	<0.01
pGlu-His-Trp-Ser-Tyr-Gly-Leu-NH_2	<0.01
pGlu-His-Trp-Ser-Tyr-Gly-NH_2	<0.01
pGlu-His-Trp-Ser-Tyr-NH_2	<0.01
pGlu-His-Trp-Ser-NH_2	<0.01
pGlu-His-Trp-NH_2	<0.01
Des-His^2-LRF	<0.01
Substitution analogs	
[$CH_3CH_2C(=O)$-Gly^1] LRF	0.2
Met^2-LRF[1]	1.0
Phe^2-LRF	4.0
Tyr^2-LRF	10.0
Trp^2-LRF	40.0
Phe^3-LRF[2]	4.0
Gly^4-LRF	1.5
Gly^5-LRF	<0.1
L-Ala^6-LRF	1.0
D-Ala^6-LRF[3]	400.0
D-Ileu^6-LRF	65.0
D-Val^6-LRF	30.0
Gly^7-LRF	0.2
Gly^8-LRF	0.1
Gly^9-LRF	0.2

1 Substitutions of Gly, Ala, D-Ala, Cys or Asp for His^2 of LRF yields LRF antagonists with less than 0.1% agonistic activity.
2 Substitution of Gly or Ala for Trp^3 of LRF yields LRF antagonists with less than 0.1% agonistic activity.
3 Des-His^2-[D-Ala^6] LRF is an antagonistic LRF analog several times more potent an inhibitor than Des-His^2-LRF [25, 35].

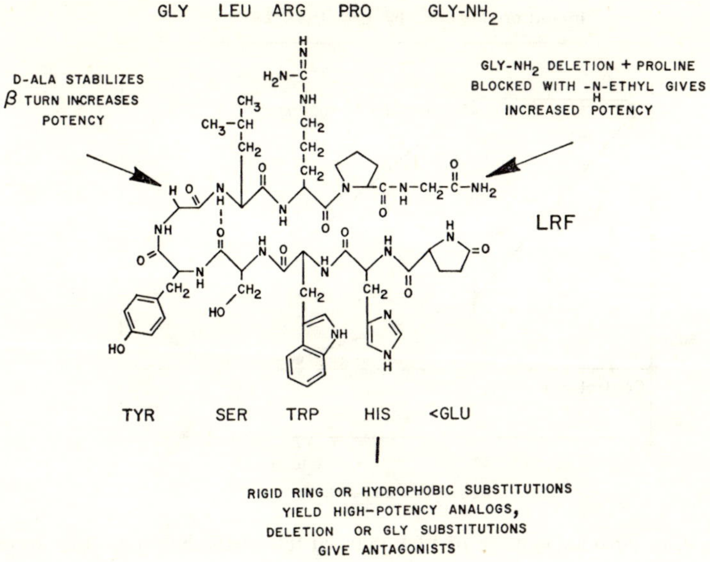

GLY LEU ARG PRO GLY-NH2

D-ALA STABILIZES β TURN INCREASES POTENCY

GLY-NH2 DELETION + PROLINE BLOCKED WITH -N(H)-ETHYL GIVES INCREASED POTENCY

LRF

TYR SER TRP HIS <GLU

RIGID RING OR HYDROPHOBIC SUBSTITUTIONS YIELD HIGH-POTENCY ANALOGS, DELETION OR GLY SUBSTITUTIONS GIVE ANTAGONISTS

Fig. 3. Diagrammed structure of LRF containing a β turn [24, 25]. The remaining backbone structures are drawn parallel to suggest that a portion of the molecule could resemble TRF [12], especially as the His² and Trp³ positions have been concluded to be part of the active site of luteinizing hormone releasing factor (LRF) [35].

LH releasing ability of the peptide, it also stimulates the secretion of FSH. All active synthetic LRF analogs and derivatives so far examined release both LH and FSH.

LRF has an apparent activity constant of ca. 5×10^{-10} M, while direct binding experiments with ^{3}H-Pro^{9}-LRF show a similar receptor affinity constant of 2×10^{-9} M [15, 34].

Table III gives the biological potencies of a number of LRF derivatives. These data, together with the report of Fujino *et al.* [10] that the alkylated amides of the des-Gly^{10}LRF have very high biological potencies, can be formulated into a functional, diagrammatic structure (fig. 3).

It appears that the histidine moiety is required to activate the LRF receptor, but this moiety is not obligatorily required for recognition. When histidine is absent the derivative becomes an antagonist of LRF; by definition such an analog must bind but not activate the LRF receptor [24, 35] (fig. 4).

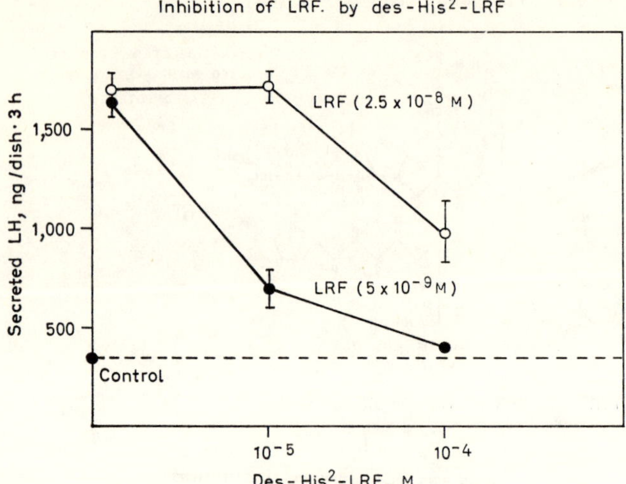

Fig. 4. In vitro antagonism of LRF-induced luteinizing hormone (LH) secretion by des-His^2-LRF.

Recent analogs made with substitutions at the Gly^6 position of LRF have been interpreted to propose a β turn in the middle of this decapeptide [24, 25]. With the β turn, the folded LRF structure may have its ends proximal to each other. The ends of the molecule form a site that resembles TRF; an idea put forward in a previous communication [12].

As it has been found that modification of the His^2 and also Trp^3 positions [35, 38] can yield antagonists, it is obvious that these residues are involved in the intrinsic activity of the molecule. Any model relating the interaction of LRF with the ultimate stimulation of the secretory process must consider both the LRF molecule, the modifications which produce antagonists and the membrane LRF receptor. RUDINGER *et al.* have recently reviewed similar phenomena and elegantly summarized such interactions with a comprehensible model [28].

IV. Somatostatin or SRIF

The final section of our discussion will be somewhat different from the two preceding ones in that we will give some information about

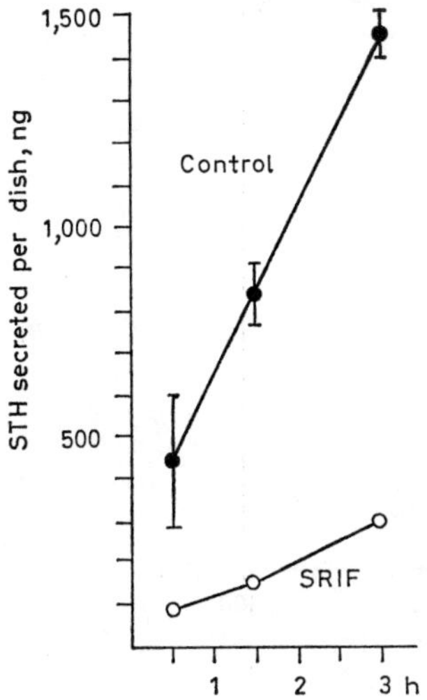

Fig. 5. Inhibition of resting growth hormone (GH) secretion by 25 nM somatotropin release inhibiting factor (SRIF) [33]. Effect of SRIF on the rate of secretion of somatotropin (STH) by rat anterior pituitary cell cultures.

somatostatin – not necessarily membrane interaction or structure-activity relationships but mainly validation of this newest of characterized hypothalamic peptides as a regulatory agent of the secretion of biologically active GH.

KRULICH *et al.* [19] had reported that hypothalamus contains a substance that could inhibit the secretion of GH *in vitro*. Later, KRULICH *et al.* [20] examined various parts of the rat hypothalamus and found GH release inhibiting activity in the vicinity of the median eminence; assays (radioimmunoassays *in vitro*) were based on the ability of crude extracts to modify the secretion of GH.

Searching for an active GH releasing factor (GH-RF) our laboratory tested extracts of sheep hypothalami and found that very low amounts of such extracts (1/1,000 of a fragment per milliliter medium) would inhibit the secretion of immunoreactive GH (rGH) by primary rat pituitary cell cultures. Based on this assay, a GH release inhibiting factor (somatostatin or SRIF) was isolated and its primary sequence [3, 7, 21, 33] determined to be:

H-Ala-Gly-Cys-Lys-Asn-Phe-Phe-Trp-Lys-Thr-Phe-Thr-Ser-Cys-OH.

This peptide was synthetized by solid phase methods [27] and the reduced linear synthetic replica was found to have activity equivalent to

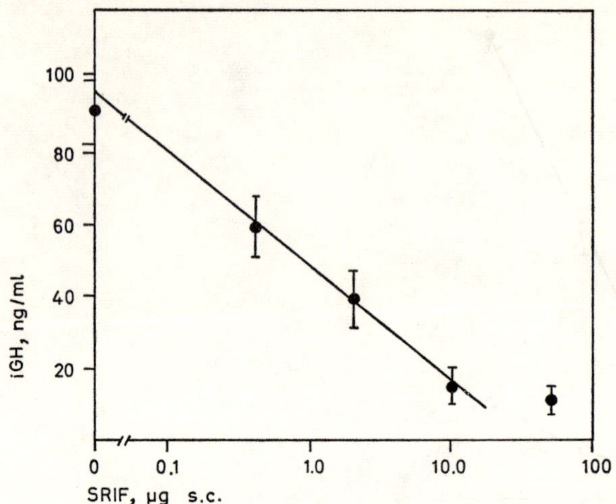

Fig. 6. In vivo effect of increasing doses of SRIF on plasma immunoassayable GH [4]. Somatostatin dose response (15 min) in gentled rats.

native SRIF. Interestingly, cyclized (disulfite-bridged) synthetic SRIF has the same potency as noncyclized linear synthetic SRIF in our *in vitro* assays.

SRIF acutely inhibits the secretion of immunoassayable somatotropin within 30 min *in vitro*. Following a change of medium, cultured pituitary cells secrete rGH at a linear rate for up to 4 h. The addition of SRIF leads to a linear rate of secretion considerably lower than that of the control (fig. 5). Synthetic SRIF is active at concentrations $\geqq$ 0.2 nM *in vitro* to inhibit spontaneous rGH secretion as well as that stimulated by high K^+ theophylline or dibutyryl 3′ 5′ cyclic AMP [33]. The activity of SRIF is not restricted to pituitary cell cultures [34] but it is also seen on hemi-pituitaries incubated in short-term experiments [33] and *in vivo* in gentled rats (fig. 6) and rats stimulated to secrete rGH by pentobarbital [3, 4].

It has been found subsequently [37, 39] that both synthetic and native SRIF inhibit the secretion of TSH mediated by TRF or by elevated [K^+] *in vitro*. Furthermore, a simultaneous injection of SRIF with TRF decreases the plasma TSH levels reached with TRF alone when blood is sampled 5 min after injection. SRIF demonstrates some specificity in that the secretion of LH and FSH in response to LRF is uninfluenced by SRIF.

The effect of SRIF on prolactin (PRL) secretion is seen as a function of other experimental variables, but we have found that SRIF dissociates the TRF-induced release of PRL from that of TSH [36, 39].

Conclusions

Our knowledge and understanding of the hypothalamic peptides regulating the hormone secretion of the anterior pituitary has expanded dramatically in the last few years. There remain many more, as yet chemically uncharacterized, hypophysiotropic hormones. Further progress in the field will concern the elucidation of the structure of these physiologically defined entities and an extension of our knowledge of structure-activity relationships of the known peptides and their mechanism of action.

The clinical experimentation and application of the hypothalamic peptides to physiological and pathological conditions is only in its infancy but shows satisfying promise.

References

1 BLAGDON, D. E.; RIVIER, J. and GOODMAN, M.: Proposed conformation of TRF. Proc. nat. Acad. Sci., Wash. *70:* 1166 (1973).

2 BOWERS, C. Y.; FRIESEN, H.; WANT, H.; GUYDA, H., and FOLKERS, K.: Prolactin and TSH release in man by synthetic TRF. Biochem. biophys. Res. Commun. *45:* 1033 (1971).

3 BRAZEAU, P.; VALE, W.; BURGUS, R.; LING, N.; BUTCHER, M.; RIVIER, J., and GUILLEMIN, R. R.: Hypothalamic polypeptide that inhibits the secretion of immunoreactive growth hormone. Science *179:* 77–79 (1973).

4 BRAZEAU, P.; RIVIER, J.; VALE, W., and GUILLEMIN, R.: Physiology of somatostatin. Endocrinology *94:* 184–188 (1974).

5 BURGUS, R.; DUNN, R. F.; DESIDERIO, D.; VALE, W., et GUILLEMIN, R.: Dérivés polypeptidiques de synthèse doués d'activité. C. R. Acad. Sci. *269:* 226 (1969).

6 BURGUS, R.; BUTCHER, M.; LING, N.; MONAHAN, M.; RIVIER, J.; FELLOWS, R.; AMOSS, M.; BLACKWELL, R.; VALE, W., et GUILLEMIN, R.: Structure moléculaire du contrôlant la sécrétion de LH. C. R. Acad. Sci. *273:* 1611 (1971).

7 BURGUS, R.; LING, N.; BUTCHER, M., and GUILLEMIN, R.: Primary structure of somatostatin. Proc. nat. Acad. Sci., Wash. *70:* 684–688 (1973).

8 CONVEY, E. M.; TUCHER, H. A.; SMITH, V. G., and ZOLMAN, J.: Bovine prolactin, growth hormone thyroxine and corticoid response to TRF. Endocrinology *92:* 471 (1973).

9 FERMANDJIAN, S.; PRADELLES, P., and FROMAGEOT, P.: PMR spectroscopy of TRF. FEBS Letters *28:* 156 (1973).

10 FUJINO, M.; KOBAYASHI, S.; OBAYASHI, M.; SHINAGAWA, S.; FUKUDA, T.; KITADA, C.; NAKAYAMA, R., and YAMAZAKI, I.: Structure activity relationships of the C-terminal sequence of LHRH. Biochem. biophys. Res. Commun. *49:* 863 (1972).

11 GRANT, G.; LING, N.; RIVIER, J., and VALE, W.: Orientation restrictions of TRF due to intramolecular H-bonding. Biochemistry *11:* 3070 (1972).

12 GRANT, G. and VALE, W.: Speculations on structural relationships of hypothalamic releasing factors. Nature New Biol. *237:* 182–183 (1972).

13 GRANT, G.; VALE, W., and GUILLEMIN, R.: Interaction of TRF with membrane receptors of pituitary cells. Biochem. biophys. Res. Commun. *46:* 28–34 (1972).

14 GRANT, G.; VALE, W., and GUILLEMIN, R.: Characteristics of pituitary binding sites for TRF. Endocrinology *92:* 1455 (1973).

15 GRANT, G.; VALE, W., and RIVIER, J.: Pituitary binding sites for H^3-LRF. Biochem. biophys. Res. Commun. *50:* 771 (1973).

16 GUILLEMIN, R. and VALE, W.: Bioassays of the hypophysiotropic hormones; in MEITES Hypophysiotropic hormones of the hypothalamus. Assay and chemistry, pp. 12–35 (Williams & Wilkins, Baltimore).

17 GUILLEMIN, R.; BURGUS, R., and VALE, W.: The hypothalamic hypophysiotropic TRF. Vitamins Hormones, N.Y. *29:* 1–39 (1971).

18 KAPLAN, S. L.; GRUMBACH, M. M.; FRIESEN, H. G., and COSTOM, B. H.: Prolactin and TSH response to TRF. J. clin. Endocrin. *35:* 825 (1972).

19 KRULICH, L.; DHARIWAL, A. P. S., and MCCANN, S. M.: Stimulating and inhibitory effects of purified hypothalamic extracts on growth hormone release from rat AP. Endocrinology *83:* 783 (1968).

20 KRULICH, L.; ILLNER, P.; FAWCETT, C. P.; QUIJADA, M., and MCCANN, S. M.: Control of growth hormone secretion; in PECILE and E. E. MULLER Growth and growth hormone, p. 306 (Excerpta Medica, Amsterdam 1972).

21 LING, N.; BURGUS, J.; RIVIER, J.; VALE, W., and BRAZEAU, P.: Use of mass spectrometry in deducing the sequence of somatostatin. Biochem. biophys. Res. Commun. *50:* 127–133 (1973).

22 MATSUO, H.; BABA, Y.; NAIR, R. M. G.; ARIMURA, A., and SCHALLY, W. V.: Structure of porcine LH and FSH releasing hormone. Biochem. biophys. Res. Commun. *43:* 1334 (1971).

23 MONAHAN, M.; RIVIER, J.; VALE, W.; LING, N.; GRANT, G.; AMOSS, M.; GUILLEMIN, R.; BURGUS, R.; NICOLAIDES, E., and REBSTOCK, M.: Structure biological activity relationships of TRF and LRF. 3rd Amer. Peptide Symp., Boston 1972. Chemistry and biology of peptides, Ann Arbor Science Publ. No. 6010608 (1972).

24 MONAHAN, M. and RIVIER, J.: Synthesis of LRF. B.B.R.C. *48:* 1100 (1973).

25 MONAHAN, M.; AMOSS, M., and VALE, W.: Synthetic analogs of LRF with increase against or antagonist properties. Biochemistry *12:* 4616–4621 (1973).

26 NAIR, R. M. G.; BARRETT, J. F.; BOWERS, C. Y., and SCHALLY, A. V.: Structure of porcine TRF. Biochemistry *9:* 1103 (1970).

27 RIVIER, J.; BRAZEAU, P.; VALE, W.; LING, N.; BURGUS, R.; GILON, C.; YARDLEY, G., and GUILLEMIN, R.: Synthesis of the polypeptide somatostatin. C. R. Acad. Sci. *276:* 2737 (1973).

28 RUDINGER, J.; PLISKA, V., and KREJCI, I.: Oxytacin analogs in the analysis of some phases of hormone action; in ASTWOOD Recent progress in hormone research, vol. 28, p. 131 (Academic Press, New York).

29 SIEVERTSSON, H.: Chemistry and structure activity studies of TRF. Acta pharm. succ. *9:* 19–46 (1972).

30 TASHJIAN, A.; BAROWSKI, N., and JENSEN, D.: TRF. Evidence for stimulation of prolactin by pituitary cells in culture. Biochem. biophys. Res. Commun. *43:* 516 (1971).

31 VALE, W.; BURGUS, R.; DUNN, T. F., and GUILLEMIN, R.: *In vitro* plasma inactivation of TRF. Hormones *2:* 193 (1971).

32 VALE, W.; RIVIER, J., and BURGUS, R.: Synthetic TRF analogs. II. pGlu N^{3im}Me-His-ProNH_2. Endocrinology *89:* 1485 (1971).

33 VALE, W.; BRAZEAU, P.; GRANT, G.; NUSSEY, A.; BURGUS, R.; RIVIER, J.; LING, N., et GUILLEMIN, R.: Premières observations sur le mode d'action de la somatostatin. C. R. Acad. Sci. *275:* 2913 (1972).

34 VALE, W.; GRANT, G.; AMOSS, M.; BLACKWELL, R., and GUILLEMIN, R.: Enzymatically dispersed AP cells. Validation of a method. Endocrinology *91:* 562–572 (1972).

35 VALE, W.; GRANT, G.; RIVIER, J.; MONAHAN, M.; AMOSS, M.; BLACKWELL, R.; BURGUS, R., and GUILLEMIN, R.: Synthetic polypeptide antagonists of LRF. Science *176:* 933 (1972).

36 VALE, W.; BLACKWELL, R.; GRANT, G., and GUILLEMIN, R.: TRF and thyroid hormones on prolactin secretion by rat AP cells *in vitro*. Endocrinology *93:* 26–33 (1973).

37 VALE, W.; BRAZEAU, P.; RIVIER, C.; RIVIER, J.; GRANT, G.; BURGUS, R., and GUILLEMIN, R.: Inhibitory hypophysiotropic activities of hypothalamic somatostatin. Fed. Proc. *32:* 211 (1973).

38 VALE, W.; GRANT, G., and GUILLEMIN, R.: Chemistry of the hypothalamic releasing factors; in MARTINI and GANONG Frontiers in neuroendocrinology, 1973, pp. 375–413 (Oxford University Press, New York).

39 VALE, W.; BRAZEAU, P., and RIVIER, C.: Effect of SRIF or prolactin and TSH (Abstract). Endocrinology *91* (1973).

Author's address: Dr. G. GRANT, The Salk Institute, P.O. Box 1809, *San Diego, CA 92112* (USA)

Recent Studies of Hypothalamic Function
Int. Symp. Calgary 1973, pp. 196–206 (Karger, Basel 1974)

Endocrine and Extra-Endocrine Studies of Hypothalamic Hormones in Man

A. J. Kastin, A. V. Schally, R. H. Ehrensing and A. Barbeau

Endocrinology Section of the Medical Service, and Endocrine and Polypeptide Laboratory, Veterans Administration Hospital; Department of Medicine, Tulane University School of Medicine, New Orleans, La.; Department of Psychiatry, Louisiana State University School of Medicine, New Orleans, La.; Department of Neurobiology, Clinical Research Institute of Montreal, and University of Montreal, Montreal

I. Introduction

During the last few years, neuroendocrinology has progressed to the stage where clinical studies with hypothalamic hormones have been performed in several medical centers. The skepticism which accompanied early reports of the existence of hypothalamic hormones has disappeared. The emphasis has shifted from the investigation of whether the hypothalamus contains hormones to the testing of the diagnostic and clinical uses of these substances.

II. Endocrine Studies with Hypothalamic Hormones

A. Thyrotropin Releasing Hormone (TRH)

Most of the early clinical studies with TRH have been reviewed recently [8]. These investigations showed that administration of TRH results in increased plasma levels of thyrotropin stimulating hormone (TSH) which reach a maximum about 20–30 min after a single, rapid intravenous (i.v.) injection. Although there are indications that women may be more responsive to TRH than men, this has not been found by all investigators. Hypothyroid patients, lacking the negative feedback control of the thyroid hormones, have elevated plasma TSH levels which show greater increases than normal after injection of TRH. The opposite condition exists in hyperthyroid patients or normal subjects treated with large doses of thyroid hormone(s). Because the principal site of the negative feedback action of thyroid hormone is exerted on the pituitary, individuals with high levels of thyroid hormone (hyperthyroidism) do not respond to TRH with a release of TSH. Likewise, a patient with a pituitary tumor involving the thyrotropic cells of the pituitary gland would not respond to TRH whereas one with a tumor not involving these cells might release TSH. Normal responses also have been found in patients with galactorrhea, primary and secondary amenorrhea, gonadal dysgenesis, hypogonadotropic hypogonadism and isolated growth hormone (GH) deficient dwarfs. The same report [8] which reviewed the above data, also discussed the evidence that TRH does not normally release GH, luteinizing hormone (LH), follicle stimulating hormone (FSH), cortisol, insulin or glucose. Many subsequent articles have described elevations of thyroid hormone(s) after administration of TRH, but these increases still have not been detected in all patients or by all investigators.

Since then, however, there have been several reports which show that TRH raises GH levels in patients with acromegaly [9, 33, 34] and severe renal failure [6]. Administration of GH may inhibit TRH-mediated TSH secretion, but the mechanism is unknown [31]. Stimulation of TSH release by TRH does not seem to be affected by either α- or β-adrenergic stimulation or blockade [39] or short-term steroid administration [40]; however, prolonged administration of glucocorticoids does appear to inhibit TSH release after TRH [22]. The prolactin release which has been described by several investigators [8] as occurring after administration of TRH can be blocked by pretreatment with one of the ergot alkaloids [35]. TRH also has been reported to cause prolactin stimulation in a patient with isolated TSH deficiency [32]. Although the clinical endocrine use of TRH has been limited to diagnosis, the ability of this hypothalamic hormone to release prolactin may have wide applications, particularly in animal husbandry.

B. LH Releasing Hormone (LH-RH)

At the same time that the clinical studies with TRH were reviewed [8], the clinical studies with LH-RH were summarized [13]. It was shown that natural LH-RH is able to release LH in normal men, women and children. The response to LH-RH in patients with pituitary tumors was variable, as was the case with TRH. Moreover, administration of both TRH and LH-RH to the same patients resulted in several patterns of release of the respective pituitary hormones. In contrast to subjects who failed to respond to TRH after pretreatment with thyroid hormones, pretreatment of men and women with sex steroids did not block the response to LH-RH. This does not mean that thyroid hormone or the sex steroids do not act at both the pituitary and hypothalamic levels, but only indicates the principal site of their negative feedback actions in man at the doses studied. Individuals with elevated basal values of gonadotropins respond to LH-RH with even greater increases. This includes postmenopausal women, Turner's syndrome, Klinefelter's syndrome, and men pretreated with clomiphene. Since synthetic LH-RH induces FSH release, this FSH releasing activity is intrinsic to LH-RH, although the rise in FSH is less dramatic than the increase in LH. Additional studies with natural, porcine LH-RH demonstrated a dose response curve, induction of ovulation and a small but significant release of LH in patients with Kallman's syndrome.

The availability of synthetic LH-RH to many investigators has permitted more studies to be performed. Somewhat surprisingly, however, relatively little additional information has been obtained so far. In some instances, of course, extension of the LH-RH studies with the synthetic material has been helpful since investigations using natural LH-RH were limited by the extreme scarcity of the material. For example, only two subjects were injected with LH-RH by the subcutaneous (s.c.) route. Although this showed that routes of administration other than i.v. were feasible, direct comparisons were not made in the same subjects. This situation was rectified in a recent study where it was found that the same subjects injected 3 weeks apart with the same dose of LH-RH gave essentially the same LH and FSH responses when injected i.v. or s.c. [1]. Different subjects injected i.v., s.c., intramuscularly (i.m.), or as a 4-hour i.v. infusion at two different doses also released LH and FSH, but not as much FSH as LH [7]. In normal subjects, porcine LH-RH does not release TSH, GH or cortisol, and several investigators have confirmed this with synthetic LH-RH. Moreover, another study with synthetic LH-RH showed that, unlike TRH, LH-RH does not alter the release of prolactin [15].

A more difficult problem has been the location of the defect in

Kallman's syndrome of hypogonadotropic hypogonadism with anosmia. Studies with synthetic LH-RH confirmed that a single i.v. dose caused a small increase in gonadotropin release. However, repeated i.m. injections of LH-RH for 5 days failed to increase the responsiveness to a second i.v. injection of LH-RH [43]. Yet, two patients with familial panhypopituitarism (asexual ateliotic dwarfism), who showed a similar small increase in gonadotropin release after i.v. LH-RH, showed a greater response after being 'primed' with an 8-hour infusion of LH-RH [19]. Although the synthetic material has permitted additional information to be obtained concerning such syndromes, the location of the defect within the hypothalamic-pituitary system is still not established.

Perhaps the greatest additional clinical information which has been obtained with the synthetic material concerns the effect of LH-RH on ovulation and spermatogenesis. It had been found earlier that natural LH-RH could induce ovulation if given as a prolonged i.v. infusion. In the single such subject studied, the ovary was primed with a material containing FSH. Since LH-RH releases FSH as well as LH, it was logical to use LH-RH early as well as late in the cycle. This regimen was tried in 3 women, and two of them ovulated [42]. In subsequent studies, synthetic LH-RH also induced ovulation in 10 of 23 anovulatory women [44]. The percentage of ovulation has been reported to be higher when clomiphene is used as the 'priming' material and then followed by LH-RH [18] or when anovulation occurs in women with normal urinary gonadotropin excretion [41]. Preliminary studies with the use of LH-RH in men with oligospermia and azoospermia showed a slight but transient improvement in spermatogenesis [2, 36]. A recent review mentions many of the current investigations with synthetic LH-RH [17]. It is reasonable to expect additional reports concerning not only the diagnostic but also the therapeutic uses of LH-RH.

C. GH Releasing Hormone (GH-RH)

Since GH is measured by radioimmunoassay (RIA) in man, it only would be possible to detect the activity of a GH-RH in the human being if it was capable of releasing a GH which could be measured by RIA. A highly purified, and synthetic substance proposed as a GH-RH is a decapeptide which releases GH in rats which can be detected only by bioassay, but not by RIA. It is not surprising, therefore, that a clinical evaluation of this substance failed to demonstrate any convincing increase in plasma levels of GH as detected by RIA [14].

D. Melanocyte Stimulating Hormone Release Inhibiting Hormone (MIF or MRIH)

A series of studies has shown that the release of melanocyte stimulating hormone (MSH) in mammals is controlled by a hypothalamic inhibitor (MIF or MRIH) [10]. Although this concept is now well established, it is not clear which of the several proposed chemical substances represents the true MRIH. Two of these have been isolated from bovine hypothalamic tissue (MRIH-I, MRIH-II) [20, 21] and two others have been proposed [3] on the basis of studies with synthetic analogs of oxytocin. Only Pro-Leu-Gly-NH_2 (MRIH-I), shown by CELIS *et al.* [4] to be cleaved from oxytocin by an enzyme present in hypothalamic tissue, has been tested for its endocrine effects in man. Preliminary clinical findings after a rapid i.v. injection of MRIH-I did not show any decrease in MSH release [13].

III. Extra-Endocrine Studies with Hypothalamic Hormones

A. MSH Release Inhibiting Hormone (MIF, MRIH)

As it became increasingly evident that naturally occurring peptides such as MSH affected the central nervous system (CNS) of mammals, including man [16], we started to examine the effects of other peptides found in the hypothalamic-pituitary system. Pro-Leu-Gly-NH_2 (MRIH-I) was studied in several animal systems routinely used for screening drugs suspected of having CNS effects. MRIH-I was found to be highly active in increasing the behavioral effects induced in mice by small doses of dopa and a monoamine oxidase inhibitor. This potentiation of the effects of dopa occurred whether the mouse was hypophysectomized or not, demonstrating for the first time that the presence of the pituitary gland was not necessary for some of the actions of MRIH-I [23]. Similarly, MRIH-I was shown to antagonize the CNS and peripheral effects of oxotremorine. It also potentiated the effects of L-dopa in reducing these effects of oxotremorine. Again, the effects were observed in both intact and hypophysectomized mice [24]. Mice and monkeys treated with a monoamine oxidase inhibitor and dopa as well as the rauwolfia alkaloid deserpidine showed marked sedation; this effect was reversed by treatment with MRIH-I [26]. These systems have been considered by many investigators to represent animal models of Parkinson's disease and mental depression.

The first clinical study investigating the extra-endocrine effects of a hypothalamic hormone reported preliminary results obtained in 16 parkinsonian patients after administration of MRIH-I [11]. Eight patients with Parkinson's disease received 20 mg MRIH-I in an i.v. infusion lasting 1 h. The main effect was a diminition of rigidity by some 20 % as compared with the placebo period. The same 8 patients were later given 30 mg of MRIH-I a day by mouth for 2 days. Again, most of the improvement was a reduction of rigidity, with a slight decrease in tremor. Three different parkinsonian patients received 50 mg MRIH-I in divided doses for 2 months. Rigidity was reduced about 40 %, temor by 38 % and akinesia about 30 %. In the same clinic, about a 15-percent improvement can be seen during placebo periods, and various double-blind trials of anticholinergic drugs result in about a 30-percent improvement. Five more patients with dyskinesias induced by L-dopa were given MRIH-I. MRIH-I caused a 48-percent reduction in the dyskinesias. Since no side-effects were noted in this clinical study, it is anticipated that higher doses of MRIH-I may result in further improvement in parkinsonism. It was intended to try MRIH-I in mental depression next. However, TRH was more readily available.

B. TRH

Although TRH was not as effective as MRIH-I in the dopa-potentiation test [23, 25], it did have some activity in animal experiments. Accordingly, TRH was administered to 5 patients with mental depression as part of a double-blind crossover study [12]; the study lasted 6 days. For 3 consecutive days the patients received TRH intravenously and for 3 consecutive days the patients received a placebo injection. The order was randomized. The investigator determining the improvement was not present at the injections to avoid clues from any possible side-effects. Patients showed only mild and transient improvement. Measurement of TSH released by TRH in these patients showed a blunted response in

4 of the 5 patients. Prolactin levels were not measured, but the diminished rise in plasma TSH levels suggested that the effects of TRH may not have been mediated by the pituitary gland, just as we had shown in the case of MRIH-I in the animal studies. Additional studies with a higher dose of TRH administered for a longer period of time are presently in progress.

Using an entirely different approach, PRANGE *et al.* also found relief of mental depression after administration of TRH [30]. Earlier, these investigators had shown [27, 28, 38] that thyroid hormone enhanced the antidepressant effect of imipramine, a tricyclic antidepressant. Other workers [5] could not confirm these findings with thyroid hormone in a double-blind controlled study using a homogeneous population of 49 patients. PRANGE *et al.* next tried TSH which also resulted in a more rapid recovery from depression, presumably by causing the secretion of thyroid hormone [29]. It was then a logical step to inject TRH, reasoning that TSH release would, in turn, augment thyroid hormone secretion. However, as pointed out earlier in this review, even in normal individuals, the release of thyroid hormone after injection of TRH is not a constant or marked finding. When it was found later that the presence of the pituitary was not required for some of the actions of TRH on the CNS [25], it became obvious that TRH was probably exerting a direct effect upon the brain. Beneficial effects of TRH in depression also have been found in preliminary studies by others [30, 37].

IV. Comments and Discussions

The diagnostic uses of hypothalamic hormones in clinical endocrinology is now well established. Their therapeutic uses in endocrinology are becoming increasingly evident. Although most of the interest in the field has been directed to endocrinological studies, the new concept of extra-endocrine uses of hypothalamic hormones has broad therapeutic implications. The issue is not whether TRH will prove to be the final answer for mental depression, as we think it may not, or whether MRIH-I may be the best treatment of parkinsonism, but rather the continued exploration of nonendocrine uses of naturally occurring substances found in the hypothalamus and elsewhere in the brain, and their analogs, in clinical medicine.

References

1 ARIMURA, A.; SAITO, M.; YAOI, Y.; KUMASAKA, T.; SATO, H.; KOYAMA, T.; NISHI, N.; KASTIN, A. J., and SCHALLY, A. V.: Comparison of the effects of subcutaneous and intravenous injection of synthetic LH-releasing hormone (LH-RH) on serum LH and FSH levels in men. J. clin. Endocrin. *36* (2): 385–388 (1973).

2 BERGADA, C.; MANCINI, R. E.; RIVAROLA, M. A.; VILAR, O.; CALAMERA, J. C.; BIANCULLI, C.; SCHALLY, A. V., y KASTIN, A. J.: Accion del LH-RH/FSH-RH sintetico en el varon prepuberal. Progr. Soc. Argent. Endocrin. Metab. abstract No. 37 (1972).

3 BOWER, A.; HADLEY, M. E., and HRUBY, V. J.: Comparative MSH release-inhibiting activities of tocinoic acid (the ring of oxytocin), and L-Pro-L-Leu-Gly-NH_2 (the side chain of oxytocin). Biochem. biophys. Res. Commun. *45:* 1185–1191 (1971).

4 CELIS, M. E.; TALEISNIK, S., and WALTER, R.: Regulation of formation and proposed structure of the factor inhibiting the release of melanocyte-stimulating hormone. Proc. nat. Acad. Sci., Wash. *68:* 1428–1433 (1971).

5 FEIGHNER, J. P.; KING, L. J.; SCHUCKIT, M. A.; CROUGHAN, J., and BRISCOE, W.: Hormonal potentiation of imipramine and ECT in primary depression. Amer. J. Psychiat. *128* (10): 1230–1238 (1972).

6 GONZALEZ-BARCENA, D.; KASTIN, A. J.; SCHALCH, D. S.; TORRES-ZAMORA, M.; PEREZ-PASTEN, E.; KATO, A., and SCHALLY, A. V.: Responses to thyrotropin-releasing hormone in patients with renal failure and after infusion in normal men. J. clin. Endocrin. *36* (1): 117–120 (1973).

7 GONZALEZ-BARCENA, D.; KASTIN, A. J.; SCHALCH, D. S.; BERMUDEZ, J. A.; LEE, D.; ARIMURA, A.; RUELAS, J.; ZEPEDA, I., and SCHALLY, A. V.: Synthetic LH-releasing hormone (LH-RH) administered to normal men by different routes. J. clin. Endocrin. *37:* 481–484 (1973).

8 GUAL, C.; KASTIN, A. J., and SCHALLY, A. V.: Clinical experience with hypothalamic releasing hormones. I. Thyrotropin releasing hormone (TRH). Recent Progr. Hormone Res. *28:* 173–200 (1972).

9 IRIE, M. and TSUSHIMA, T.: Increase of serum growth hormone concentration following thyrotropin-releasing hormone injection in patients with acromegaly or gigantism. J. clin. Endocrin. *35* (1): 97–100 (1972).

10 KASTIN, A. J. and SCHALLY, A. V.: Control of MSH release in mammals. Excerpta Med. Int. Congr. Series No. 238, pp. 311–317 (Excerpta Medica, Amsterdam 1970).

11 KASTIN, A. J. and BARBEAU, A.: Preliminary clinical studies with L-prolyl-L-leucyl-glycine amide in Parkinson's disease. Canad. med. Ass. J. *107:* 1079–1081 (1972).

12 KASTIN, A. J.; EHRENSING, R. H.; SCHALCH, D. S., and ANDERSON, M. S.: Improvement in mental depression with decreased thyrotropin response after administration of thyrotropin-releasing hormone. Lancet *ii:* 740–742 (1972).

13 KASTIN, A. J.; GUAL, C., and SCHALLY, A. V.: Clinical experience with hypothalamic releasing hormones. II. Luteinizing hormone-releasing hormone and other hypophysiotropic hormones. Recent Progr. Hormone Res. *28:* 201–227 (1972).

14 KASTIN, A. J.; SCHALLY, A. V.; GUAL, C.; GLICK, S., and ARIMURA, A.: Clinical evaluation in men of a substance with growth-hormone releasing activity in rats. J. clin. Endocrin. *35:* 326–329 (1972).

15 KASTIN, A. J.; GONZALES-BARCENA, D.; FRIESEN, H.; JACOBS, L. S.; SCHALCH, D. S.; ARIMURA, A.; DAUGHADAY, W. H., and SCHALLY, A. V.: Unaltered plasma prolactin levels in men after administration of synthetic LH-releasing hormone. J. clin. Endocrin. *36* (2): 375–377 (1973).

16 KASTIN, A. J.; MILLER, L. H.; NOCKTON, R.; SANDMAN, C. A.; SCHALLY, A. V., and STRATTON, L. O.: Behavioral aspects of MSH. In: Progress in brain research, vol. 93, pp. 461–470 (Elsevier, Amsterdam 1973).

17 KASTIN, A. J.; SCHALLY, A. V.; ZARATE, A.; ARIMURA, A.; GONZALEZ-BARCENA, D.; MEDEIROS-NETO, G., and SCHALCH, D. S.: Analysis of clinical studies with natural and synthetic LH-RH in man. Israel J. med. Sci. (in press).

18 KELLER, P. J.: Induction of ovulation by synthetic luteinizing-hormone releasing factor in infertile women. Lancet *ii:* 570–572 (1972).

19 MEDEIROS-NETO, G. A.; TOLEDO, S. P. A.; SUCUPIRA, M.; FRAIGE-FILHO, F.; MATTAR, E.; KASTIN, A. J., and SCHALLY, A. V.: Characterization of the LH response to luteinizing hormone-releasing hormone (LH-RH) in asexual and sexual ateliotic dwarfism. J. clin. Endocrin. Metab. *37:* 972–976 (1973).

20 NAIR, R. M. G.; KASTIN, A. J., and SCHALLY, A. V.: Isolation and structure of hypothalamic MSH release-inhibiting hormone. Biochem. biophys. Res. Commun. *43:* 1376–1381 (1971).

21 NAIR, R. M. G.; KASTIN, A. J., and SCHALLY, A. V.: Isolation and structure of another hypothalamic peptide possessing MSH-releasing inhibiting activity. Biochem. biophys. Res. Commun. *47:* 1420–1425 (1972).

22 OTSUKI, M.; DAKODA, M., and BABA, S.: Influence of glucocorticoids on TRF-induced TSH response in man. J. clin. Endocrin. *36* (1): 95–102 (1973).

23 PLOTNIKOFF, N. P.; KASTIN, A. J.; ANDERSON, M. S., and SCHALLY, A. V.: Dopa potentiation by MIF. Life Sci. *10:* 1279–1283 (1971).

24 PLOTNIKOFF, N. P.; KASTIN, A. J.; ANDERSON, M. S., and SCHALLY, A. V.: Oxotremorine antagonism by a hypothalamic hormone, melanocyte-stimulating hormone release-inhibiting factor, MIF. Proc. Soc. exp. Biol. Med. *140:* 811–814 (1972).

25 PLOTNIKOFF, N. P.; PRANGE, A. J.; BREESE, G. R.; ANDERSON, M. S., and WILSON, I. C.: Thyrotropin releasing hormone. Enhancement of dopa activity by a hypothalamic hormone. Science *178:* 417–418 (1972).

26 PLOTNIKOFF, N. P.; KASTIN, A. J.; ANDERSON, M. S., and SCHALLY, A. V.: Deserpidine antagonism by a tripeptide, L-prolyl-L-leucylglycinamide. Neuroendocrinology *11:* 67–71 (1973).

27 PRANGE, A. J.; WILSON, I. C.; RABON, A. M., and LIPTON, M. A.: Enhancement of imipramine by triiodothyronine in unselected depressed patients. Excerpta Med. Int. Congr. Ser. No. 180, pp. 532–535 (Excerpta Media, Amsterdam 1969).

28 PRANGE, A. J.; WILSON, I. C.; RABON, A. M., and LIPTON, M. A.: Enhancement of imipramine antidepressant activity by thyroid hormone. Amer. J. Psychiat. *126:* 457–469 (1969).

29 PRANGE, A. J.; WILSON, I. C.; KNOX, A.; MCCLANE, T. K., and LIPTON, M. A.: Enhancement of imipramine by thyroid stimulating hormone. Clinical and theoretical implications. Amer. J. Psychiat. *127* (2): 109–199 (1970).

30 PRANGE, A. J.; WILSON, I. C.; LARA, P. P.; ALLTOP, L. B., and BREESE, G. R.: Effects of thyrotropin-releasing hormone in depression. Lancet *ii:* 999–1002 (1972).

31 ROOT, A. W.; SNYDER, P. J.; REZVANI, I.; DIGEORGE, A. M., and UTIGER, R. D.: Inhibition of thyrotropin-releasing hormone-mediated secretion of thyrotropin by human growth hormone. J. clin. Endocrin. *36* (1): 103–106 (1973).

32 SACHSON, R.; ROSEN, S. W.; CUATRECASAS, P.; ROTH, J., and FRANTZ, A. G.: Prolactin stimulation by thyrotropin-releasing hormone in a patient with isolated thyrotropin deficiency. New Engl. J. Med. *287* (19): 972–973 (1972).

33 SAITO, S.; ABE, K.; YOSHIDA, H.; KANEKO, T.; NAKAMURA, E.; SHIMIZU, N., and YANAIHARA, N.: Effects of synthetic thyrotropin-releasing hormone on plasma thyrotropin, growth hormone and insulin levels in man. Endocrinology, Jap. *18* (1): 101–108 (1971).

34 SCHALCH, D. S.; GONZALEZ-BARCENA, D.; KASTIN, A. J.; SCHALLY, A. V., and LEE, L. A.: Abnormalities in the release of TSH in response to thyrotropin-releasing hormone (TRH) in patients with disorders of the pituitary, hypothalamus and basal ganglia. J. clin. Endocrin. *35* (4): 609–615 (1972).

35 SCHAMS, D.: Prolactin releasing effects of TRH in the bovine and their depression by a prolactin inhibitor. Hormone Metab. Res. *4:* 405–406 (1972).

36 VALDES-LAVALLINA, F.; ZARATE, A.; SCHALLY, A. V.; CANALES, E. S.; GONZALEZ, A.; PEREZ-UBIERNA, C., and KASTIN, A. J.: Treatment of idiopathic oligospermia and azoospermia with synthetic luteinizing hormone-releasing hormone (LH-RH). Preliminary results. Progr. Fertility and Sterility Meeting, 1973 (in press).

37 VIS-MELSEN, M. J. E. VAN DER and WIENER, J. D.: Improvement in mental depression with decreased thyrotropin response after administration of thyrotropin-releasing hormone. Letter to the editor. Lancet *ii:* 1415 (1972).

38 WILSON, I. C.; PRANGE, A. J.; MCCLANE, T. K.; RABON, A. M., and LIPTON, M. A.: Thyroid hormone enhancement of imipramine in nonretarded depression. New Engl. J. Med. *282:* 1063–1067 (1970).

39 WOOLF, P. D.; LEE, L. A., and SCHALCH, D. S.: Adrenergic manipulation and thyrotropin-releasing hormone (TRH)-induced thyrotropin (TSH) release. J. clin. Endocrin. *35* (4): 616–618 (1972).

40 WOOLF, P. D.; GONZALEZ-BARCENA, D.; SCHALCH, D. S.; LEE, L. A.; PEDRO ARZAC, J.; SCHALLY, A. V., and KASTIN, A. J.: Lack of effect of steroids on thyrotropin-releasing hormone (TRH) mediated thyrotropin (TSH) release in man. Neuroendocrinology *13:* 56–62 (1973).

41 ZANARTU, J.; DABANCENS, A.; KASTIN, A. J., and SCHALLY, A. V.: Effect of synthetic hypothalamic gonadotropin releasing hormone in anovulatory sterility. Fertil. Steril. (in press).

42 ZARATE, A.; CANALES, E.; SCHALLY, A. V.; AYALA-VALDES, L., and KASTIN, A. J.: Successful induction of ovulation with synthetic luteinizing hormone-releasing hormone in anovulatory infertility. Fertil. Steril. *23:* 672–674 (1972).
43 ZARATE, A.; KASTIN, A. J.; SORIA, J.; SCHALLY, A. V., and CANALES, E. S.: Effect of synthetic luteinizing hormone-releasing hormone (LH-RH) in two brothers with hypogonadotropic hypogonadism and anosmia. J. clin. Endocrin. *36:* 612–614 (1973).
44 ZARATE, A.; CANALES, E. S.; SORIA, J.; GONZALEZ, A.; SCHALLY, A. V., and KASTIN, A. J.: Further observations on the therapy of anovulatory sterility with synthetic luteinizing hormone-releasing hormone. Assessment of follicular maturation by estrogen determination. Fertil. Steril. *25:* 3–10 (1974).

Author's address: Dr. A. J. KASTIN, Endocrinology Section, Veterans Administration Hospital, 1601 Perdido Street, *New Orleans, LA 70146* (USA)

Recent Studies of Hypothalamic Function
Int. Symp. Calgary 1973, pp. 207–215 (Karger, Basel 1974)

Mechanism of Action of Hypothalamic Releasing Hormones

Role of Ions in the Release of Adenohypophysial Hormones

J. KRAICER

Department of Physiology, Queen's University at Kingston, Kingston, Ontario

This review will concern itself with the mechanisms by which the hypothalamic releasing hormones stimulate the secretion of hormones from the adenohypophysis, but will be restricted to the role of the inorganic cations – Ca^{++}, Na^{+} and K^{+} – in the release process. A more comprehensive review has recently been carried out [7].

The term 'releasing hormone' suggests that these substances stimulate the release of preformed hormone. The fundamental questions then are: (1) Do the releasing hormones act primarily to stimulate release of preformed hormone, with synthesis of new hormone being secondary to the release process; (2) Do they act primarily to stimulate synthesis, with release being secondary, or as a consequence of new synthesis, or (3) do the releasing hormones have direct and specific actions on both synthesis and release? These fundamental questions are under active investigation in a number of laboratories. It seems reasonable to conclude at this time that there is a primary effect on the release mechanism, for if synthesis is blocked, release of preformed hormone will still occur [9].

This review will deal with the ionic mechanisms by which the hypothalamic releasing hormones stimulate release of preformed hormone. It is assumed for this discussion that their mechanism of action is in all cases the same. As far as we know, this is a valid assumption.

We began with the model proposed by DOUGLAS [3] and his colleagues, a model derived from their studies of hormone release from the neurohypophysis and adrenal medulla – the stimulus secretion coupling hypothesis. According to this model:

1. Release is initiated by a secretagogue interacting with the plasma membrane [this interaction is discussed in detail by Dr. GRANT in this volume].

2. This results in an alteration in plasma membrane chemistry or conformation.

3. This, in turn, results in an increase in the permeability of the plasma membrane to various ions, in particular to Ca^{++}, and possibly also to Na^{+}, with a resultant decrease in transmembrane potential (TMP).

4. But more important, there is an increased Ca^{++} influx into the cells.

5. This increased influx of Ca^{++} then, in some way, triggers the release process.

Because of the obvious analogy to the excitation-coupling mechanism in muscle, DOUGLAS [3] and colleagues termed the process 'stimulus-secretion' coupling with Ca^{++} as the link, or coupling, between stimulus and secretion.

The Role of Na^{+} and K^{+} in the Release Process

There is some evidence that Na^{+} and K^{+} may be involved in the release process [4, 14–16]. We have recently carried out a series of *in vitro* studies [6] to determine if changes in plasma membrane permeability or intracellular uptake of Na^{+} or K^{+} occur under conditions of augmented hormone release, and if such changes could be related to the release mechanism. Glands were incubated in Krebs-Ringer bicarbonate solution (KRB) containing both the labeled cation and ^{3}H or ^{14}C-mannitol, the latter as an extracellular fluid volume marker [6]. If the uptake of the labeled cation is greater than the mannitol uptake, then the cation is presumed to have entered the cells. The uptake, with time, gives a measure of the one-way flux of the ion. ^{24}Na was distributed rapidly and completely in a compartment slightly greater than the volume of distribution and mannitol; however, as was the case with mannitol, the ^{24}Na remained constant after 6 min, suggesting that the adenohypophysial cells are essentially impermeable to Na^{+}. On the other hand, the cells are quite permeable to K^{+} since ^{42}K was taken up steadily for at least 96 min, by which time the uptake reached 58 mM/kg wet weight.

We also examined the effect of a number of secretagogues on both the uptakes and the intracellular concentrations of Na^{+} and K^{+}. These included 30 mM $[K^+]_0$, synthetic lysine-vasopressin, dibutyryl cyclic AMP, theophylline, crude hypothalamic stalk-median eminence extract and a purified ovine GH releasing hormone (GRH) preparation [6]. None of these substances except the elevated $[K^+]_0$ altered either the uptake or the intracellular concentrations of Na^+ or K^+. Increasing the external $[K^+]_0$ concentration produced a slight decrease in ^{22}Na uptake with no change in the intracellular concentration of Na^+. The uptake of ^{42}K was increased approximately 3-fold with no change in the intracellular concentration of K^+. Thus, the effect was to increase the permeability of the cells to K^+ itself with no change in net transport.

Since none of the secretagogues tested except for elevated $[K^+]_0$ altered the permeability or net transport of Na^+ and K^+, these two major cations appear not to be involved in the mechanisms of hormone release.

The Role of Ca^{++} in the Release Process

We can ask two interrelated questions based on the [3] hypothesis: (1) Is hormone release Ca^{++}-dependent, and (2) Is there an influx of Ca^{++} into the cells associated with the release process?

The Ca^{++} dependence of the release process has been extensively studied. The rationale for most studies was that if an influx of Ca^{++} is essential for release, then the removal of Ca^{++} from the incubation medium should prevent release. A number of studies have reported that the augmented release of hormone induced by a whole host of secretagogues, including elevated K^+ and releasing hormone preparations, is suppressed when Ca^{++} is removed from the incubation medium and that this effect is reversible and repeatable [7]. However, in most of these studies very rigorous procedures were used to remove Ca^{++} from the system. These included prolonged and repeated washing of tissues in Ca^{++}-free media and/or the use of Ca^{++} chelating agents. One cannot conclude from such studies that release was prevented solely by the removal of Ca^{++} from the extracellular compartment, thus preventing influx into the cells. Such rigorous procedures could have prevented release, not only by preventing influx, but also by the removal or leeching out of an essential intracellular Ca^{++} compartment.

We have recently attempted to define more closely the Ca^{++} requirement for ACTH release, comparing a quick Ca^{++}-free rinse procedure with a more rigorous multiple Ca^{++}-free wash procedure [12]. The augmented release of ACTH induced by elevated $[K^+]_0$ and by vasopressin was abolished by the quick rinse procedure, while that induced by theophylline and by a crude extract of hypothalamus stalk-median eminence was not. It was only after the vigorous Ca^{++} wash procedure that the augmented release of hormone induced by the latter two secretagogues was depressed. The quick Ca^{++}-free rinse procedure decreased $[Ca^{++}]_i$ only slightly, while the multiple Ca^{++}-free wash procedure reduced $[Ca^{++}]_i$ to about 66 % of the *in vivo* value.

These results clearly demonstrate that Ca^{++} is essential for the release process, and that different intracellular pools of Ca^{++} may be involved, depending on the secretagogue. Some secretagogues appear to require a loosely bound or easily accessible Ca^{++} pool, others a more tightly bound or less accessible pool.

Of more direct relevance is the second question asked above. Is there an influx of Ca^{++} into the cells of the pars distalis associated with hormone release?

We have studied this in an *in vitro* system with glands incubated in KRB containing ^{45}Ca with ^{3}H-mannitol as the extracellular space marker [11, 13]. If ^{45}Ca is distributed in a volume larger than the mannitol space, then Ca^{++} is presumed to have entered the cells. This then gives a measure of the volume of distribution of ^{45}Ca or, more precisely, the one-way flux of Ca^{++}. The Ca^{++} space was found to increase rapidly, reaching a volume of distribution equal to the total gland volume within 6 min [11]. There followed a gradual increase over a subsequent 90-min period, with the volume of distribution reaching about 160 % of the total gland volume.

The augmented release of hormone induced by elevated $[K^+]_0$ was associated with a concurrent increase in Ca^{++} influx [11]. Of more physiological significance was our finding that the augmented release of growth hormone (GH) induced by a partially purified ovine growth hormone releasing hormone preparation was accompanied by a substantially increased Ca^{++} influx [13].

The Ca^{++} flux studies have not yet been correlated with concurrent measurements of $[Ca^{++}]_i$. Therefore, we do not know whether the changes in Ca^{++} influx represent a change in permeability of the cells to Ca^{++} or a change in the net transport of Ca^{++}, or both.

These data establish a pivotal role for Ca^{++} in the release process. Further studies are required to identify the intracellular Ca^{++} compartments and to explore the intracellular events associated with the role of

Ca^{++} in the release process. We have recently speculated on the possible steps in the release process where Ca^{++} may act [7].

Transmembrane Potentials in the Adenohypophysis

In order to investigate the role of electrical changes in the plasma membrane in the action of the hypothalamic releasing hormones, two interrelated questions are posed.

1. Will a decrease in TMP stimulate hormone release?

2. Do the releasing hormones alter the electrical characteristics of the plasma membrane?

To answer the first question, the standard electrophysiological procedure used to decrease TMP in an *in vitro* system was employed; namely, elevation of the $[K^+]$ in the incubation medium. The rationale for this is the Nernst equation [2] in which the equilibrium potential for K^+ is given by the equation:

$$E_k = \frac{RT}{FZ_k} \log_e \frac{[K^+]_0}{[K^+]_i}.$$

This can be simplified to:

$$E_k = 61 \log_{10} \frac{[K^+]_0}{[K^+]_i},$$

where E_k = equilibrium potential for K^+, in millivolts, $[K^+]_0$ = concentration of extracellular free K^+ in millimoles per litre, $[K^+]_i$ = concentration of free intracellular K^+ in millimoles per litre.

In nerve and muscle, the TMP can be described by this Nernst equilibrium potential for K^+, since K^+ is the most permeable ion species [5]. Thus, by increasing $[K^+]_0$, one can decrease TMP in a predicted way. According to the Douglas hypothesis, this would result in hormone release.

As predicted, it has been repeatedly and consistently reported that incubating adenohypophyses in high $[K^+]_0$ will increase the release of all of the hormones of the adenohypophysis except for prolactin [7]. This release is reversible and repeatable.

However, the interpretation of these findings is not straightforward. We have recorded TMP from the cells of the rat adenohypophysis using conventional intracellular recording techniques both *in vivo* [17, 18] and *in vitro* [8, 10]. TMP recorded *in vitro* are low, averaging about –12 mV. We originally reported the presence of positive TMP *in vitro* [10] but this was not confirmed in a subsequent study [8]. *In vivo* TMP averages about –20 mV [17, 18]. If the Nernst potential for K^+ were to hold for the cells of the adenohypophysis, the low resting TMP would indicate an extremely low $[K^+]_i$ as compared to other tissues. This is not the case, since the intracellular $[K^+]_i$ of the rat adenohypophysis is about 112 mM/kg wet weight, comparable to that found in other tissues. If the cells were selectively permeable to K^+, the TMP derived from the Nernst relation would be about –75 mV and not the –12 to –24 mV actually observed.

Is the increased hormone release induced by high $[K^+]_0$ associated with a depolarization? In an *in vitro* study [8] TMP were continuously recorded from individual cells when the medium was changed from KRB to KRB containing 30 mM $[K^+]_0$. All cells showed a depolarization. But the mean depolarization was only 4 mV and not the 40 mV predicted.

How can we account for the low TMP observed? The intracellular concentrations of K^+ and Na^+ in adenohypophysial cells are in the same range as those found in other tissues ($[K^+]_i = 112$ mM, $[Na^+]_i = 25$ mM) [6]. The cells of the adenohypophysis are relatively impermeable to Na^+ and quite permeable to K^+ and Ca^{++}. If we assume that the free $[Ca^{++}]_j$ is very low, as in most other tissues [1], then the low TMP observed could be due to the opposing concentration gradients of these two permeable ions – K^+ and Ca^{++}. The small but significant depolarization observed in elevated $[K^+]_0$ is associated with an increased influx of both K^+ and Ca^{++} [6, 11] with no increase in Na^+ permeability [6]. Thus, the depolarization in response to elevated $[K^+]_0$ might be due to the increased Ca^{++} uptake as well as the altered K^+ gradient. A significant, but smaller depolarization still occurs during stimulation with elevated $[K^+]_0$ in a Ca^{++}-free medium [8] when augmented hormone release has been abolished. The depolarization is smaller in Ca^{++}-free media since there is no increased Ca^{++} influx, and is therefore primarily due to the altered K^+ gradient.

A small but significant depolarization is also observed *in vitro* when KRB is replaced with Ca^{++}-free KRB [8]. Under these conditions there is an increase in Na^+ influx with no change in K^+ influx [6]. This would suggest that the depolarization observed under these conditions might be due to an increased Na^+ permeability.

Thus, although an increase in $[K^+]_0$ does stimulate hormone release, the mechanism by which it does this is by no means clear-cut. It does not cause release in the simple way predicted by the Nernst relationship. The absolute values of TMP observed, and the magnitude of the depolarization found, are not those predicted from the Nernst relation.

The second question raised above was whether the releasing hormones alter the electrical characteristics of the plasma membrane. More specifically, is there a resultant decrease in TMP and decrease in electrical resistance of the plasma membrane? We have carried out two series of experiments using standard intracellular recording techniques in an attempt to answer this question; the first indirect, and the second less indirect. In the first study [17] the rationale was to produce a situation in which there are a large number of pars distalis cells secreting increased amounts of hormone and then to see whether this would alter the TMP distribution. Following the chronic administration of propylthiouracil (PTU) in the rat there is, through feedback control, a massive increase in thyroid stimulating hormone (TSH) secretion, with a marked hypertrophy and hyperplasia of hypersecreting thyrotrophs. We therefore sampled the TMP of cells in the pars distalis in controls and in rats receiving PTU for prolonged periods. We then looked for a change in the frequency distribution of TMP; more specifically, for a shift in the frequency distribution to less negative values. When the frequency distributions were subjected to statistical analysis, no differences were found in the control and PTU treated groups. Thus, the response predicted by the Douglas hypothesis was not seen; no decrease in TMP was observed.

In the second series of *in vivo* studies [18], a more direct approach was used. TMP and electrical resistance were measured following the intracarotid injection of a crude extract of rat hypothalamus stalk-median eminence, rich in releasing hormone activity. We did not find the simple response pattern expected. The most consistent change was not a decrease in TMP and a decrease in electrical resistance, but no change in TMP and an increase in electrical resistance. Of the 51 cells sampled, 74 % demonstrated an increase in electrical resistance. A depolarization was observed in 9 cells, a hyperpolarization in 9 cells and no change in TMP in 33 cells. Control extracts were without effect.

The results of this limited series of studies do not lead to a consistent pattern. The direct *in vitro* experiment to test the effect of a pure or synthetic releasing hormone directly on its target cell has not yet been reported. One can conclude only that the releasing hormones can alter the electrical characteristics of the plasma membrane, but not in a consistent way. They do not appear to do so in the simple way predicted by the stimulus-secretion coupling hypothesis.

Conclusions

1. The hypothalamic releasing hormones alter the electrical characteristics of plasma membrane of adenohypophysial cells, but not in a simple predictable way.

2. There is an increased movement of Ca^{++} into the cells associated with releasing hormone induced release. This influx is unique, since there is no concurrent influx of the other two common cations, Na^{+} and K^{+}.

3. Ca^{+} is essential for the release process. However, its site of action has yet to be elucidated. It would appear to be involved at some specific cytoplasmic site(s) where translocation of Ca^{++} within the cell is the critical event.

4. Changes in Na^{+} and K^{+} uptake and permeability appear to play no functional role in the mechanisms governing the release of adenohypophysial hormones.

5. The low TMP of adenohypophysial cells may be due to their relatively high permeability to both K^{+} and Ca^{++}. Hormone release induced by high $[K^{+}]_0$ is likely due to an increase in the permeability of the cell membranes to Ca^{++} resulting in an increased influx of Ca^{++}, and perhaps also a redistribution of intracellular Ca^{++}.

Acknowledgements

I wish to thank Dr. J. V. MILLIGAN and Dr. D. H. YORK for their assistance in the preparation of this manuscript. The studies reported from our laboratories were supported by grants from the Medical Research Council of Canada, Associated Medical Services Inc., and Queen's University.

References

1 BORLE, A. B.: Calcium transport in kidney cells and its regulation; in NICHOLS and WASSERMAN Cellular mechanisms for calcium transfer and homostasis, pp. 151–174 (Academic Press, New York 1971).

2 DAVSON, H.: A textbook of general physiology (Churchill, London 1970).

3 DOUGLAS, W. W.: Stimulus-secretion coupling. The concept and clues from chromaffin and other cells. Brit. J. Pharmacol. *34:* 451–474 (1968).

4 FLEISCHER, N.; ZIMMERMAN, G.; SCHINDLER, W., and HUTCHINS, M.: Stimulation of adrenocorticotropin (ACTH) and growth hormone (GH) release by ouabain. Relationship to calcium. Endocrinology *91:* 1436–1441 (1972).

5 KATZ, B.: Nerve, muscle and synapse, 1st ed., p. 62 (McGraw-Hill, London 1966).

6 KRAICER, J. and MILLIGAN, J. V.: Effects of various secretagogues upon ^{42}K and ^{22}Na uptake during *in vitro* hormone release from the rat adenohypophysis. J. Physiol., Lond. *232:* 221–237 (1973).

7 KRAICER, J.: Mechanisms involved in the release of adenohypophysial hormones; in FARQUHAR and TIXIER-VIDAL Ultrastructure in biological systems. The anterior pituitary (in press).

8 MARTIN, S.; YORK, D. H., and KRAICER, J.: Alterations in transmembrane potential of adenohypophysial cells in elevated potassium and calcium-free media. Endocrinology *92:* 1084–1088 (1973).

9 MCCANN, S. M.: Mechanism of action of hypothalamic-hypophyseal stimulating and inhibiting hormones; in MARTINI and GANONG Frontiers in neuroendocrinology, pp. 209–235 (Oxford University Press, New York 1971).

10 MILLIGAN, J. V. and KRAICER, J.: Adenohypophysial transmembrane potentials: Polarity reversal by elevated external potassium ion concentration. Science *167:* 182–184 (1970).

11 MILLIGAN, J. V. and KRAICER, J.: ^{45}Ca uptake during the *in vitro* release of hormones from the rat adenohypophysis. Endocrinology *89:* 766–773 (1971).

12 MILLIGAN, J. V. and KRAICER, J.: Physical characteristics of the Ca^{++} compartments associated with *in vitro* ACTH release. Endocrinology *94:* 435–443 (1974).

13 MILLIGAN, J. V.; KRAICER, J.; FAWCETT, C. P., and ILLNER, P.: Purified growth hormone releasing factor increases ^{45}Ca uptake into pituitary cells. Canad. J. Physiol. Pharmacol. *50:* 613–617 (1972).

14 PARSONS, J. A.: Effects of cations on prolactin and growth hormone secretion by rat adenohypophyses *in vitro*. J. Physiol., Lond. *210:* 973–987 (1970).

15 SCHOFIELD, J. G. and COLE, E. N.: Behaviour of systems releasing growth hormone *in vitro*. Mem. Soc. Endocrin. No. 19, 185–199 (1971).

16 WAKABAYASHI, K.; KAMBERI, I. A., and MCCANN, S. M.: *In vitro* response of the rat pituitary to gonadotrophin-releasing factors and to ions. Endocrinology *85:* 1046–1056 (1969).

17 YORK, D. H.; BAKER, F. L., and KRAICER, J.: Electrical properties of cells in the adenohypophysis. An *in vivo* study. Neuroendocrinology *8:* 10–16 (1971).

18 YORK, D. H.; BAKER, F. L., and KRAICER, J.: Electrical changes induced in rat adenohypophysial cells, *in vivo*, with hypothalamic extract. Neuroendocrinology *11:* 212–228 (1973).

Author's address: Dr. J. KRAICER, Department of Physiology, Queen's University at Kingston, *Kingston, Ontario* (Canada)

Metabolic and Behavioral Aspects of
Hypothalamic Function

Recent Studies of Hypothalamic Function
Int. Symp. Calgary 1973, pp. 216–231 (Karger, Basel 1974)

Influence of Limbic Cortex on Hypothalamus

New Anatomic and Microelectrode Findings

P. D. MacLean

Laboratory of Brain Evolution and Behavior, National Institute of Mental Health, Bethesda, Md.

In evolution, the primate forebrain has expanded along the lines of 3 basic patterns which, in figure 1, I have labeled reptilian, paleomammalian and neomammalian. Radically different in chemistry and structure and, in an evolutionary sense, countless generations apart, the 3 cerebrotypes comprise a hierarchy of three brains in one or, what I call for short, a *triune* brain [27, 28].

Of the 3 basic formations only one – the paleomammalian – has thus far been shown to be extensively interconnected with the hypothalamus. As I shall point out, however, future research may require some revision of present anatomical views regarding the reptilian-type brain. The matter is one of great importance because of the recognized pivotal role of the hypothalamus in the regulation of neuroendocrine and somatovisceral functions, as well as emotional behavior. After a brief introduction I will review some recent microelectrode and neuroanatomic studies concerned with the influence of the cortex of paleomammalian origin on hypothalamic function.

The stain for cholinesterase and the histofluorescence technique for revealing dopamine have been most useful in identifying corresponding structures in the brains of reptiles, birds and mammals. In mammals, the major counterpart of the reptilian forebrain is represented by the striatopallidal complex. Nauta and Mehler have presented evidence that peripallidal gray matter projects to the hypothalamus, but they find no indication that the so-called pallidohypothalamic tract terminates in the ventromedial nucleus or elsewhere in the hypothalamus [37]. The ansa lenticularis has both a compact and diffuse trajectory through the lateral and dorsolateral hypothalamus. Some of the diffuse fibers become entangled with those of the medial forebrain bundle, and improved anatomical techniques may eventually reveal hypothalamic connections.

In the lost transitional animals between reptiles and mammals (the so-called mammal-like reptiles) the rudimentary reptilian cortex is presumed to have undergone expansion and differentiation. In all existing mammals, most of the phylogenetically old cortex occupies a large convolution which BROCA, in 1878, called the great limbic lobe because it surrounds the brain stem [5]. In 1952, I suggested the term 'limbic system' as a designation for the limbic cortex and structures of the brain stem with which it has primary connections [23]. The limbic system represents an inheritance from lower mammals.

As illustrated in figure 2, the limbic system comprises 3 main subdivisions, each centered around a particular group of nuclei. The two subdivisions connected with the amygdala and septum are involved, respectively, in functions required for self-preservation and the procreation of the species. In the evolution of higher primates and man, the third subdivision undergoes great enlargement. It should be emphasized that its main afferent pathway, the mammillothalamic tract, bypasses the olfactory apparatus. According to HOPF one of its main nuclei, the anterior

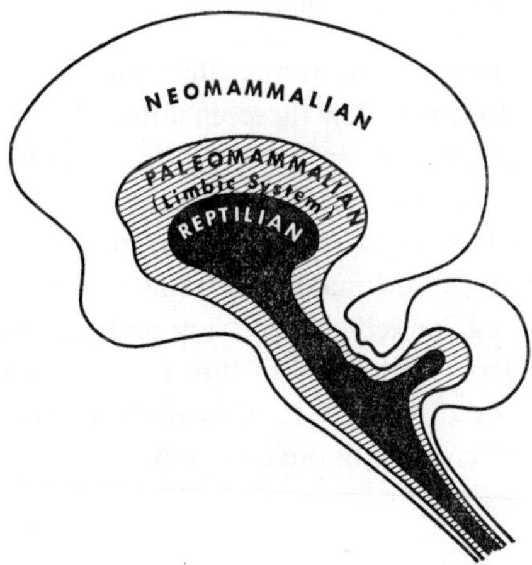

Fig. 1. In evolution the primate forebrain expands in hierarchic fashion along the lines of 3 basic patterns named in the diagram. The 3 main evolutionary formations (which are radically different in chemistry and structure) comprise, so to speak, three brains in one – a *triune* brain. The limbic system represents the counterpart of the paleomammalian brain [25b].

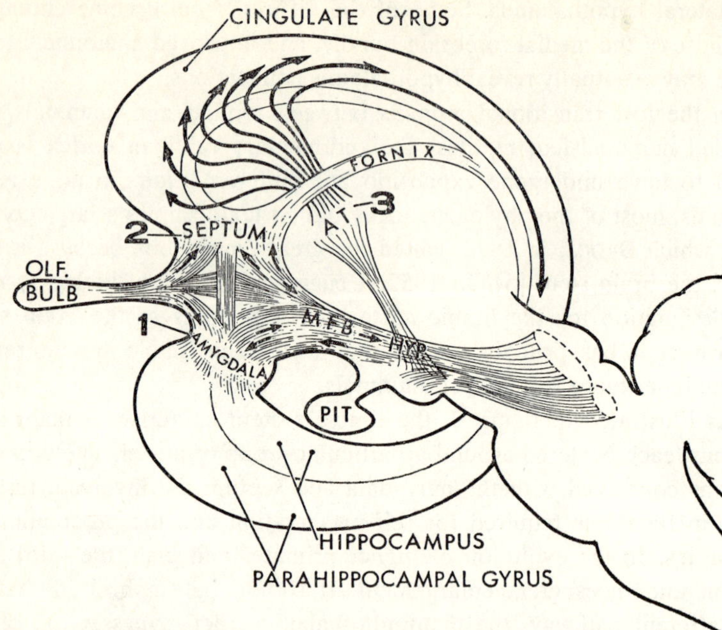

Fig. 2. Diagram of the 3 main subdivisions of the limbic system and their major pathways. AT = anterior thalamic nuclei; Hyp. = hypothalamus; MFB = medial forebrain bundle; Olf. = olfactory [adapted from 25a].

ventral nucleus, enlarges proportionally more in man than the rest of the thalamus [16]. I have suggested elsewhere that these anatomical features, together with functional considerations, may reflect a shifting in emphasis from olfactory to visual influences in sociosexual behavior [28].

An assortment of clinical and experimental evidence indicates that the limbic system, as a whole, derives information in terms of emotional feelings that guide behavior required for self-preservation and the preservation of the species. The strongest evidence for this is derived from patients with limbic epilepsy. Seizure discharges in or near the limbic cortex may trigger a broad spectrum of vivid emotional feelings.

The Question of Sensory Inputs to Limbic Cortex

In 1949, I proposed that emotional feelings generated by the limbic cortex depend upon a blending of internal and external experience [22]. There are also indications that a melding of such experience is essential for a feeling of individuality, a sense of reality and memory [cf. 27]. Except for known olfactory connections and some indirect evidence of gustatory and visceral projections, there existed at that time no information as to whether or not other sensory systems are specifically related to the limbic cortex. Such knowledge was desirable as a first step in understanding the influence of the limbic cortex on hypothalamic function.

In that same paper, I suggested that information from all the intero- and exteroceptive systems reaches the hippocampus via the hippocampal gyrus [22]. Subsequently, our neuronographic findings supported the inference of transcortical links from the primary visual, auditory and somatic areas [31, 43]. Three years ago, JONES and POWELL reported an anatomical study in macaques that also provided evidence of stepwise connections from these areas to the hippocampal gyrus [19]. As I will now explain in a brief summary of our published studies, further investigation has indicated that there are also subcortical inputs from the exteroceptive systems. Then I shall proceed to the question of limbic cortical influences of the hypothalamus.

Microelectrode and anatomical findings. In continuing the study of inputs to the limbic cortex, we have used a stereotaxic, closed system technique for exploring the brain and recording from single cells in awake, sitting squirrel monkeys. Such experiments have the advantage of avoiding the depressive effects of anesthesia on neural transmission and, as opposed to those employing macroelectrodes, are able to pinpoint the locus of a neural response. We have tested more than 7,500 cerebral units, of which about one-third was located in various limbic cortical areas.

Cells responsive to photic stimulation were found only in the posterior parahippocampal cortex [32]. About one-half the responsive cells in the posterior hippocampal gyrus showed a sustained tonic on-response to illumination of the eye, suggesting the possibility that they play a role in neuroendocrine activity affected by diurnal or seasonal changes in light, or are possibly involved in states of alerting, wakefulness and attention.

Response latencies of the photically activated units suggested a subcortical pathway. Accordingly, we placed lesions in various parts of the geniculopulvinar complex and traced degeneration with improved silver techniques for demonstrating fine fibers. A continuous band of degeneration could be followed from the ventrolateral part of the lateral geniculate body into the so-called Meyer's loop which makes a temporal detour into the core of the hippocampal gyrus [29]. Some fibers entered the posterior parahippocampal cortex. A coarser, more conspicuous degeneration resulted when lesions involved the inferior pulvinar. The pulvinar projections form a band just lateral to the optic radiations [29].

Units responding to auditory stimulation, some with latencies as short as 7 msec, were found in the limbic cortex overlying the claustrum [45]. Both anterograde and retrograde anatomical studies have shown connections of the insular cortex with the medial geniculate body [see 45 for references]. Units responding to somatic stimulation were found somewhat more forward of the auditory representation in the insula. The receptive fields were usually large and bilateral. Anatomical evidence about somatic projections to the insula is conflicting.

While exploring the insula, we also tested units for gustatory responses [45]. A number of responsive cells was found rostrally in the same region from which Benjamin and Burton recorded evoked slow potentials with stimulation of the chorda tympani [4].

So far, we have explored all of the limbic cortex except the piriform and posterior orbital areas. Somewhat surprisingly, the cells of the cingulate gyrus proved to be virtually unresponsive to visual, auditory, or somatic stimulation [3]. In a further attempt to discover the nature of the inputs to this cortex, we tested the effects of vagal stimulation. In these experiments, stimulating electrodes were chronically implanted on the cervical vagus nerve or at the site where it enters the jugular foramen. To date, we have tested about 300 units in the cingulate cortex of which a little more than 20 % were responsive to triple shocks applied to the vagus nerve [1, 12]. The ratio of initially excited to initially inhibited units was approximately 2:3. Most of these units were located in the mid-portion of the cingulate gyrus. In an attempt to control and implement these findings we injected microamounts of serotonin through a catheter into the superior vena cava as a means of exciting pulmonary receptors [2]. Of 80 cingulate cells tested with repeated injections, 18 % showed excitatory or inhibitory effects, with the ratio of excited to inhibited units being 3:1. Most of these units were recorded in the same part of the gyrus as the cells responding to vagal volleys.

The finding that cingulate units responded to vagal volleys with latencies as short as 15 msec suggested a rather direct afferent pathway. Morest has provided anatomical evidence of connections from nucleus solitarius to the dorsal tegmental nucleus of Gudden [36]. From this nucleus impulses might be transmitted to the cingulate gyrus via the mammillary peduncle [35] and mammillothalamic tract. Recent histofluorescence studies on aminergic systems [17, 38] leave open the possibility of ascending noradrenergic fibers from the nucleus solitarius. The anterior ventral nucleus which projects to the cingulate gyrus has numerous fine terminals containing norepinephrine [9].

The units of the various responsive limbic areas identified in the awake, sitting monkey appeared to be modality-specific. This raises the question of where sensory information is integrated and processed in the limbic system. Stated metaphorically, where do the 'viewers' reside in the limbic system? One likely place is in the entorhinal cortex of the hippocampal gyrus which receives connections from all the areas mentioned and which, in turn, projects to the hippocampus. The hippocampus itself also receives connections from the various responsive areas and the Schaffer collateral system would provide a means of interrelating information of multisensory origin.

According to MOREST's anatomical findings [36], the septal region would also be a recipient of the presumed visceral projections from Gudden's nucleus [35]. The septum in turn projects principally to stratum oriens of the hippocampus where the fibers may end on the basal dendrites (fig. 3). Projections from the hippocampal gyrus, on the other

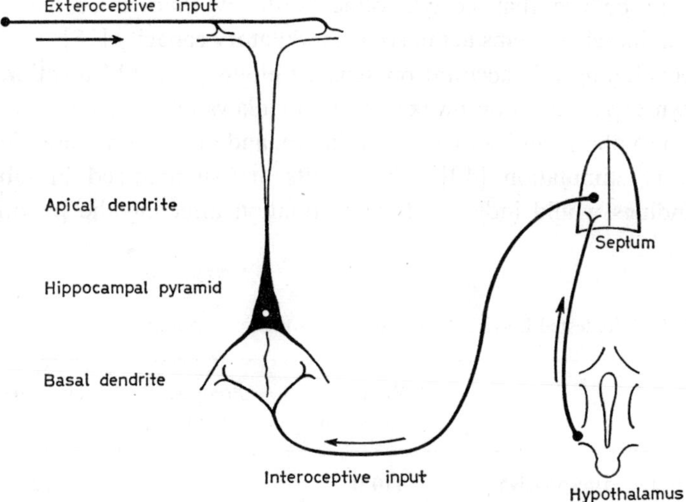

Fig. 3. Sketch of the essential neural circuity discussed in text with reference to the differential action of interoceptive and exteroceptive pathways on hippocampal pyramids.

hand, terminate on the apical dendrites of the hippocampal pyramids. They convey impulses from olfactory and other exteroceptive systems. In an intracellular study of hippocampal neurons in the awake, sitting monkey we found that septal stimulation resulted in excitatory postsynaptic potentials (EPSP) associated with neuronal discharge, whereas stimulation of the olfactory bulb generated only EPSP without spikes [49]. In terms of classical conditioning, the impulses from the respective intero- and exteroceptive systems would correspond to unconditional and conditional stimuli [cf. 27].

In this respect it is of interest that the afferents from the septum to stratum oriens are cholinergic [20] whereas the radiate layer with the apical dendrites is seeded with terminals containing norepinephrine [9]. The latter are presumed to stem from an ascending pontomedullary noradrenergic system that joins the medial forebrain bundle and provides an extensive innervation of certain hypothalamic and limbic structures [6, 17, 38]. In a histofluorescence study including observations on the pygmy marmoset and squirrel monkey, Jacobowitz and I [18] have found that the pattern of organization of all the recognized aminergic systems seems to have been preserved with remarkable consistency in the evolution of primates. My own experience pertaining to the behavioral and electrophysiological effects of chemical stimulation [24, 25] would lead me to believe that acetylcholine is the workhorse of the brain, whereas aminergic systems act in some modulatory capacity [28].

In concluding this account on sensory inputs, I should mention that in our own experiments on awake, sitting monkeys no hypothalamic units responded to photic and somatic stimulation, and only a few were affected by auditory stimulation [40]. The results are summarized in table I. These findings would indicate that information affecting the hypothala-

Table I. Units tested by visual, auditory and somatic stimulation

	Basal forebrain	Preoptic region	Hypothalamus
Visual, tested/responsive	110/0	78/0	321/2
Auditory, tested/responsive	109/0	77/0	315/8
Somatic, tested/responsive	43/0	27/0	168/0

mus is first integrated and processed in related structures such as the limbic cortex.

Influence of Limbic Cortex on Hypothalamus

Microelectrode findings. Having given this background, I take up next the question of the influence of the limbic cortex on the hypothalamus. Firstly, I will describe work in which we tested the effects of hippocampal stimulation on units of the hypothalamus, preoptic region and basal forebrain [40]. Many workers include the preoptic region as part of the hypothalamus. As shown in figure 4, stimulating electrodes were placed in the anterior and posterior hippocampus. The experiments were conducted on squirrel monkeys adapted to sit quietly without medication. In addition to testing the effects of hippocampal volleys, we also observed changes induced by hippocampal after-discharges.

As evident in table II, more than 30 % of the units in the basal forebrain and preoptic areas responded to hippocampal volleys whereas only 14 % of hypothalamic units were affected. Specific hypothalamic areas, however (namely, the perifornical area, mammillary region and posterior hypothalamic area), contained a large percentage of responsive units. There were only 3 hypothalamic structures (the supraoptic, arcuate and premammillary nuclei) that had no responsive units, but in each case the sample was very small. As I shall comment upon in the Discussion, table II lists the significant finding that more than 83 % of the responsive units in all regions were initially excited while the rest showed initial inhibition. Figure 5 shows an example of an oscillographic record and dot display of a unit in the medial preoptic area showing initial excitation,

Table II. Units responsive to hippocampal volleys

	Basal forebrain	Preoptic region	Hypothalamus
Number tested	177	99	390
Number and percent responding	60 (34%)	30 (30%)	56 (14%)
Initially excited, %	87	83	86
Initially inhibited, %	13	17	14

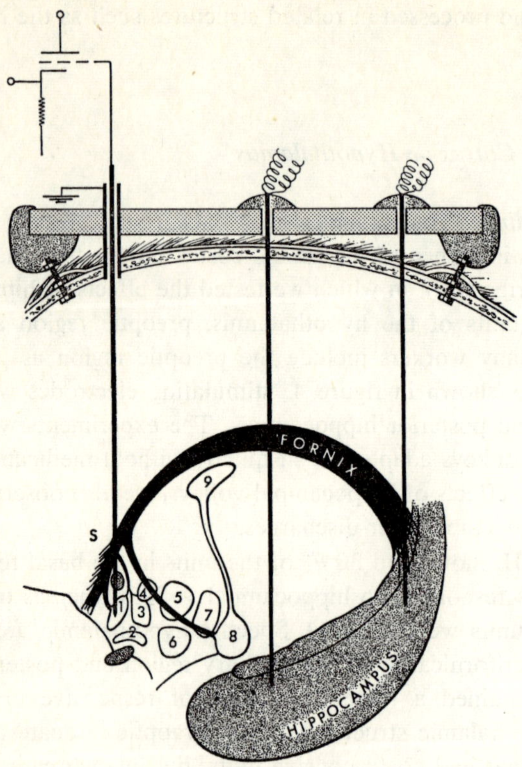

Fig. 4. Diagram of electrode array in testing effects of hippocampal volleys on unit activity of the hypothalamus, preoptic region and basal forebrain. The stereotaxic platform provides a closed system technique for microelectrode exploration of the brain in awake, sitting, squirrel monkeys and recording from single cells. Diagram shows bipolar, stimulating electrodes in the anterior and posterior hippocampus and a recording electrode in the preoptic region. S = septal area; 1 = preoptic region; 2 = supraoptic nucleus; 3 = anterior hypothalamus; 4 = paraventricular nucleus; 5 = dorsomedial nucleus; 6 = ventromedial nucleus; 7 = posterior hypothalamus; 8 = mammillary body; 9 = anterior thalamic nuclei [40].

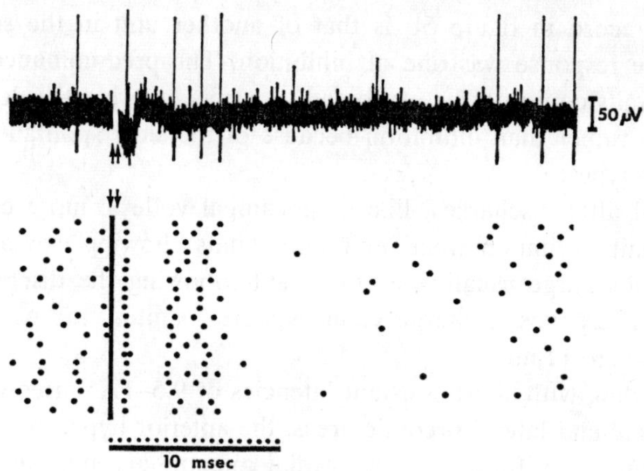

a

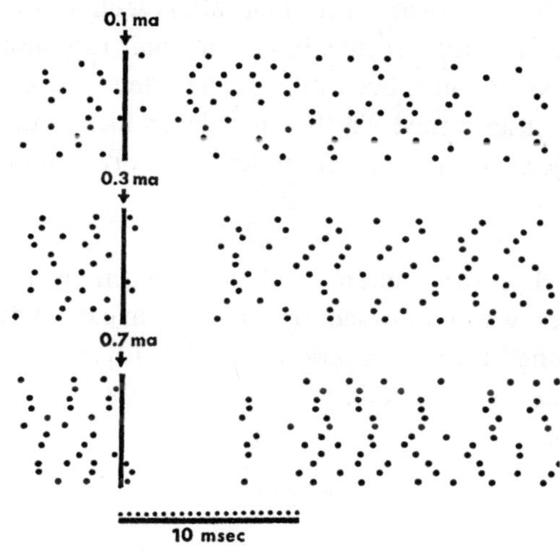

b

Fig. 5. a Illustration of oscillographic record of unit in medial preoptic area responding to hippocampal volleys. See original paper [40] for discussion of different types of response patterns of units in various areas. *b* Dot display for another medial preoptic unit showing prolongation of an inhibitory effect by increasing the shock intensity.

while the lower record in figure 5b is that of another unit in the same area in which the response was one of inhibition. The predominance of initially excited units could not be attributed to a bias favoring the detection of excitation rather than inhibition because of the slow spontaneous firing rate of both types.

Hippocampal after-discharges, like hippocampal volleys, more commonly elicited unit excitation than inhibition. Units showing excitation during the after-discharge usually became silent following the discharge and *vice versa*. These postdischarge changes were manifest for periods ranging from 50 sec to 11 min.

Units responding with short constant latencies of 9.5–12.5 msec were found in the medial and lateral preoptic areas, the anterior hypothalamus, dorsal hypothalamus, perifornical area, medial mammillary nucleus and posterolateral hypothalamus. These and other considerations suggested direct orthodromic conduction.

Poletti *et al.* [41] have shown since that after section of the fornix, hippocampal volleys are still effective in eliciting unit responses in the preoptic region and hypothalamus, but with a longer latency. It is probable that the impulses are transmitted via the amygdala which is one of the major avenues for projections from the frontotemporal division of the limbic system.

Anatomical findings. The supposition of direct projections to the above-mentioned structures was confirmed by an anatomical study in which Poletti *et al.* sectioned the fornix on one side [cf. 40] and degeneration was traced by Voneida's silver stain [47].

Discussion

Two anatomical findings were of particular interest because they provided evidence for the first time in a primate that there are direct projections from the fornix to the medial preoptic area and perifornical region [cf. 40]. Long ago, the work of Hess and Brügger implicated the perifornical region in the expression of angry behavior [14]. The medial preoptic area is believed to be involved in the control of body temperature, cardiovascular function, water balance, food intake, sexual functions and sleep [see 13 for review]. In recent years, the medial preoptic area has aroused increasing interest because of the finding in rodents that circulating testosterone during the first 2 weeks of life determines sexual differentiation by its action on this area and adjoining parts of the septum and anterior hypothalamus [see 10 for review]. It has been shown by radioautography that this area has an affinity for estradiol [34, 44] and L-testosterone [39] and that the direct application of estrogen brings about changes in sexual receptivity [21]. Electrical stimulation of the medial preoptic area results in ovulation [8] and induces genital tumescence [30]. The new findings regarding a direct hippocampal influence on the medial preoptic area call for a revision of restrictive views that limbic influences on this area are mediated by the amygdala.

On the basis of diverse evidence, some workers have proposed that the hippocampus exerts its influence on behavior [11] and neuroendocrine functions largely through inhibitory mechanisms [see 40 for example references]. There are a few physiological studies that would support such an interpretation. Hippocampal stimulation, for example, has been observed to suppress the release of adrenocorticotrophic hormone (ACTH) [7, 33, 42] and to inhibit cortically-induced extensor reflexes [46]. We found, in a previous microelectrode study, that fornix volleys inhibited, but did not augment, the responses of caudal intralaminar units to fifth-nerve stimulation [48]. There is evidence from other studies, however, that hippocampal stimulation, depending on the physiological state of the organism at the time, may have a facilitatory or inhibitory effect with respect to ACTH release [19a], cardiovascular reflexes [15] and visceral responsiveness [25]. The accumulated microelectrode findings in the awake, sitting monkey would indicate that there is 'leeway for attributing a range of inhibitory, excitatory, and modulatory functions to the hippocampus' [40].

Finally, the prolonged changes in unit activity observed following hippocampal after-discharges would seem to correlate with the correspondingly long 'rebound' autonomic and behavioral changes seen in animals following hippocampal seizures [26]. Such changes possibly reflect an imbalance in the production and inactivation of transmitter substances.

Summary

Of the 3 main evolutionary developments of the forebrain, only the limbic convolution has been shown to have extensive connections with the hypothalamus. Except for known olfactory connections and some evidence of gustatory and visceral projections, there has existed no information as to whether or not other sensory systems are specifically related to the limbic cortex. Such information is essential for an analysis of how intero- and exteroceptive systems participate in the regulation of somatovisceral, endocrine and emotional functions of the hypothalamus and related structures.

By microelectrode recording in awake, sitting monkeys, we have found that photic stimulation specifically activates single nerve cells in the posterior parahippocampal cortex, and that gustatory, auditory and somatic stimulation evokes unit potentials in respective parts of the limbic cortex of the insula. The existence of direct visual connections has been confirmed by a neuroanatomical study.

Various limbic areas of the cingulate gyrus are virtually unresponsive to stimulation of exteroceptive systems, but vagal volleys activate cells in the midportion of the gyrus. Neurons in this same area are also excited by vena cava injections of microamounts of serotonin, a known activator of visceral receptors.

Since all of the above limbic areas project to the hippocampus, the findings suggest one mechanism by which limbic integration of internally and externally derived information could influence the hypothalamus. Microelectrode studies have shown that hippocampal volleys excite a large percentage of neurons in basal forebrain structures, preoptic area and parts of the hypothalamus. Hippocampal afterdischarges may induce prolonged changes in the firing rate of cells. A parallel neuroanatomic study has revealed for the first time in a primate that the fornix projects to the perifornical area and to the medial preoptic area. The implications of these respective findings are discussed with respect to limbic cortical influences on emotional behavior and on genital and gonadal aspects of sexual function.

The presentation of results includes a brief comment on the possible modulatory role of ascending aminergic systems on limbic and hypothalamic functions. A current histofluorescence study, including comparative observations on the pygmy marmoset and squirrel monkey, has revealed that the organizational pattern of recognized aminergic systems has been preserved with remarkable consistency in the evolution of primates.

References

1 Bachman, D. S.; Katz, H. M., and MacLean, P. D.: Vagal influence on units of cingulate cortex in the awake, sitting squirrel monkey. Electroenceph. clin. Neurophysiol. *33:* 350–351 (1972).

2 Bachman, D. S.; Katz, H. M., and MacLean, P. D.: Effect of intravenous injections of 5-hydroxytryptamine (serotonin) on unit activity of cingulate cortex of awake squirrel monkeys. Fed. Proc. *31:* 303 (1972).

3 Bachman, D. S. and MacLean, P. D.: Unit analysis of inputs to cingulate cortex in awake, sitting squirrel monkeys. I. Exteroceptive systems. Int. J. Neurosci. *2:* 109–113 (1971).

4 Benjamin, R. M. and Burton, H.: Projection of taste nerve afferents to anterior opercular-insular cortex in squirrel monkey *(Saimiri sciureus).* Brain Res. *7:* 221–231 (1968).

5 Broca, P.: Anatomie comparée des circonvolutions cérébrales. Le grand lobe limbique et la scissure limbique dans la série des mammifères. Rev. Anthrop. Ser. 2 *1:* 385–498 (1878).

6 Dahlström, A. and Fuxe, K.: Evidence for the existence of monoamine neurons in the central nervous system. Acta physiol. scand. *64:* 1–85 (1965).

7 Endroczi, E. and Lissák, K.: The role of the mesencephalon and archicortex in the activation and inhibition of the pituitary-adrenocortical system. Acta physiol. hung. *15:* 25 (1959).

8 Everett, J. W.: Ovulation in rats from preoptic stimulation through platinum electrodes. Importance of duration and spread of stimulus. Endocrinology *76:* 1195–1201 (1965).

9 Fuxe, K.: Evidence for the existence of monoamine neurons in the central nervous system. IV. Distribution of monoamine nerve terminals in the central nervous system. Acta physiol. scand. *64:* 37–84 (1965).

10 Gorski, R. A.: Sexual differentiation of the hypothalamus; in Mack The neuroendocrinology of human reproduction (Thomas, Springfield 1971).

11 Grastyan, E.: The hippocampus and higher nervous activity. In: 2nd Conf. on the Central Nervous System and Behavior. Transactions, pp. 119–205 (Josiah Macy, jr., Foundation, New York 1959).

12 Hallowitz, R. A. and MacLean, P. D.: Unpublished observations (1973).

13 Haymaker, W.; Anderson, E., and Nauta, W. J. H.: The hypothalamus (Thomas, Springfield 1969).

14 Hess, W. R. und Brügger, M.: Das subkortikale Zentrum der affektiven Abwehrreaktion. Helv. physiol. Acta *1:* 33 (1943).

15 Hockman, C. H.; Talesnik, J., and Livingston, K. E.: Central nervous system modulation of baroceptor reflexes. Amer. J. Physiol. *217:* 1681–1689 (1969).

16 Hopf, A.: Volumetrische Untersuchungen zur vergleichenden Anatomie des Thalamus. J. Hirnforsch. *8:* 25–38 (1965).

17 Jacobowitz, D. M. and Kostrzewa, R.: Selective action of 6-hydroxydopa on noradrenergic terminals. Mapping of preterminal axons of the brain. Life Sci. *10:* 1329–1342 (1971).

18 Jacobowitz, D. M. and MacLean, P. D.: Unpublished observations (1973).

19 Jones, E. G. and Powell, T. P. S.: An anatomical study of converging sensory pathways within the cerebral cortex of the monkey. Brain *93:* 793–820 (1970).

19a Kawakami, M.; Seto, K.; Terasawa, E.; Yoshida, K.; Miyamoto, T.; Sekiguchi, M., and Hattori, Y.: Influence of electrical stimulation and lesion in limbic structure upon biosynthesis of adrenocorticoid in the rabbit. Neuroendocrinology *3:* 337–348 (1968).

20 Lewis, P. R. and Shute, C. C. D.: The cholinergic limbic system. Projections to hippocampal formation, medial cortex, nuclei of the ascending cholinergic reticular system, and the subfornical organ and supra-optic crest. Brain *90:* 521–540 (1967).

21 Lisk, R. D.: Diencephalic placement of estradiol and sexual receptivity in the female rat. Amer. J. Physiol. *203:* 493–496 (1962).

22 MacLean, P. D.: Psychosomatic disease and the 'visceral brain'. Recent developments bearing on the Papez theory of emotion. Psychosom. Med. *11:* 338–353 (1949).

23 MacLean, P. D.: Some psychiatric implications of physiological studies on frontotemporal portion of limbic system (visceral brain). Electroenceph. clin. Neurophysiol. *4:* 407–418 (1952).

24 MacLean, P. D.: Chemical and electrical stimulation of hippocampus in unrestrained animals. I. Methods and electroencephalographic findings. Amer. med. Ass. Arch. Neurol. Psychiat. *78:* 113–127 (1957).

25 MacLean, P. D.: Chemical and electrical stimulation of hippocampus in unrestrained animals. II. Behavioral findings. Amer. med. Ass. Arch. Neurol. Psychiat. *78:* 128–142 (1957).

25a MacLean, P. D.: Contrasting functions of limbic and neocortical systems of the brain and their relevance to psychophysiological aspects of medicine. Amer. J. Physiol. *25:* 611–626 (1958).

25b MacLean, P. D.: The brain in relation to empathy and medical education. J. Nerv. Ment. Dis. *144:* 374–382 (1967).

26 MacLean, P. D.: Ammon's horn. A continuing dilemma; in Cajal The structure of Ammon's horn (Thomas, Springfield 1968).

27 MacLean, P. D.: The triune brain, emotion, and scientific bias; in Schmitt The neurosciences second study program (Rockefeller University Press, New York 1970).

28 MacLean, P. D.: A triune concept of the brain and behaviour. I. Man's reptilian and limbic inheritance. II. Man's limbic brain and the psychoses. III. New trends in man's evolution; in Boag and Campbell The Hincks Memorial Lectures (University of Toronto Press, Toronto 1973).

29 MacLean, P. D. and Creswell, G.: Anatomical connections of visual system with limbic cortex of monkey. J. comp. Neurol. *138:* 265–278 (1970).

30 MacLean, P. D. and Ploog, D. W.: Cerebral representation of penile erection. J. Neurophysiol. *25:* 29–55 (1962).

31 MacLean, P. D. and Pribram, K. H.: Neuronographic analysis of medial and basal cerebral cortex. I. Cat. J. Neurophysiol. *16:* 312–323 (1953).

32 MacLean, P. D.; Yokota, T., and Kinnard, M. A.: Photically sustained on-responses of units in posterior hippocampal gyrus of awake monkey. J. Neurophysiol. *31:* 870–883 (1968).

33 Mason, J. W.: The central nervous system regulation of ACTH secretion. In: Reticular formation of the Brain (Little, Brown, Boston 1958).

34 Michael, R. P.: Oestrogens in the central nervous system. Brit. med. Bull. *21:* 87–90 (1965).

35 Morest, D. K.: Connexions of dorsal tegmental nucleus in rat and rabbit. J. Anat., Lond. *95:* 1–18 (1961).

36 Morest, D. K.: Experimental study of the projections of the nucleus of the tractus solitarius and the area postrema in the cat. J. comp. Neurol. *130:* 277–299 (1967).

37 Nauta, W. J. H. and Mehler, W. R.: Projections of the lentiform nucleus in the monkey. Brain Res. *1:* 3–42 (1966).

38 OLSON, L. and FUXE, K.: Further mapping out of central noradrenaline neuron systems. Projections of the 'subcoeruleus' area. Brain Res. *43:* 289–295 (1972).

39 PFAFF, D. W.: Autoradiographic localization of radioactivity in rat brain after injection of tritiated sex hormones. Science *161:* 1355–1356 (1968).

40 POLETTI, C. E.; KINNARD, M. A., and MACLEAN, P. D.: Hippocampal influence on unit activity of hypothalamus, preoptic region, and basal forebrain in awake, sitting squirrel monkeys. J. Neurophysiol. *36:* 308–324 (1973).

41 POLETTI, C. E.; SUJATANOND, M., and SWEET, W. H.: Hypothalamic, preoptic, and basal forebrain unit responses to hippocampal stimulation in awake, sitting squirrel monkeys with fornix lesions. Fed. Proc. *31:* 404 (1972).

42 PORTER, R. W.: The central nervous system and stress-induced eosinopenia. Recent Progr. Hormone Res. *10:* 1–27 (1954).

43 PRIBRAM, K. H. and MACLEAN, P. D.: Neuronographic analysis of medial and basal cerebral cortex. II. Monkey. J. Neurophysiol. *16:* 324–340 (1953).

44 STUMPF, W. E.: Estradiol-concentrating neurons. Topography in the hypothalamus by dry-mount autoradiography. Science *162:* 1001–1003 (1968).

45 SUDAKOV, K.; MACLEAN, P. D.; REEVES, A. G., and MARINO, R.: Unit study of exteroceptive inputs to claustrocortex in awake, sitting squirrel monkey. Brain Res. *28:* 19–34 (1971).

46 VANEGAS, H. and FLYNN, J. P.: Inhibition of cortically-elicited movement by electrical stimulation of the hippocampus. Brain Res. *11:* 489–506 (1968).

47 VONEIDA, T. J. and TREVARTHEN, C. B.: An experimental study of transcallosal connections between the proreus gyri of the cat. Brain Res. *12:* 384–395 (1969).

48 YOKOTA, T. and MACLEAN, P. D.: Fornix and fifth-nerve interaction on thalamic units in awake, sitting squirrel monkeys. J. Neurophysiol. *31:* 358–370 (1968).

49 YOKOTA, T.; REEVES, A. G., and MACLEAN, P. D.: Differential effects of septal and olfactory volleys on intracellular responses of hippocampal neurons in awake, sitting monkeys. J. Neurophysiol. *33:* 96–107 (1970).

Author's address: Dr. P. D. MACLEAN, Laboratory of Brain Evolution and Behavior, National Institute of Mental Health, *Bethesda, MD 20014* (USA)

Recent Studies of Hypothalamic Function
Int. Symp. Calgary 1973, pp. 232–250 (Karger, Basel 1974)

Effects of Ethyl Alcohol, Angiotensin and Several Essential Amino Acids on the Lateral Hypothalamus[1]

M. J. WAYNER, T. ONO, D. NOLLEY and A. DE YOUNG
Brain Research Laboratory, Syracuse University, Syracuse, N.Y.

Introduction

It is generally accepted that the hypothalamus plays a primary role in the neural control of mammalian ingestive behavior. In many ways, the hypothalamus is involved in the regulation of vital physiological functions and serves to coodinate these activities with those of the interactions of the organism with its external environment. Nerve cells of certain vascularized portions of this general region of the brain appear to be affected directly by changes in the temperature and composition of the blood which are indirectly related to ambient temperature and physiological imbalances due to dehydration and caloric depletion. Sensory signals which arise from the oropharyngeal tissue, gastrointestinal tract and other organs associated with fluid regulation and the metabolism of food can also modulate neuronal activity within these regions of the hypothalamus. Cells of the perifornical, lateral *(LH)* and far lateral regions of the hypothalamus are concerned with the initiation of action involved in eating and drinking whereas the ventromedical nucleus is intimately involved in the mediation of activity which results in the reduction and termination of eating and drinking [9, 26]. Recently the validity of using such concepts as drinking and eating was challenged and the specificity of so-called regulatory behavior was questioned [40, 41]. An explanation of drinking and eating as neuromuscular processes where the lateral hypothalamus is involved in

1 This research was supported by NSF Grants GB-18414X and GB-35506, and NIMH Grant 15473 and Training Grant MH-06969.

the control of the motor reflex excitability and effectiveness of the appropriate sensory feedback to sustain these motor activities has been developed [39, 40, 41]. The ability of cells within the lateral preoptic-lateral hypothalamic region to respond differentially to changes in extra-cellular ion concentrations, blood glucose, free fatty acids and amino acid ratios, and blood temperature is essential to the theory. Apparently these cells display a great deal of convergence within the reciprocating connections of the *LH* [21] between the limbic forebrain and limbic midbrain and brain stem motor control systems and consequently considerable plasticity which is usually reduced in the presence of relatively constant environmental stimuli [40]. The purpose of the present report will be to review some of our research on the sensitivities of lateral hypothalamic neurons in the rat to extracellular Na and glucose, ethyl alcohol, angiotensin and several of the essential amino acids.

Studies on Single Neurons of the LH

The anatomy of the *LH* is relatively complex and represents the confluence of many neural pathways of diverse origins and functions [23, 26]. Although the activity of the *LH* is modulated by many other parts of the brain such as limbic structures, one of its most intimate relations seems to be with the VMH. In the control of ingestive behavior the *LH* and ventromedial hypothalamus (VMH) are reciprocally related. Lesions in the *LH* disrupt drinking and eating and usually result in a permanent adipsia. Lesions in the VMH usually result in hyperphagia without any pronounced effects on water drinking. Some of the symptoms which result from *LH* and VMH lesions can be reproduced by small knife cuts between the two regions [14, 17, 33] and indicate the possible existence of direct anatomical connections between them. Results of electrical and chemical stimulation in the *LH* are more difficult to understand. Cholinergic stimulation in the *LH* produces drinking whereas adrenergic substances inserted through the same implanted cannula result in eating [16]. The behavior elicited during electrical stimulation of the *LH* depends primarily upon the availability and variety of environmental stimuli [37, 39]. Considerable circumstantial evidence has accumulated to indicate the presence of VMH chemoreceptors sensitive to blood glucose [22] or receptors somewhere in the brain sensitive to reduced glucose utilization [10]. Whether or not the *LH* contains glucosensitive neurons has been a moot question for many years. The most convincing recent evidence demonstrates clearly that rats with bilateral *LH* lesions do not eat in response to insulin [11] or 2-deoxy-D-glucose [45], a specific glucoprivic drug [34]. The fact that the hypothalamus contains osmosensitive neurons involved in the regulation of drinking is well known [5, 20, 28, 30, 42]. Also there is some evidence which implicates the *LH* and the fact that destruction of the *LH* produces a relatively permanent adipsia is well established. Because of the reciprocal relationship between cells of the *LH* and VMH it has been difficult to ascertain whether or not either or both of these regions contain neurons sensitive to glucose and/or sodium, the prevalent extracellular ion which would be involved in any meaningful osmotic pressure change. A more precise approach to the problem requires the electrophoretic application of minute quantities of chemically active substances to single brain cells and an analysis of results in terms of the modulation of ongoing discharge frequency [29].

Methods

The methods and procedures employed in these experiments varied considerably and have all been described previously [44, 46]. In brief, acute rat preparations lightly anesthetized with ether, urethan or urethan plus chlorase were used, subjected to a minimum amount of surgery, fixed in a stereotaxic instrument, and multibarreled glass capillary microelectrodes were driven through a hole in the skull to the *LH* according to predetermined stereotaxic coordinates by means of micromanipulators. The coordinates were calculated according to the atlas of DE GROOT [15]; A 5.8, L 1.9, H -2.5, and a lateral angle of 20–25° to the vertical. Electrode tip positions were verified by standard histological methods. One capillary filled with concentrated NaCl served as the recording electrode and the others were filled with various substances such as NaCl, glucose plus NaCl, valine^{5}-angiotensin II and ethyl alcohol. The DC resistances of the capillaries varied from 5–150 mΩ. Chemical substances were electrophoretically applied to nerve cells by a constant current source. Action potentials were recorded extracellularly, amplified by a high input impedance preamplifier, monitored visually on an oscilloscope, and recorded on magnetic tape. When a spontaneously active neuron was located in a predetermined site, action potentials were inspected carefully for waveform and amplitude and only stable cells were retained for further study. The length of baseline depended upon the spontaneous discharge rate. Data were discarded if there were any evidence of electrode induced membrane damage or change in spike frequency due to current. Discharge frequencies were counted on line, recorded by means of a printer and displayed graphically on an X-Y plotter. Apparent changes in discharge frequency were checked by appropriate statistical tests. All of the effects reported here are statistically significant.

Effects of Glucose and Sodium on LH Neurons

The *LH* contains glucosensitive cells. Results on a typical glucosensitive neuron are illustrated in figure 1. The cell was not sensitive to Na for stimulating currents up to 40 nA. A typical osmosensitive *LH* neuron is illustrated in figure 2. This type of cell is sensitive to both Na and glucose. In contrast to the VMH the *LH* glucosensitive neurons are of two types, those which increase and those which decrease in discharge frequency when glucose is applied. This latter type is illustrated in figure 3. Of the 64 *LH* neurons tested with glucose, 21 increased, 15 decreased and 28

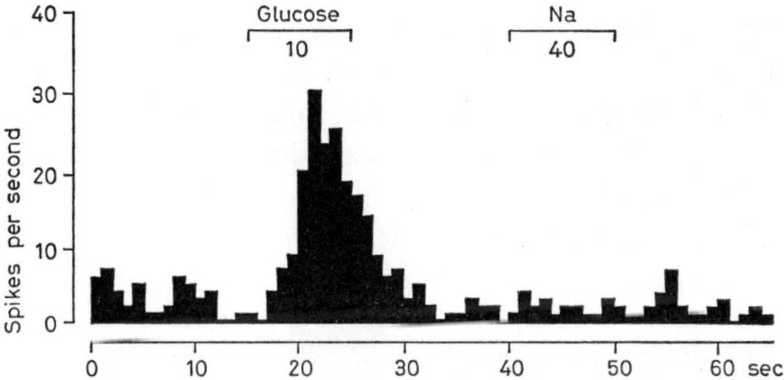

Fig. 1. An X-Y plot of the discharge frequencies of a lateral hypothalamic *(LH)* neuron sensitive to glucose but not Na. The electrophoretic ejection current indicated in nanoamperes and the duration by the horizontal solid line.

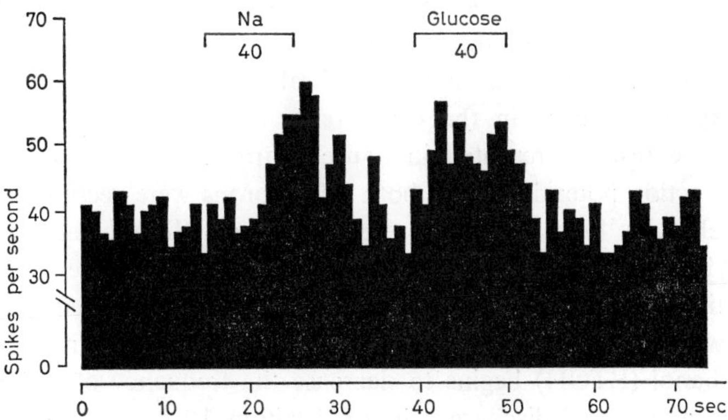

Fig. 2. Same as figure 1 except for an *LH* neuron sensitive to both Na and glucose.

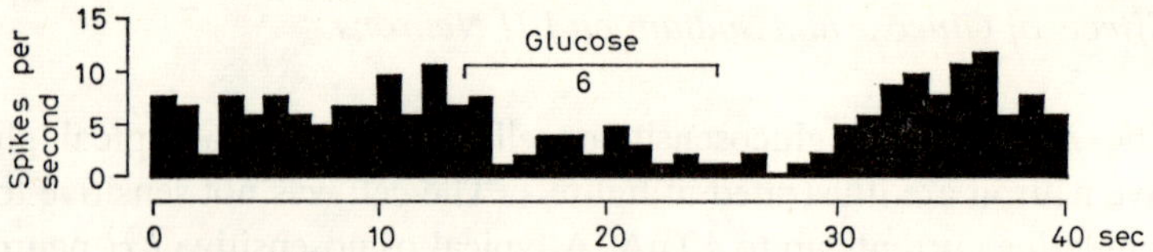

Fig. 3. Same as figure 1 except for an *LH* neuron sensitive to glucose but in this case the ejected glucose resulted in a decrease in spontaneous discharge rate.

were not affected. Of the 21 which increased, 7 were tested with Na and 4 increased and 3 were not affected. Of the 15 which decreased, 8 were tested with Na and one decreased; the other 7 were not affected. Of the 28 which were not affected by glucose, 9 were tested with Na and 1 increased and the other 8 were not affected. These results indicate that the *LH* contains both glucosensitive cells, some of which increase in discharge frequency and others which decrease during glucose application, and osmosensitive cells which in general increase in discharge rate during Na ejection, only one cell decreased in rate.

Effects of Ethyl Alcohol on LH Neurons

We have reported previously that 85 % of the neurons examined in this part of the *LH* of rat are sensitive to sodium and ethyl alcohol; 53 % increased in discharge rate and 32 % decreased and the remainder were not affected [43]. Cells which have been studied in other parts of the brain do not appear to the affected at these low doses, 40 mg/kg, administered intravenously through the lateral tail vein. At higher doses, 1,000 mg/kg, many cells in the other parts of the brain seem to be depressed. The results presented in figure 4 are particularly significant because the action potentials from both *LH* neurons were recorded from the same relatively high impedance electrode and indicate that the two cells are lying close together. The two cells appear to be reciprocally related in time. *LH1* has a relatively continuous and high frequency of discharge; whereas *LH2* is slow and intermittent. Administration of 0.2 cc of 10 % ethanol (ETOH) begins to decrease the frequency of *LH1* and increase the number of discharges in *LH2* within 10 sec. Although the correlation is not perfect, *LH1* is discharging at a low rate, close to zero, when *LH2* is discharging at a very high rate. Sodium chloride produces a very obvious decrease in frequency of *LH2*. The negative correlation between the two cells seems to decrease from 1,700 to 2,400 sec. At 2,380 sec *LH2* began to discharge at an increasing rate reaching 79 spikes/sec at 2,390 sec. When the infusion of Ringer's lactate solution began at 2,400 sec there was a decrease in the frequency of both cells within 10 sec which endured for at least 10 min. These results and additional similar data indicate that the *LH* cells with high spontaneous discharge frequencies are interneurons and that they normally inhibit, as part of a small functional aggregate of cells, the neurons with relatively low spontaneous frequencies of discharge (lower part of fig. 4). Therefore, ethyl alcohol in the brain might have a selective effect on interneurons just as it does in the spinal cord; except that these interneurons in the *LH* have a very sensitive inward Na pump which makes them particularly vulnerable to ethanol because of its effect on sodium conductance and the excitability of the cell membrane. Consequently, these Na sensitive interneurons are turned off by ethanol and the other cell is released from inhibition and increases in frequency or, in other words, is turned on. The lack of a perfect negative cross-correlation between the cells in figure 4 should be expected because the activity of each cell is undoubtedly affected by other neurons which continually modulate their discharge patterns.

The effects of ethyl alcohol applied electrophoretically to a neuron located more posteriorly in the *LH* are presented in figure 5. This cell, although its spontaneous discharge rate is relatively high, increased in frequency during the ejection of ethyl alcohol. A typical dose related effect can be observed with little increase for ejection currents above 100 nA. Only a relatively small increase in frequency occurred during the ejection of chloride ions by 240 nA of current.

As only the sodium sensitive *LH* neurons seem to display a low threshold to the administration of ethyl alcohol, a Na transport mechanism in the membrane appears to be involved. Results on squid giant axon indicate that membrane excitability is depressed by ethyl alcohol through suppression of the increase in Na conductance which normally accompanies adequate stimulation [3, 24, 25]. The ethyl alcohol is more effective when applied externally and can reduce the efflux of ^{22}Na by about 30 % without any noticeable effect on the amplitude of the action potential [19]. Evidence has also accumulated which indicates that ethyl alcohol usually inhibits brain (Na + K) ATPase but that active ion transport (Na + K) ATPase activity increases after chronic administration

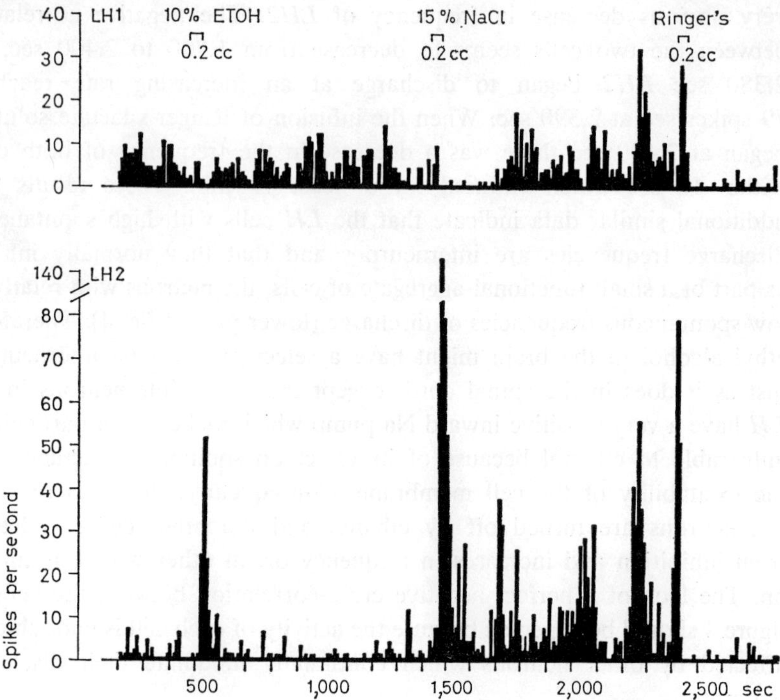

Fig. 4. The discharge frequencies of two *LH* cells recorded simultaneously through the same high impedance electrode. The administration of 0.2 cc vol. of 10 % ethyl alcohol, 15 % NaCl and Ringer's solution via the lateral tail vein at the rate of 0.54 ml/min indicated by the short horizontal solid line. *LH1* had a higher spontaneous discharge rate and decreased whereas *LH2* increased in discharge frequency. ETOH = ethanol.

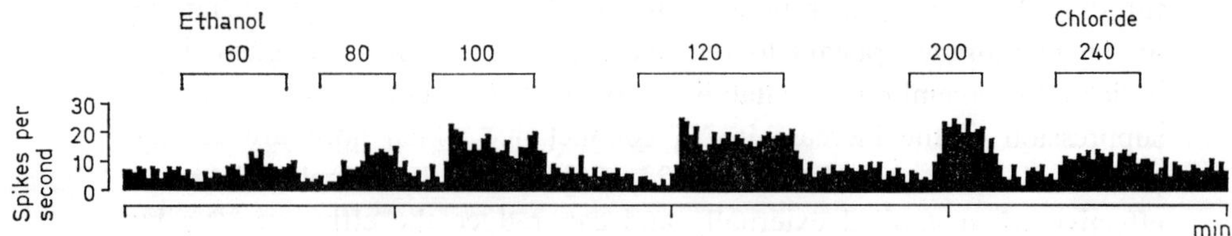

Fig. 5. A dose related increase in the spontaneous discharge frequency of a lateral hypothalamus-medial forebrain bundle (*LH*-MFB) neuron during the electrophoretic ejection of ethanol. Reversed current and the ejection of chloride at 240 nA had no appreciable effect.

of ethanol [19]. Apparently, ethyl alcohol has a profound effect on excitable cell membrane which involves molecular processes that precede the action potential and on the subsequent active transport of Na and K. As ethyl alcohol permeates the blood brain barrier readily it should be expected to produce many and varied effects within the central nervous system. In a recent study of ethyl alcohol on cat spinal reflexes [7], dorsal horn interneuron spontaneous discharges were depressed whereas spinal motoneurons were more resistant and higher doses produced a decrease of membrane conductance to both inward and outward constant current. In the cerebellum, ethyl alcohol tended to accelerate the discharge of interneurons and depress what appear to be Purkinje cells. Single cell activity recorded from the lateral vestibular nuclei was also depressed [8]. Therefore, the results on *LH* neurons are unusual only in the low threshold and high sensitivity to ethyl alcohol.

Effects of Angiotensin on LH Neurons

Angiotensin is a substance of considerable current interest because it appears to be involved in normal body fluid regulation by stimulating the release of aldosterone and antidiuretic hormone [13] and by a direct action on hypothalamic neurons implicated in the control of drinking [12]. Since angiotensin applied intraventricularly elicits drinking [6] and there is no evidence that it crosses the blood-brain barrier [32], an indirect route to central neurons via the choroid plexuses, cerebrospinal fluid and ventricles, particularly into the walls of the third ventricle [38], seems very likely. In addition, both angiotensin and carbachol when injected in the midline region of the rat brain produce drinking [35]. These results raise some serious doubts concerning the functional significance of the renin-angiotensin system in the elicitation of drinking and raise the important question of specific sensitivity of central neurons to angiotensin. Two brief reports have appeared recently in which the sensitivity of hypothalamic neurons to angiotensin was determined by more direct means [27, 31]. The purpose of the present experiment was to examine in greater detail the effects of valine^{5}-angiotensin II on the discharge frequency of lateral hypothalamic neurons of the rat when applied microelectrophoretically. Results indicate two types of neurons within the lateral hypothalamus which are affected differently by angiotensin and potentiated by sodium.

Table I. A summary of the effects of angiotensin II and Na on the neurons of 5 different brain sites

Site	N	Angiotensin			N	Na		
		E	I	O		E	I	O
LH	17	0	9	8	2/9	0	1	1
LH-MFB	29	19	0	10	14/19	3	0	11
					10/10	0	0	10
Zona incerta	23	16	1	6	16/16	4	0	12
					1/1	0	1	0
					3/6	0	0	3
Thalamus	66	18	7	41	18/18	0	0	18
					7/7	0	0	7
Cortex	13	0	1	12	2/12	0	0	2

LH = lateral hypothalamus; *LH*-MFB = lateral hypothalamus-medial forebrain bundle.

The data on 148 cells from 5 different brain sites are summarized in table I. The number of cells, N, studied in each site is also included. Two different types of neuron were found in the *LH*; those definitely within the medial forebrain bundle (*LH*-MFB), and a different variety located about 0.5 mm more ventral and slightly more medial (LH) from the center of the *LH*. Of 17 *LH* cells studied, 9 were inhibited (I), none were excited (E) and 8 were not affected (O) by the ejected angiotensin. Of the 9 which were inhibited, 2 were tested with Na and in one the discharge frequency decreased (I) and the other was not affected. These cells of the lower *LH* have a relatively low spontaneous discharge frequency which make them particularly difficult to locate. Searching was facilitated by the electrophoretic application of glutamate through one of the capillaries. An example of such a cell with a high spontaneous discharge frequency maintained by the electrophoretic application of glutamate is illustrated in figure 6. The neuron was tested with 5 μg of angiotensin applied intravenously via the lateral tail vein and then twice with angiotensin ejected iontophoretically by outward currents of 50 and 100 nA. The latencies and time to attain a maximum effect differ considerably for the two techniques and indicate a slower route of entry to the *LH* cells when the angiotensin is administered intravenously. Since 3

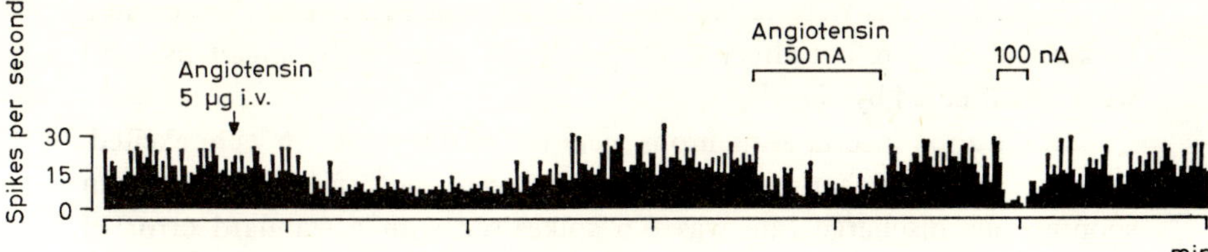

Fig. 6. An X-Y plot of the discharge frequency of a lower *LH* neuron in which the spontaneous discharge rate is continuously enhanced by the electrophoretic ejection of glutamate. Discharge frequency is inhibited by the simultaneous intravenous administration of 5 μg of angiotensin and by the electrophoretic ejection of angiotensin with currents of 50 and 100 nA, as indicated by the horizontal solid lines.

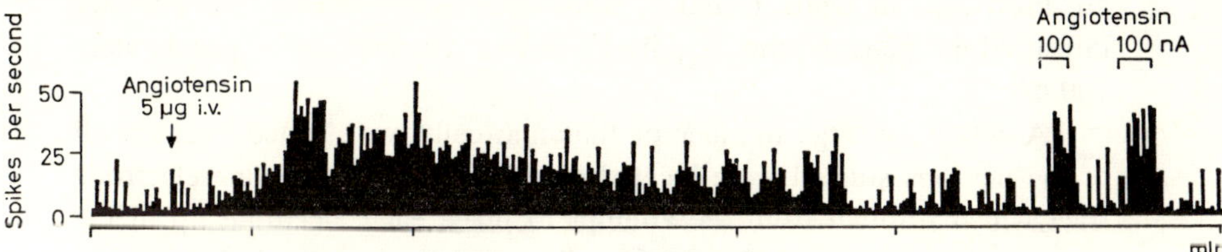

Fig. 7. An *LH*-MFB neuron with a low sporadic spontaneous discharge rate enhanced by the intravenous administration of 5 μg of angiotensin and the electrophoretic ejection of angiotensin with currents of 100 nA.

barreled electrodes were used and one capillary was filled with glutamate, it was impossible to test any one cell with both angiotensin and Na. The two cells tested for Na therefore did not have enhanced discharge frequencies due to glutamate.

29 *LH*-MFB neurons were tested. 19 were excited by angiotensin, increased in discharge frequency, none were inhibited and 10 were not affected. An example of such a cell is illustrated in figure 7 where the spontaneous discharge rate is definitely enhanced by the intravenous administration of 5 μg of angiotensin and by the iontophoretic ejection of angiotensin by outward currents of 100 nA. Again the differences in latency and time to attain a maximum effect are obvious. Dose related increases in discharge rate and after effect were also observed but not illustrated. As these cells had a relatively high spontaneous discharge rate as compared to the lower *LH,* mean of 7.3 spikes/sec, with a standard error of 1.4 spikes/sec, glutamate facilitation was not necessary and 14 of

the 19 excited cells were tested with Na. Three of these cells were excited by Na and 11 were not affected. Of the 10 not affected by angiotensin, all were not affected by Na.

In the zona incerta 23 neurons were tested. Of these, 16 were excited, 1 decreased in discharge frequency and 6 were not affected. The mean spontaneous discharge rate was 6.6 spikes/sec with a standard error of 2.1 spikes/sec. Of the 16 which were excited by angiotensin, 4 increased in discharge frequency due to Na and 12 were not affected. The one which was inhibited by angiotensin was also inhibited by Na. Three of the 6 cells not affected by angiotensin were also tested by Na and were not affected. The neurons of the *LH, LH*-MFB and zona incerta usually have a low threshold to angiotensin with a required ejection current of less than 20 nA. Both an increase and decrease in discharge frequency are illustrated in figure 6 and 7. The Na sensitive effects had a higher threshold in general and required ejection currents of approximately 100 nA.

A relatively large number of thalamic cells were tested because all of the neurons studied were located along essentially the same electrode tract which resulted from the attempt to place each electrode tip in the same predetermined site. Of the 66 cells tested, 18 increased, 7 decreased in discharge frequency and 41 were not affected. All of these neurons displayed high thresholds and usually required more than 50 nA of ejection current to produce an observable increase in frequency. All of the 18 which were increased by angiotensin were tested with Na and were not affected. The 7 which decreased were also tested with Na and were not affected. In the cerebral cortex 13 cells were tested by angiotensin, 1 decreased in discharge frequency and 12 were not affected. The decrease in discharge frequency by the one cell required an ejection current of over 200 nA and it is unlikely that the effect can be attributed only to the angiotensin. Two cells were tested with Na and were not affected.

Effects of Combined Angiotensin and Na on LH Neurons

As angiotensin and ethyl alcohol both affect sodium transport mechanisms [3, 7, 8, 19, 24, 25, 36], it seemed reasonable to test their combined effects on LH sodium sensitive neurons. Results are illustrated on the *LH*-MFB unit in figure 8 where angiotensin was ejected

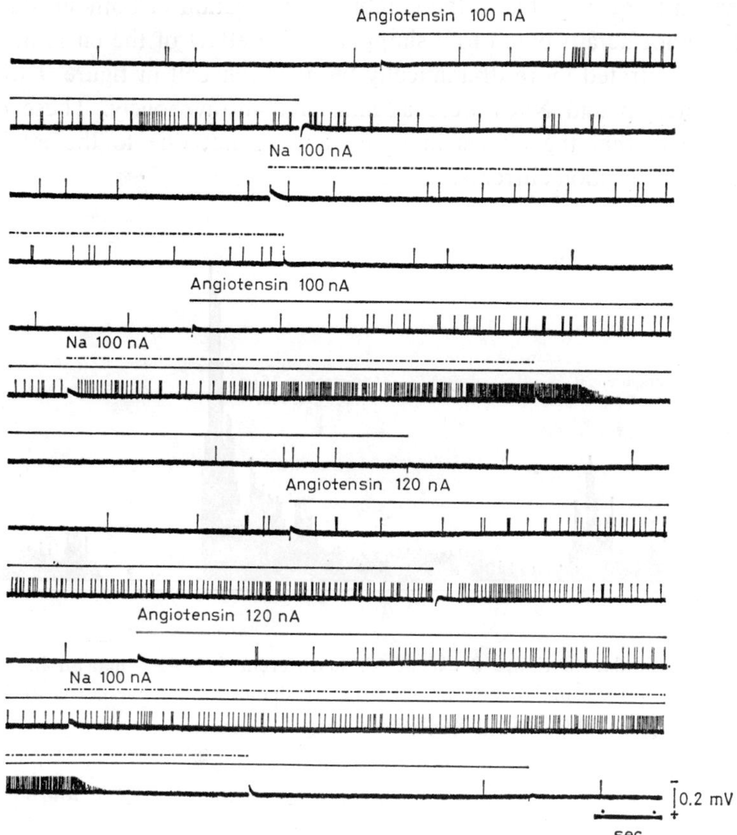

Fig. 8. Continuous unit recording in an *LH*-MFB neuron. The electrophoretic ejection of angiotensin indicated by the solid horizontal line and the ejection of Na by the broken line. Simultaneous ejection indicated by both solid and broken lines. The decrease in spike amplitude is characteristic of only relatively intense combined applications of Na with angiotensin. Voltage and time calibrations as indicated.

first, as indicated by the horizontal solid line; then Na, as indicated by the horizontal broken line, and then the simultaneous ejection of both substances. The sequence was repeated employing a larger ejection current for the angiotensin. The results are obvious and indicate a multiplicative increase in frequency due to the combined application of both substances which cannot be attributed to a simple additive effect of the current. The effect is illustrated more dramatically by a similar cell in figure 9 where the angiotensin and NaCl were administered intravenously. These data also indicate that the effects in figure 8 were not due to the additive effects of the ejection currents.

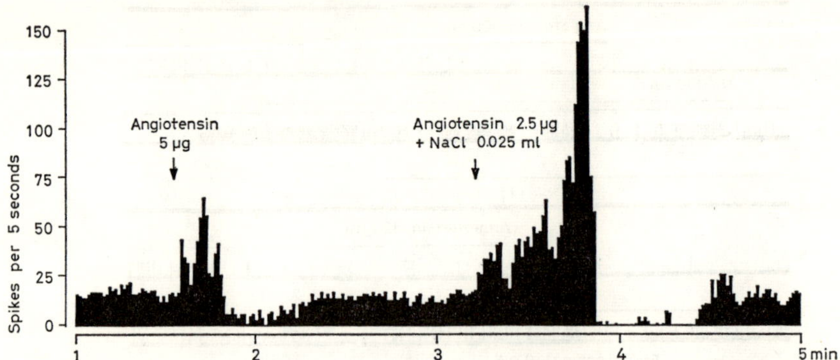

Fig. 9. Similar to figure 8 except for a different *LH*-MFB neuron in which the combined effects of Na and angiotensin are demonstrated by means of intravenous administration of both substances.

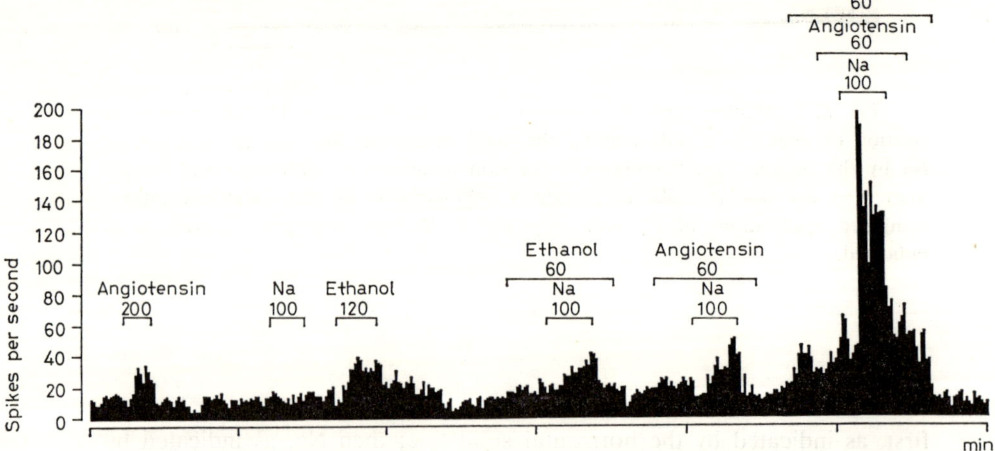

Fig. 10. An *LH*-MFB neuron. The electrophoretic ejection of angiotensin, Na and ethanol and several simultaneous ejection combinations are indicated by the horizontal solid lines. The numbers refer to the ejection currents in nanoamperes.

Effects of Combined Angiotensin, Na and Ethyl Alcohol on LH Neurons

Because of the interesting data obtained with angiotensin and Na, it seemed logical to study the combined effects of angiotensin, Na and ethyl alcohol on several *LH* neurons. The data obtained from an *LH*-MFB neuron are illustrated in figure 10 where the combined ejection of ethyl alcohol, angiotensin and Na produced the most dramatic effect and confirm the notion of a synergistic action of angiotensin, ethyl alcohol and Na on common ionic membrane mechanisms.

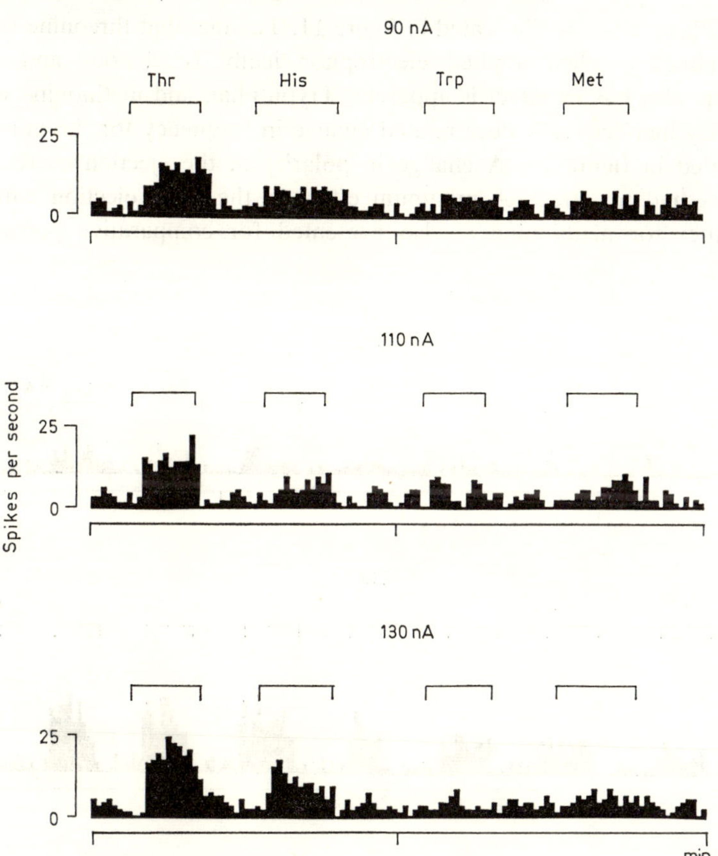

Fig. 11. An illustration of the effects of the electrophoretic ejection of threonine, histidine, tryptophan and methionine with currents of 90, 110 and 130 nA on an *LH*-MFB neuron.

Effects of Several Essential Amino Acids on LH Neurons

Although the hypothalamus might not be the primary structure involved in the anorexia which develops when rats are restricted to a diet marginally deficient in an amino acid, the effects of the essential amino acids on *LH* neurons have not been assessed. Consequently, an attempt to gather such information is in progress and only a few preliminary data will be reported here. Of the 11 indispensable amino acids for the rat, the following 4 have been tested: methionine, tryptophan, histidine and threonine. Results obtained with all 4 amino acids using 5 barreled electrodes (0.05 M/liter, pH = 8.0) and negative current on the same *LH*-MFB neuron are illustrated in figure 11. The fact that threonine is the most effective when applied electrophoretically is obvious and that histidine also has an effect is apparent. Tryptophan and methionine were relatively ineffective. A dose related change in frequency for threonine is illustrated in figure 12. A change in polarity in the ejection current at 200 nA had no effect. A minimum effect at the same ejection current intensities for methionine is also presented for comparative purposes.

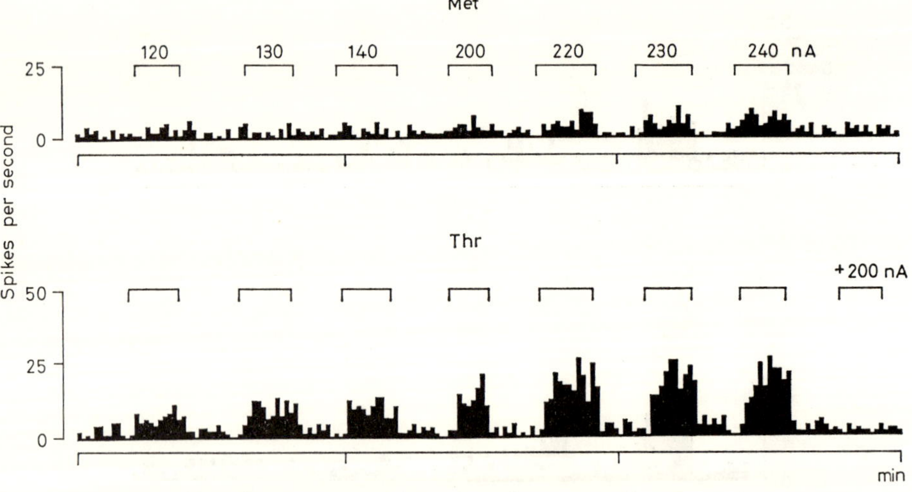

Fig. 12. Similar to figure 11 except for a different *LH*-MFB neuron. The obvious dose related effect of threonine in comparison to a relatively small effect of methionine is illustrated. Reversing the direction of the current at 200 nA had no effect.

These preliminary data are very interesting and indicate that cells differentially sensitive to some of the essential amino acids reside in the *LH*.

Discussion

The fact that the *LH* contains neurons sensitive to glucose, Na and ethyl alcohol had been established previously. Present results indicate that many cells of the *LH* and zona incerta are relatively more sensitive to angiotensin and Na than neurons of the thalamus and cerebral cortex. Many more cells are sensitive to angiotensin than Na and all Na sensitive cells are also sensitive to angiotensin. The low spontaneously active ventral *LH* neurons decrease in discharge frequency when angiotensin is applied; whereas, the cells of the *LH*-MFB increase in discharge rate. The neurons of the zona incerta are very similar in this respect to the *LH*-MFB region and possibly represent cells of the same population. Therefore, angiotensin seems to have a specific effect on two types of neurons in the ventral *LH* and *LH*-MFB-zona incerta regions. With larger doses the effect is more nonspecific and many cells of the thalamus are also affected. All Na sensitive neurons of the *LH*-MFB-zona incerta were also sensitive to angiotensin. Therefore, angiotensin in small quantities might influence drinking because of a specific effect on the *LH* Na sensitive neurons [29]. Although these neurons appear to be in the same general region where hypertonic NaCl and acetylcholine elicit drinking, the effects of angiotensin appear to be independent of the effects of acetylcholine [31]. Angiotensin also seems to have a predominantly excitatory action on neurons of the supraoptic nucleus which suggests a direct effect in the release of antidiuretic hormone [27].

The multiplicative interaction between angiotensin and Na on *LH*-MFB neuron frequency confirms a similar effect on drinking and blood pressure induced by intraventricular administration of the same substances [1, 2]. Angiotensin definitely increases Na efflux from smooth muscle by enhancing the operation of the Na pump [36]. Because of the possibility that Na complexes with valine^5-angiotensin II to form a more biologically active molecule by changing the steric conformation of the peptide chain [4], it is not possible to determine if the enhancement under present conditions is due to independent effects of both treatments on the membrane. However, the fact that ethyl alcohol, angiotensin, and Na all combine to produce a multiplicative interaction which results in a pronounced increase in lateral hypothalamic neuronal activity suggests more than a physical change in a single active molecule. The relatively tenuous data on several essential amino acids indicate a dose dependent specificity for threonine on *LH* neurons which at present is difficult to reconcile with the data on dietary amino acid imbalances and resultant decreases in eating [18].

References

1 ANDERSON, B. and WESTBYE, O.: Synergistic action of sodium and angiotensin on brain mechanisms controlling fluid balance. Life Sci. *9:* 601–608 (1970).

2 ANDERSON, B.; ERIKSSON, L., and FERNANDEZ, O.: Reinforcement by Na+ of centrally mediated hypertensive response to angiotensin II. Life Sci. *10:* 633–638 (1971).

3 ARMSTRONG, C. M. and BINSTOCK, L.: The effects of several alcohols on the properties of the squid giant axon. J. gen. Physiol. *48:* 265–277 (1964).

4 BERGMANN, P.; OEHME, J.; JELINEK, J., and CORT, J. H.: Potentiation of angiotensin and eledoisin activities by sodium chloride. Life Sci. *10:* 969–975 (1971).

5 CROSS, B. A. and GREEN, J. D.: Activity of single neurons in the hypothalamus. Effect of osmotic and other stimuli. J. Physiol., Lond. *148:* 554–569 (1959).

6 DANIELS-SEVERS, A.; OGDEN, E., and VERNIKOS-DANELLIS, J.: Centrally mediated effects of angiotensin II in the unanesthetized rat. Physiol. Behav. *7:* 785–787 (1971).

7 EIDELBERG, E. and WOOLEY, D. F.: Effects of ethyl alcohol upon spinal cord neurons. Arch. int. Pharmacodyn. *185:* 388–396 (1970).

8 EIDELBERG, E.; BOND, M. L., and KELTER, A.: Effect of alcohol on cerebellar and vestibular neurons. Arch. int. Pharmacodyn. *192:* 213–219 (1971).

9 EPSTEIN, A. N.: The lateral hypothalamic syndrome. Its implications for the physiological psychology of hunger and thirst; in STELLAR and SPRAGUE Progress in physiological psychology, vol. 4, pp. 263–317 (Academic Press, New York 1971).

10 EPSTEIN, A. N.: The glucoprivic control of food intake and the glucostatic theory of feeding behavior; in MOGENSON Regulation of food and water intake (in press).

11 EPSTEIN, A. N. and TEITELBAUM, P.: Specific loss of the hypoglycemic control of feeding in recovered lateral rats. Amer. J. Physiol. *213:* 1159–1167 (1967).

12 EPSTEIN, A. N.; FITZSIMONS, J. T., and ROLLS, B. J.: Drinking induced by injection of angiotensin into the brain of the rat. J. Physiol., Lond. *210:* 457–474 (1970).

13 FITZSIMONS, J. T.: The renin-angiotensin system in the control of drinking; in MARTINI, MOTTA and FRASCHINI The hypothalamus, pp. 195–212 (Academic Press, New York 1970).

14 GOLD, R. M.: Hypothalamic hyperphagia produced by parasagittal knife cuts. Physiol. Behav. *5:* 23–25 (1970).

15 Groot, J. de: The rat forebrain in stereotaxic coordinates. Verh. K. Ned. Akad. Wet. *52:* 1–40 (1959).

16 Grossman, S. P.: Direct adrenergic and cholinergic stimulation of hypothalamic mechanisms. Amer. J. Physiol. *202:* 872–882 (1962).

17 Grossman, S. P. and Grossman, L.: Food and water intake in rats with parasagittal knife cuts medial or lateral to the lateral hypothalamus. J. comp. physiol. Psychol. *74:* 148–156 (1971).

18 Harper, A. E.; Benevenga, N. J., and Wohlhueter, R. M.: Effects of ingestion of disproportionate amounts of amino acids. Physiol. Rev. *50:* 428–558 (1970).

19 Israel, Y.; Rosenmann, E.; Hein, S.; Colombo, G., and Canessa-Fischer, M.: Effects of alcohol on the nerve cell; in Israel and Mardones Biological basis of alcoholism (Wiley-Interscience, New York 1971).

20 Joynt, R. J.: Functional significance of osmosensitive units in the anterior hypothalamus. Neurology, Minneap. *14:* 584–590 (1964).

21 Kotlyar, B. I. and Yeroshenka, T.: Hypothalamic glucoreceptors. The phenomenon of plasticity. Physiol. Behav. *7:* 609–615 (1971).

22 Mayer, J. and Thomas, D.: Regulation of food intake and obesity. Science *156:* 328–337 (1967).

23 Millhouse, O. E.: A Golgi study of the descending medial forebrain bundle. Brain Res. *15:* 341–363 (1969).

24 Moore, J. W.: Effects of ethanol on ionic conductances in the squid axon membrane. Psychosom. Med. *28:* 450–457 (1966).

25 Moore, J. W.; Ulbricht, W., and Takata, M.: Effect of ethanol on the sodium and potassium conductances of the squid axon membrane. J. gen. Physiol. *48:* 279–295 (1964).

26 Morgane, P. J.: The function of the limbic and rhinic forebrain. Limbic midbrain systems and reticular formation in the regulation of food and water intake. Ann. N. Y. Acad. Sci. *157:* 806–848 (1969).

27 Nicoll, R. A. and Barker, J. L.: Excitation of supraoptic neurosecretory cells by angiotensin II. Nature, Lond. *233:* 172–173 (1971).

28 Novin, D. and Durham, R.: Unit and D-C potential studies of the supraoptic nucleus. Ann. N. Y. Acad. Sci. *157:* 740–754 (1969).

29 Oomura, Y.; Ono, T.; Ooyama, H., and Wayner, M. J.: Glucose and osmosensitive neurons of the rat hypothalamus. Nature, Lond. *222:* 282–284 (1969).

30 Oomura, Y.; Ooyama, H.; Yamamoto, T., and Naka, F.: Reciprocal relationship of the lateral and ventromedial hypothalamus in the regulation of food intake. Physiol. Behav. *2:* 97–115 (1967).

31 Oomura, Y.; Sugimori, M.; Nakamura, T.; Gawronski, D., and Fukuda, R.: Catecholamine effects on angiotensin receptive neurons in the rat hypothalamus. Abstract IUPS Regional Meeting, Sydney 1972.

32 Osborne, M. J.; Pooters, N.; d'Auriac, G. A.; Epstein, A. N.; Worcel, M., and Meyer, P.: Metabolism of tritiated angiotensin II in anesthetized rats. Pflügers Arch. ges. Physiol. *326:* 101–114 (1971).

33 Sclafani, A. and Grossman, S. P.: Hyperphagia produced by knife cuts between the medial and lateral hypothalamus in rats. Physiol. Behav. *4:* 533–538 (1969).

34 SMITH, G. P. and EPSTEIN, A. N.: Increased feeding in response to decreased glucose utilization in the rat and monkey. Amer. J. Physiol. *217:* 1083–1087 (1969).

35 SWANSON, L. W.; SHARPE, L. G., and GRIFFIN, D.: Drinking to intracerebral angiotensin II and carbachol. Dose-response relationships and ionic involvement. Physiol. Behav. (in press).

36 TURKER, R. K.; PAGE, I. H., and KHAIRALLAH, P. A.: Angiotensin alteration of sodium fluxes in smooth muscle. Arch. int. Pharmacodyn. *165:* 394–404 (1967).

37 VALENSTEIN, E. S.; COX, V. C., and KAKOLEWSKI, J. W.: The hypothalamus and motivated behavior; in TAPP Reinforcement and behavior, pp. 242–283 (Academic Press, New York 1969).

38 VOLICER, L. and LOEW, C. G.: Penetration of angiotensin II into the brain. Neuropharmacology *10:* 631–636 (1971).

39 WAYNER, M. J.: Motor control functions of the lateral hypothalamus and adjunctive behavior. Physiol. Behav. *5:* 1319–1325 (1970).

40 WAYNER, M. J.: Specificity of behavioral regulation; in MOGENSON Regulation of food and water intake (in press).

41 WAYNER, M. J. and CAREY, R. J.: Basic drives. Annu. Rev. Psychol. *24:* 53–80 (1973).

42 WAYNER, M. J. and KAHAN, S.: Central pathways involved during the salt arousal of drinking. Ann. N. Y. Acad. Sci. *157:* 701–722 (1969).

43 WAYNER, M. J.; GAWRONSKI, D., and ROUBIE, C.: Effects of ethyl alcohol on lateral hypothalamic neurons. Physiol. Behav. *6:* 747–749 (1971).

44 WAYNER, M. J.; ONO, T., and NOLLEY, D.: Effects of angiotensin applied electrophoretically on lateral hypothalamic neurons. Pharmacol. Biochem. Behav. *1:* 223–226 (1973).

45 WAYNER, M. J.; COTT, A.; MILLNER, J., and TARTAGLIONE, R.: Loss of 2-deoxy-D-glucose induced eating in recovered lateral rats. Physiol. Behav. *7:* 881–884 (1971).

46 WAYNER, M. J.; GAWRONSKI, D.; ROUBIE, C., and GREENBERG, I.: Effects of ethyl alcohol on lateral hypothalamic neurons; in MELLO and MENDELSON Recent advances in studies of alcoholism, pp. 219–273 (US Government Printing Office, Washington 1971).

Authors' address: Dr. M. J. WAYNER, Dr. T. ONO, D. NOLLEY and A. DE YOUNG, Brain Research Laboratory, Syracuse University, 601 University Avenue, *Syracuse, NY 13210* (USA)

Recent Studies of Hypothalamic Function
Int. Symp. Calgary 1973, pp. 251–267 (Karger, Basel 1974)

Natural and Experimental Hypothalamic Changes in Hibernators

N. MROSOVSKY

Departments of Zoology and Psychology, University of Toronto, Toronto

Introduction

Over the years there has been an increasing sophistication in the experimental manipulations of the hypothalamus. Miniaturised multiple electrode assemblies, precisely controlled thermodes and push-pull cannulae are a few examples. Much remains to be learned with these methods, or even with lesions. A general limitation, however, is that hypothalamic manipulations often affect different cell types or fibres from different areas, and this sometimes greatly complicates analysis [see MOGENSON, this volume]. One possible escape from this problem is to use drugs that affect specific hypothalamic elements. For instance, gold thioglucose is initially taken up by oligodendrocytes in the ventromedial hypothalamus (VMH); these cells are, therefore, favoured candidates for glucoreceptors [8]. For the thermoregulatory system there is capsaicin. This selectively damages mitochondrial elements in certain neurones in the preoptic-anterior hypothalamic (PO-AH) region, but not other cell types there. Since these changes are correlated with initial heat loss and later desensitisation of the response to warmth, the damaged cells may well be the warmth detectors [18, 39]. Such drugs are not common and unfortunately the specificity of a drug is generally inversely correlated with how much it has been studied.

Another approach is to go to the phylogenetic supermarket and discover whether nature is offering us any of these manipulations ready-prepared. This could be a useful auxilliary to chemical and other experimental hypothalamic manipulations. Much ground-work needs to be done, but for the study of body weight and of thermoregulation the hibernators already look promising.

Regulation of Body Weight

Many mammalian hibernators have an annual cycle of weight with a period of fattening in the autumn. At this time they look quite like animals made obese by VMH lesions [see 28 for photographs]. The similarity extends to their responses to certain experimental procedures. For instance, if a hypothalamic hyperphagic rat in the static phase of obesity is forced to reduce weight by food restriction, on return to *ad libitum* feeding its weight returns to the static level. If an already obese animal is made still fatter by forced feeding and then left to eat on its own, its weight drops back to the static level [14]. These findings can be conveniently described in terms of changes in a negative feedback system with a set point for body weight or body fat. The effects of VMH lesions might be due to direct elevation of the set point or to some other alteration such as decreased sensitivity to signals associated with satiety [see 13 for a quantitative treatment]. In either case, operationally, the lesioned system works as if it were regulating about higher levels.

Similar tests made on hibernators, such as ground squirrels, suggest that there are similar changes in their regulatory systems. If food is removed for 2–3 days from an animal during the weight gain phase, then

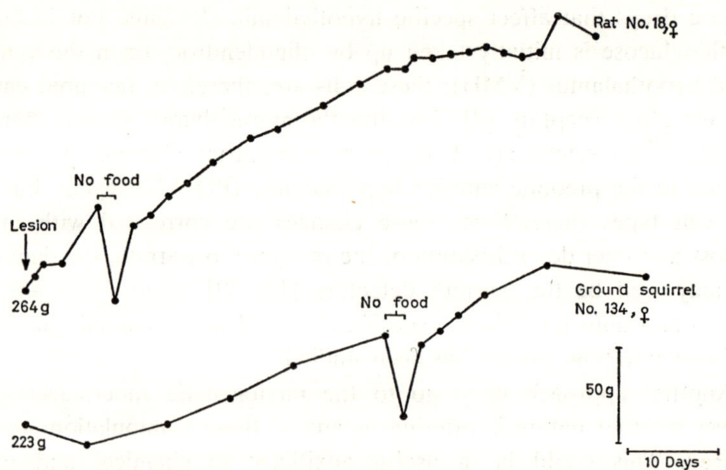

Fig. 1. Body weights of a rat with a ventromedial hypothalamic (VMH) lesion and of a golden-mantled ground squirrel during natural fattening showing the reactions to temporary withdrawal of food [data from 1].

on return of food the animal compensates for the set-back by extra-fast weight gains (fig. 1). When it reaches a weight that would have been expected for that time of year had no temporary food deprivation occurred, rates of weight gain slacken. This shows that before the deprivation test the animal was not gaining weight as fast as it was capable of doing. The implication is that natural changes in the regulatory system are relatively gradual. The animal can meet slowly climbing set points in this phase of its cycle without gaining weight as fast as it can. When forced to fall behind its set point, it makes up by increasing the rate of weight gain.

This point can be demonstrated in another way. If a VMH lesion is made during a fattening phase the rate of gain is much increased (fig. 2). This shows that before the lesion the squirrel was not gaining weight as fast as it might have. Since the deprivation experiments point to regulation of body weight, one must again infer that in a natural weight gain phase the set point rises only gradually.

Gradual natural set point changes and sudden lesion-induced changes are associated with different shapes in the curves of weight gain. Weight gain in golden-mantled and thirteen-lined ground squirrels is often relatively uniform over large portions of the curve [1, 32]. There must, of course, be some flattening off as the animals reach peak weights and

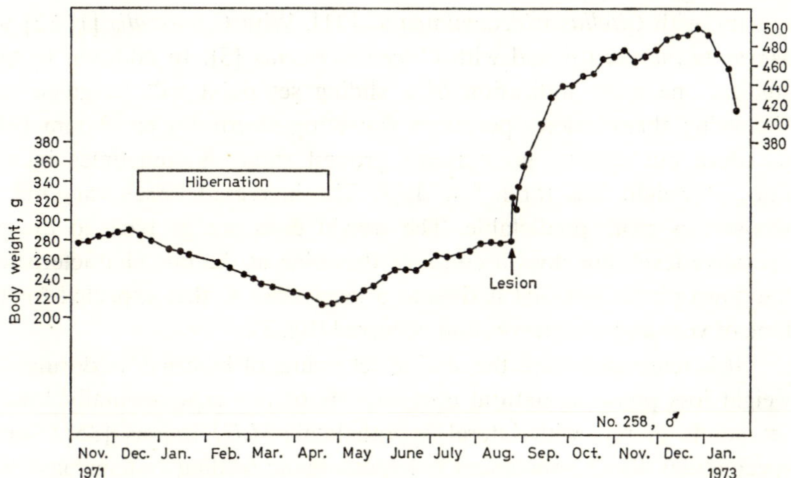

Fig. 2. Body weight of a golden-mantled ground squirrel kept in constant conditions (12–12 h light-dark cycle and 12.5 ± 3 °C) showing natural and lesion-induced weight gains. The lesion (2 mA for 25 sec) damaging the VMH was made by methods similar to those described by MROSOVSKY [29].

go into a phase of weight loss. In contrast, for hypothalamic hyperphagic rats, the literature shows many examples of exponential-looking curves, with levelling off as the animals approach plateau weights. The same probably occurs with ground squirrels after VMH lesions (compare the natural and lesion-induced weight gains in fig. 2), but further analysis is required since lesions may not eliminate the continuation of natural changes in the system.

The exponential weight curve in the rats with VMH damage is compatible with the operation of a proportional controller: the greater the difference between the actual and the set weight, the greater the rate of weight gain. If this is a correct interpretation, a hypothalamic hyperphagic rat should not react to a few days food deprivation during its dynamic phase in the same way as a ground squirrel during natural fattening. Figure 1 illustrates what happens: on re-feeding after a few days food deprivation the rat continues to gain weight at a rate similar to his pre-deprivation rate for a specific body weight. Beyond the last pre-deprivation weight, the rat traces a post-deprivation curve similar to a projection of his pre-deprivation weight curve, but delayed by 3 days [1, 34].

Turning to the weight loss phase of the annual cycle, conceptually similar experiments indicate that there is a slowly declining set point for body weight in hibernators. Supportive data come from studies on several species, with *Citellus tridecemlineatus* [31]. With *C. lateralis* [1, 12] with *Marmota monax* [6] and with *Cricetus cricetus* [3]. In addition to these studies, one more indication of a sliding set point will be given here. Following sham-lesion operations (lowering electrodes ca. 3 mm below the dura but passing no current) ground squirrels sometimes increase rates of weight loss for a few days. The increased losses vary but the recovery is more predictable. The weight does not go back to the pre-operative level, nor does it continue declining at the rate characteristic of the down phase, but first it rises to a level close to that expected for that time of year had no intervention occurred (fig. 3).

It is tempting to see the sliding set points of hibernators during their weight loss phase as natural counterparts of the experimentally lowered set points in rats with lateral hypothalamic *(LH)* lesions [35]. Similar speculations about imbalances in hypothalamic feeding systems have been made for pre-migratory hyperphagia in sparrows [22]. Such comparisons are not negated but rather made more precise if it turns out that the critical aspect of *LH* lesions is damage to fibres of passage rather than to local cell bodies [MOGENSON, this volume].

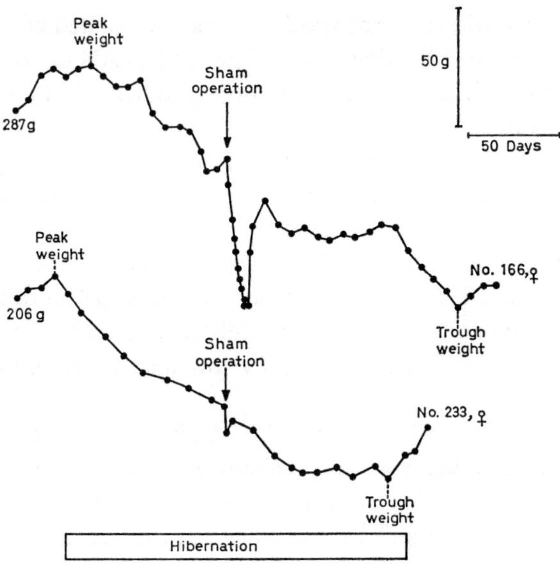

Fig. 3. Body weights of 2 golden-mantled ground squirrels during the weight loss phase of the cycle showing changes following sham operations (see text). No. 166 (top) was kept at 22.5 ± 3 °C and was sometimes lethargic but did not hibernate deeply. No. 233 (bottom) was kept at 12.5 ± 3 °C and showed bouts of hibernation over the period marked. Note that the rates of weight loss in the 2 animals were fairly similar despite these differences in energy expenditure. The light-dark cycle was 12–12 h for both animals.

Thermoregulation and Hibernation

If set points for body weight in hibernators are adjustable, perhaps the same applies for body temperature. The idea of a turning down of the thermostat has been in the hibernation literature for some time, but is only recently being tested. The situation is far from clear and there are different results from different species.

Hibernators often have bouts of shivering and increased breathing rates while becoming torpid, and this suggests that entry to hibernation is carefully regulated and allowed to proceed at a certain rate only, just as is weight loss over the winter. Moreover, temperature falls less rapidly during entry to hibernation than it does in a dead animal cooling. These points apply to edible dormice, golden hamsters and several species of marmots and ground squirrels [see 28 for a review]. HELLER and HAMMEL

[11] have provided direct evidence for hypothalamically regulated entry to hibernation. They reported that heating and cooling the PO-AH of golden-mantled ground squirrels, *C. lateralis,* during entry to hibernation altered the rate of heat production.

However, in the same series of experiments, HELLER and HAMMEL [11] found that with animals already in deep hibernation PO-AH displacements of a few degrees Celsius did not affect the oxygen consumption. If temperature was cooled below a value of about 1 °C, arousal from hibernation could be initiated. The authors suggested that set points decline during entry to hibernation and that during deep hibernation the thermostat is inactive, but can be re-activated by critically low alarm temperatures that initate arousal.

Working with yellow-bellied marmots, *M. flaviventris,* MILLS and SOUTH [26] found that PO-AH cooling of about 1 °C led to increased heart rate without producing a full arousal. Warming this region several degrees led to somewhat slower and less variable heart rates; in addition the animals sometimes left their nests and uncurled.

Given this apparent species difference, it is reasonable to ask whether the PO-AH in *C. lateralis* is perhaps a relatively unimportant part of the larger system for thermoregulation. For instance, it appears that for pigeons, *Columba livia,* and for bats, *Eptesicus fuscus,* with certain measures at least, extra-hypothalamic systems are more important than the hypothalamus [20, 36]. However, since *C. lateralis* react to PO-AH temperature changes when normothermic, and to very low alarm temperatures when hibernating [11], there is no reason to think this area is unimportant for their thermoregulation. Moreover, BECKMAN and SATINOFF [2] have found that norepinephrine and 5-hydroxytryptamine (5-HT) injections to the PO-AH arouse *C. lateralis* from hibernation while similar injections to the caudate-putamen do not, though, as they point out, mimicking of sensory stimuli rather than direct activation of thermoregulatory systems cannot be ruled out in this effect. Work with a closely related ground squirrel, *C. tridecemlineatus,* has shown that they also react to thermal changes in the PO-AH when normothermic [41]. In addition, *C. tridecemlineatus* with lesions in this region are unable to arouse from hibernation [37].

These findings leave one with the tentative conclusion that while deeply hibernating *M. flaviventris* remain relatively reponsive to changes in PO-AH temperature, deeply hibernating *C. lateralis* do not, although this area is of great importance in the normothermic state and although they still react during hibernation when their temperature falls to a critically low alarm level [11, 24].

The thirteen-lined ground squirrel, *C. tridecemlineatus,* may lie somewhere between *M. flaviventris* and *C. lateralis* in that when its brain is cooled, either by an external spoon-shaped thermode placed under its head or by decreasing the ambient temperature, sometimes it reacts with increased heart and breathing rates in the 9.7–5.0 °C range, while at other times it allows its temperature to fall to near-lethal levels (1.1–2.4 °C) before reacting. There may be changes in the critically low alarm temperature for initiating arousal [24]. However, since the animals did not always arouse fully in these experiments, there is no compelling reason for accepting this interpretation. Thermoregulation and changes in the set point about which that regulation is occurring might also be involved. In any case, further study of the conditions which determine whether *C. tridecemlineatus* responds or not to changes of brain temperature could be of great help in solving the problem of whether the thermostat remains in action during deep hibernation.

Another puzzling matter are the reports that little brown bats, *Myotis lucifugus,* with PO-AH lesions lose thermoregulatory ability and cannot arouse from hibernation, while big brown bats, *E. fuscus,* can arouse [21, 20]. However, since in the latter case it is not possible to tell from the diagrams whether parts of the PO-AH may have been spared, and since rates of arousal were somewhat slower in the lesioned bats, it is not yet clear whether this is a real species difference.

It would be premature to try to make coherence out of the present data on thermoregulation and hibernation. There is a need for more studies that use the same indices of thermoregulation on different species. It is possible that strategies for thermoregulation are different in different species or in the same species at different times of the year. For instance, Mills and South [26] indicate that vasomotor reactions to hypothalamic heating are well developed in the spring marmot but not in the pre-hibernating animal [see 30, for changes in strategies for regulation of body weight].

Tests with Hypothermic Drugs on Hibernators

Another approach to the problem of whether hibernators exhibit extreme plasticity in hypothalamic thermoregulatory systems has been to use chemicals known to affect these systems in other mammals. If extra-large temperature drops occur, or even something resembling hibernation, then suspicions that hibernation depended on natural use of such chemical mechanisms would be strengthened. There have been several experiments on the reactions of hibernators to hypothermic drugs.

In work in the author's laboratory, thirteen-lined ground squirrels were given large doses of crystalline drugs via implanted hypothalamic cannulae. Temperatures were measured 1 and 2 h later.

Subjects and housing. C. tridecemlineatus, born in the laboratory 18–20 May, 1968, from pregnant females trapped near Lawrence, Kansas, were kept at about 22 °C after weaning in cages 37 × 22 × 19 cm, provided with nesting material and Purina laboratory chow and water *ad libitum*. One animal (No. 68) was exposed to natural photoperiods 7 August to 31 October, 1968; the rest were on a 12–12 h light-dark cycle throughout. Temperature in the room over the period of testing was 23 ± 2.5 °C. During the tests the animals weighed 128–275 g; all except one were in the weight loss phase of their cycle. The exceptional animal (No. 82) showed a gradual weight gain that had started in the middle of January, 1969.

Surgery and Cannulae. The animals were implanted with cannulae by methods similar to those described by MROSOVSKY [27]. The outer guide cannula was made out of a 21-gauge hypodermic needle fitted inside a threaded piece of plastic into which the inner cannula, made of 26-gauge needle, could be screwed down flush with the tip of the guide cannula. At least 13 post-operative days elapsed before any testing.

Drugs and loading of cannulae. Spare inner cannulae were cleaned in alcohol and loaded by tamping them down 10 times into crystalline drugs. With this procedure approximate weights of drugs, as assessed with a UM6 Mettler balance, taken into the inner cannula, were 45 ± 30 μg with carbachol, and 25 ± 15 μg with L-adrenaline bitartrate, L-norepinephrine bitartrate and serotonin creatine sulphate complex. In the latter case the drug was not always completely dissolved away from the cannula tip by the end of the test. In addition to the 4 drugs tested, a sham drug procedure was used in which an empty cannula was screwed down into place.

Test procedure. Starting about 10.00 h for all tests, rectal temperatures were taken with a Yellowsprings TD telethermometer and 402 probe (4 cm immersion). Loaded inner cannulae were substituted for those previously in place. The animal was then returned to its home cage and temperatures taken again close to 1 and 2 h later when the cannula with drug was removed, inspected and a clean empty cannula replaced. There were at least 2 tests with each drug, and at least 3 days between successive tests. Treatments occurred in varying orders. Tests took place between 20 January and 15 April, 1969. Later, the brains were embedded in paraffin, sectioned at 10 μm and stained with cresyl violet.

With carbachol, drops as large as 8 °C occurred at room temperature, 23 °C (fig. 4), but the squirrels did not assume characteristic hibernating postures. These temperature changes may seem large but is should be pointed out that with 5 μg crystalline carbachol, almost one-tenth of the doses used here for squirrels, HULST and DE WEID [15] produced drops of up to 3.5 °C in rats kept at 24 °C. On the other hand, much smaller temperature decreases, or even increases, have also been found in rats after central injections of cholinergic drugs [5].

With the noradrenaline injections at some cannula placements, drops in body temperature of about 2 °C occurred in squirrels (fig. 4). Working on marmots, JACOBS *et al.* [16] found that with 144 μg epinephrine injected intraventricularly, the maximum drop was 5.4 °C; this was at 5 °C ambient. They state also that preoptic injections of catecholamines in the autumn elicited drops in core temperature similar to those reported by FELDBERG and MYERS [9] for cats, but required larger doses.

MYERS and BUCKMAN [33] have injected Ca^{++} ions intraventricularly into golden hamsters. Maximum temperature drops were 4.2 °C at 25 °C ambient temperature, 16.1 °C at 5 °C and 23.7 °C at –10 °C. Although the hamsters sometimes re-arranged nesting material at the start of the infusion, they did not maintain a hibernating position but often lay prone.

In another experiment in the author's laboratory, capsaicin (methyl-nonenoyl-vanillylamide) was injected into golden-mantled ground squirrels (see table I below).

Ten *C. lateralis* from Oregon, obtained through a dealer on 2 October, 1970, were kept at 12 ± 2.5 °C and otherwise in the same way as the animals described in detail above. Five animals received i.p. or s.c. capsaicin injections between 28 July, 1971, and 4 February, 1972, and 5 animals matched for sex and weight cycles received control injections. Rectal temperatures, measured as described above, were taken approximately every 30 min during temperature drops and less often thereafter. Ambient temperature during the tests was 12 ± 1.5 °C. Control animals were measured at the same times. For tests occurring during the hibernation season, made only during natural periodic arousals from hibernation, the control animals could not always be tested on the same day. The capsaicin was dissolved in ethanol, Tween 80 and sterile saline in amounts given by JANCSÓ *et al.* [17]. For control tests equal volumes of the same solution without capsaicin were used.

There is evidence that this drug stimulates hypothalamic warmth detectors in non-hibernators [18]. If entry into hibernation involves an alteration in the balance between high and low q10 neurones, as HAMMEL [10] has suggested, it seemed possible that capsaicin might induce especially large temperature drops in hibernators. If anything, in our limited tests, its effect on temperature was less pronounced (even though doses were

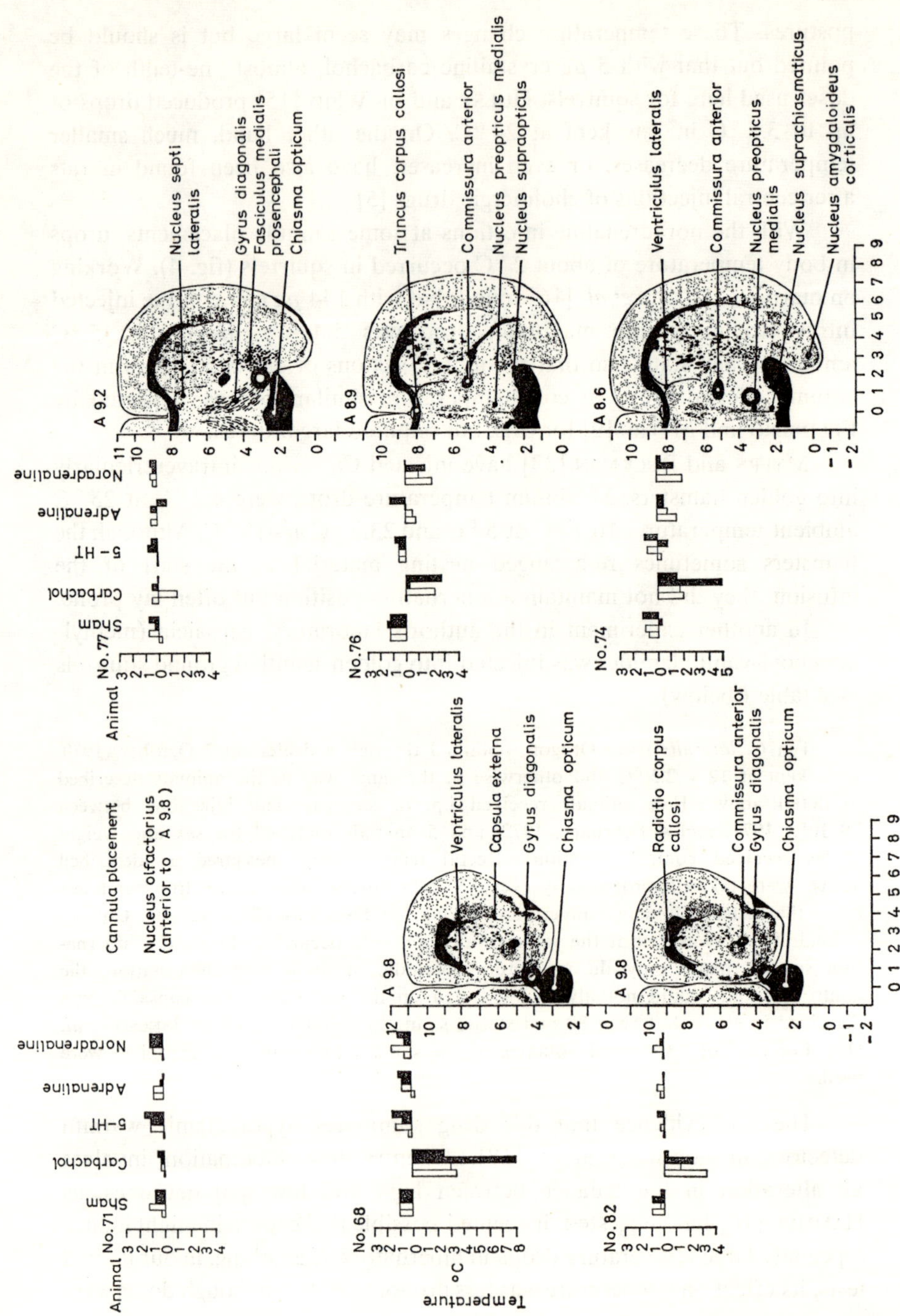
Animal
No. 71
Sham
Carbachol
5-HT
Adrenaline
Noradrenaline
Cannula placement
Nucleus olfactorius (anterior to A 9.8)
Temperature
°C
No. 68
A 9.8
Ventriculus lateralis
Capsula externa
Gyrus diagonalis
Chiasma opticum
No. 82
A 9.8
Radiatio corpus callosi
Commissura anterior
Gyrus diagonalis
Chiasma opticum
Animal
No. 77
Sham
Carbachol
5-HT
Adrenaline
Noradrenaline
A 9.2
Nucleus septi lateralis
Gyrus diagonalis
Fasciculus medialis prosencephali
Chiasma opticum
No.76
A 8.9
Truncus corpus callosi
Commissura anterior
Nucleus preopticus medialis
Nucleus supraopticus
No.74
A 8.6
Ventriculus lateralis
Commissura anterior
Nucleus preopticus medialis
Nucleus suprachiasmaticus
Nucleus amygdaloideus corticalis

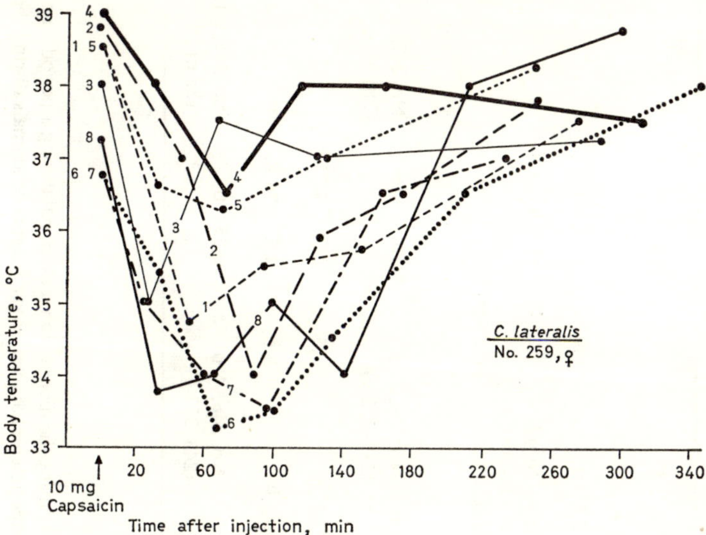

Fig. 5. Changes in body temperature of a golden-mantled ground squirrel given repeated capsaicin injections. Numbers within circles beside different lines show the order of the injections. The initial test with a different dose is ommitted. For full details see table I. The lower starting levels for 6, 7 and 8 are usual for the hibernating season.

near the lethal level, see table I). With 10 mg capsaicin per squirrel, drops of 3 °C were common (fig. 5 and table I). Commenting on this Jancsó-Gabor [17a] writes: 'It is most striking that this species is almost insensitive to capsaicin. In rats a fall of 3 °C in rectal temperature can be induced by doses as low as 0.2 mg/rat and 10 mg causes a drop of 6–7 °C.' No consistent postures were noted in these experiments; the animals were often uncurled or out of their nests.

These various tests with hypothermic drugs give no indication of an easy chemical way of triggering hibernation. The marked dependence of hypothermia on the ambient temperature in some of these experiments is

---

Fig. 4. Changes in body temperature of thirteen-lined ground squirrels following intracranial injections. The open bars show changes recorded 1 h after the injections; the shaded bars show changes 2 h post-injection. The 2 bars at each of the time-intervals represent the results for the 2 tests separately. Cannula placements are marked on sections from the atlas of Joseph *et al.* [19]. 5-HT = 5-hydroxytryptamine.

Table I. Maximum temperature drops, to the nearest 0.25 °C, of golden-mantled ground squirrels given capsaicin

Capsaicin, mg	No. 253, ♀ (233–264 g)	2*	10	5	10	5	10	No. 257, ♀ (196–264 g)	5*	10	5*	10	10	10		
Drop, °C		3.0	0	0	0.25	0.75	2.5		3.0	2.5	0	2.0	3.25	3.0		
Date		28.7.	5.8.	14.9.	27.9.	26.1.[h]	28.1.[h]		30.7.	4.8.	30.9.	19.1.[h]	25.1.[h]	4.2.[h]		
Drop for control, °C	No. 249, ♀ (201–264 g)	0	0.25	0	0	0.25	1.25 (NT)	No. 254, ♀ (198–233 g)	0.25	1.0	0	0.25 (NT)	–	–		
Capsaicin, mg	No. 259, ♀ (191–224 g)	5	10	10	10	10	10	10	10	10	No. 251, ♂ (212–222 g)	5	10	10	No. 247, ♀ (210 g)	5*
Drop, °C		1.25	3.75	4.75	3.0	2.5	2.25	3.5	3.25	3.5		1.5	3.75	4.75 (D)		3.5 (D)
Date		3.8.	5.8.	26.8.	14.9.	27.9.	30.9.	14.1.[h]	27.1.[h]	1.2.[h]		3.8.	4.8.	26.8.		30.7.
Drop for control, °C	No. 250, ♀ (184–205 g)	0	1.25	1.25	0.25	0.5	0	–	–	–	No. 246, ♂ (243–257 g)	0	0.25	0	No. 245, ♀ (188 g)	1.5

All injections were s.c. unless indicated by an asterisk for i.p. injections. [h]test occurred during the hibernation season, but during a periodic arousal; (D) animal died shortly after injection; (NT) = control not tested on same day as experimental animal. Body weights of animals during the tests are shown in brackets.

as compatible with some blocking or depression of neurones important in maintaining heat as with a regulated alteration of set point. And, although the temperature drops were large, without further data it cannot be said definitely that they were larger than in comparably sized non-hibernating species treated the same way. On the other hand, it should be pointed out that these tests were not specially designed for maximising the chance of getting hibernation. Various factors may have militated against inducing hibernation: irritant effects of capsaicin [17], too-frequent disturbance of the ground squirrels for measurements in MROSOVSKY'S experiments (above) and taping of rectal thermistors onto the tail in MYERS and BUCKMAN'S work [33] on golden hamsters. The latter species are often poor laboratory hibernators and would be unlikely to have hibernated on their own in a –10 °C environment [23, 24].

This leaves unanswered questions as to whether hibernation is the opposite to fever and whether the sensor and effector systems are the same as those used in thermoregulation in the normothermic range.

Preventing Hibernation

Experimental prevention of hibernation has also been tried. In general, difficulties of interpretation are liable to arise with this approach because hibernation is not a predictable occurrence in all animals [28] and there are non-specific ways in which it can be disrupted. For instance, LYMAN and O'BRIEN [24] found that after implantation of thermocouples many animals did not return to hibernation even though previously hibernating well. Slight disturbances, such as vibration in a thermode, led to cardiac accelerations in their experiments. Thus, it is hard to interpret MALAN'S finding [25] that European hamsters with posterior hypothalamic lesions did not build nests or hibernate. Similarly, with loss of hibernation after depletion of whole brain 5-HT by p-chlorophenylalanine treatments, it is hard to say whether thermoregulatory, sleep or other systems have been disrupted [38]. The 5-HT levels in untreated ground squirrels in these experiments decreased when the animals entered hibernation, but it is uncertain whether this was important in induction of torpor or was a result of altered metabolism. In other work on 5-HT, in hedgehogs, increases in the winter season have been found, but only slightly higher in hibernating animals than in animals in the active state at that time [42; see also 4 for changes in cerebral monoamines in bats].

In the experiments with capsaicin (table I), it had been hoped that repeated injections would lead to eventual desensitisation of warmth receptors and that hibernation would be reduced or abolished. However, no marked desensitisation was evident (fig. 5) and there were no obvious long-lasting effects on the annual weight cycle or hibernation patterns.

Different Types of Changes in Control Systems and Conclusion

At present it remains an assumption that the hypothalamus is involved in the natural weight and temperature cycles of hibernators. It is possible, for instance, that annual mechanisms affecting activity and eating bypass the VMH and *LH*. However, this does seem rather unlikely, given that the integrity of the VMH is necessary for the normal progression of the weight gain and weight loss phases (fig. 2) [9a, 29], and given the knowledge that, in terms of magnitude of effect following disruption, this is one of the major systems in mammals determining weight. Similarly, with the PO-AH, the evidence cited above suggests that this region is involved in the control of torpor rather than being inoperative or bypassed. Stronger evidence that these hypothalamic areas are involved would be the discovery of structural or biochemical changes that are tightly correlated with the weight and thermal changes.

In attempting such correlations, it is valuable to have a clear designation of the changes in the control systems. There are several different possibilities. In the case of body weight, the available evidence indicates set points continuously adjusting over many months. With temperature, it looks as if there are rapid adjustments over large ranges, with the thermostat turned down or off in a few hours. With reproductive changes, at least in some seasonally breeding animals, negative feedback systems may be inoperative altogether. Thus, snowshoe hares, non-hibernators, do not respond out of their breeding season to castration with increased pituitary *LH* and follicle stimulating hormone (FSH) [7]. The matter has not been studied sufficiently in hibernators: WELLS [40] reported that unilaterally castrated *C. tridecemlineatus* show compensatory hypertrophy even out of the breeding season, but full data were not given.

In general, more work needs to be done on hibernators in defining the essential characteristics of the preparation. It is hoped that, in the long run, just as the study of natural changes at puberty and of sexual dimorphism may be useful in discovering the neural elements controlling reproductive behaviour [see RAISMAN, this volume], so analysis of natural changes in hibernators may give insight into the hypothalamic and other brain elements controlling body weight and temperature.

Acknowledgements

This work was supported by grants from the National Research Council of Canada and the Medical Research Council of Canada. The author thanks Mrs. KIRSTEEN LANG, Mr. HUGH CRASKE, Mr. JOHN HALLONQUIST and Dr. PAT LANG for help. Dr. AURELIA JANCSÓ-GABOR kindly supplied the capsaicin.

References

1 BARNES, D. S. and MROSOVSKY, N.: Body weight regulation in ground squirrels and hypothalamically lesioned rats: slow and sudden set point changes. Physiol. Behav. (in press).

2 BECKMAN, A. L. and SATINOFF, E.: Arousal from hibernation by intrahypothalamic injections of biogenic amines in ground squirrels. Amer. J. Physiol. *224:* 875–879 (1972).

3 CANGUILHEM, B. and MARX, C.: Regulation of the body weight of the European hamster during the annual cycle. Pflügers Arch. ges. Physiol. *338:* 169–175 (1973).

4 CONSTANTINIDIS, J.; TORRE, J. C. DE LA; TISSOT, R., et HUGGEL, H.: Les monoamines cérébrales lors de l'hibernation chez la chauve-souris. Rev. suisse zool. *77:* 345–352 (1970).

5 CRAWSHAW, L. I.: Effect of intracranial acetylcholine injection on thermoregulatory responses in the rat. J. comp. physiol. Psychol. *83:* 32–35 (1973).

6 DAVIS, D. E.: Failure of schedule of torpor to alter annual rhythm of appetite of woodchucks *(Marmota monax)*. Mammalia *34:* 542–544 (1970).

7 DAVIS, G. J. and MEYER, R. K.: Seasonal variation in LH and FSH of bilaterally castrated snowshoe hares. Gener. comp. Endocr. *20:* 61–68 (1973).

8 DEBONS, A. F.; KRIMSKY, I.; FROM, A., and CLOUTIER, R. J.: Site of action of gold thioglucose in the hypothalamic satiety center. Amer. J. Physiol. *219:* 1397–1402 (1970).

9 FELDBERG, W. and MYERS, R. D.: Changes in temperature produced by microinjections of amines into the anterior hypothalamus of cats. J. Physiol., Lond. *177:* 239–245 (1965).

9a HALLONQUIST, J.: Personal commun. (1972).

10 HAMMEL, H. T.: Temperature regulation and hibernation; in FISHER, DAWE, LYMAN, SCHÖNBAUM and SOUTH Mammalian hibernation, vol. III (Oliver & Boyd, Edinburgh 1967).

11 HELLER, H. C. and HAMMEL, H. T.: CNS control of body temperature during hibernation. Comp. Biochem. Physiol. *41A:* 349–359 (1972).

12 HELLER, H. C. and POULSON, T. L.: Circannian rhythms. II. Endogenous and exogenous factors controlling reproduction and hibernation in chipmunks *(Eutamias)* and ground squirrels *(Spermophilus)*. Comp. Biochem. Physiol. *33:* 357–383 (1970).

13 HIRSCH, J.: Discussion. Adv. psychosom. Med. *7:* 229–242 (1972).

14 HOEBEL, B. G. and TEITELBAUM, P.: Weight regulation in normal and hypothalamic hyperphagic rats. J. comp. physiol. Psychol. *61:* 189–193 (1966).

15 HULST, S. G. T. and WEID, D. DE: Changes in body temperature and water intake following intracerebral implantation of carbachol in rats. Physiol. Behav. *2:* 367–371 (1967).

16 JACOBS, H. K.; SOUTH, F. E.; HARTNER, W. C. and ZATZMAN, M. L.: Thermoregulatory effects of administration of biogenic amines in to the third ventricle and preoptic area of the marmot *(M. flaviventris)*. Cryobiology *8:* 313–314 (1971).

17 JANCSÓ, N.; JANCSÓ-GABOR, A., and SZOLCSÁNYI, J.: Direct evidence for neurogenic inflammation and its prevention by denervation and pretreatment with capsaicin. Brit. J. Pharmacol. *31:* 138–151 (1967).

17a JANCSÓ-GABOR, A.: Personal commun. (1972).

18 JANCSÓ-GABOR, A.; SZOLCSÁNYI, J., and JANCSÓ, N.: Stimulation and desensitization of the hypothalamic heat-sensitive structures by capsaicin in rats. J. Physiol., Lond. *208:* 449–459 (1970).

19 JOSEPH, S. A.; KNIGGE, K. A.; KALEJS, L. M.; HOFFMAN, R. A., and REID, P.: A stereotaxic atlas of the brain of the 13-line ground squirrel *(Citellus tridecemlineatus)* (US Army Edgewood Arsenal, Maryland 1966).

20 KLUGER, M. J. and HEATH, J. E.: Effect of preoptic anterior hypothalamic lesions on thermoregulation in the bat. Amer. J. Physiol. *221:* 144–149 (1971).

21 KOSKI, J. and CONOVER, G.: The effects of hypothalamic and hippocampal lesions on the entry into hibernation in the bat. Abstracts 4th Symp. Hibernation-Hypothermia, Snowmass, Aspen 1971.

22 KUENZEL, W. J.: Dual hypothalamic feeding system in a migratory bird, *Zonotrichia albicollis*. Amer. J. Physiol. *223:* 1138–1142 (1972).

23 LYMAN, C. P.: The oxygen consumption and temperature regulation of hibernating hamsters. J. exp. Zool. *109:* 55–78 (1948).

24 LYMAN, C. P. and O'BRIEN, R. C.: Sensitivity to low temperature in hibernating rodents. Amer. J. Physiol. *222:* 864–869 (1972).

25 MALAN, A.: Contrôle hypothalamique de la thermorégulation et de l'hibernation chez le hamster d'Europe *(Cricetus cricetus)*. Arch. Sci. physiol. *23:* 47–87 (1969).

26 MILLS, S. H. and SOUTH, F. E.: Central regulation of temperature in hibernation and normothermia. Cryobiology *9:* 393–403 (1972).

27 MROSOVSKY, N.: Self-stimulation in hypothermic hibernators. Cryobiology *2:* 229–239 (1966).

28 MROSOVSKY, N.: Hibernation and the hypothalamus (Appleton-Century-Crofts, New York 1971).

29 MROSOVSKY, N.: Hypothalamic hyperphagia without plateau in ground squirrels. Physiol. Behav. (in press).

30 MROSOVSKY, N. and BARNES, D. S.: Anorexia, food deprivation and hibernation. Physiol. Behav. (in press).
31 MROSOVSKY, N. and FISHER, K. C.: Sliding set points for body weight in ground squirrels during the hibernation season. Canad. J. Zool. *48:* 241–247 (1970).
32 MROSOVSKY, N. and LANG, K.: Disturbances in the annual weight and hibernation cycles of thirteen-lined ground squirrels kept in constant conditions and the effects of temperature changes. J. interdisc. Cycle Res. *2:* 79–90 (1971).
33 MYERS, R. D. and BUCKMAN, J. E.: Deep hypothermia induced in the golden hamster by altering cerebral calcium levels. Amer. J. Physiol. *223:* 1313–1318 (1972).
34 PANKSEPP, J.: A re-examination of the role of the ventromedial hypothalamus in feeding behavior. Physiol. Behav. *7:* 385–394 (1971).
35 POWLEY, T. L. and KEESEY, R. C.: Relationship of body weight to the lateral hypothalamic feeding syndrome. J. comp. physiol. Psychol. *70:* 25–36 (1970).
36 RAUTENBERG, W.; NECKER, R., and MAY, B.: Thermoregulatory responses of the pigeon to changes of the brain and the spinal cord temperatures. Pflügers Arch. ges. Physiol. *338:* 31–42 (1972).
37 SATINOFF, E.: Disruption of hibernation caused by hypothalamic lesions. Science *155:* 1031–1033 (1967).
38 SPAFFORD, D. C. and PENGELLEY, E. T.: The influence of the neurohumor serotonin on hibernation in the golden-mantled ground squirrel, *Citellus lateralis.* Comp. Biochem. Physiol. *38A:* 239–250 (1971).
39 SZOLCSÁNYI, J.; JOÓ, F., and JANCSÓ-GABOR, A.: Mitochondrial changes in preoptic neurones after capsaicin desensitization of the hypothalamic thermodetectors in rats. Nature, Lond. *229:* 116–117 (1971).
40 WELLS, L. J.: Seasonal sexual rhythm and its experimental modification in the male of the thirteen-lined ground squirrel *(Citellus tridecemlineatus).* Anat. Rec. *62:* 409–447 (1935).
41 WILLIAMS, B. A. and HEATH, J. E.: Responses to preoptic heating and cooling in a hibernator, *Citellus tridecemlineatus.* Amer. J. Physiol. *218:* 1654–1660 (1970).
42 UUSPÄÄ, V. J.: The 5-hydroxytryptamine content of the brain and some other organs of the hedgehog *(Erinaceus europaeus)* during activity and hibernation. Experientia *19:* 156–158 (1963).

Author's address: Dr. N. MROSOVSKY, Departments of Zoology and Psychology, University of Toronto, *Toronto, Ontario* (Canada)

Recent Studies of Hypothalamic Function
Int. Symp. Calgary 1973, pp. 268–293 (Karger, Basel 1974)

Changing Views of the Role of the Hypothalamus in the Control of Ingestive Behaviors

G. J. MOGENSON

Departments of Physiology and Psychology, University of Western Ontario, London, Ontario

According to most textbooks which deal with the central control of ingestive behaviors, the lateral hypothalamus is concerned with appetite and the ventromedial hypothalamus with satiety. Lesions of the ventromedial hypothalamus (VMH) cause hyperphagia and obesity, and lesions of the lateral hypothalamus *(LH)* cause aphagia, adipsia and weight loss [2, 19, 79, 84]. Electrical stimulation of the VMH reduces feeding whereas stimulation of the *LH* elicits feeding and drinking [6, 53, 73, 78, 80]. The effects of lesioning and stimulating the hypothalamus have been explained by the dual mechanism model shown in figure 1. This model has provided a useful theoretical framework and has stimulated a good deal of interest in the central control of ingestive behaviors.

This article begins with a discussion of recent studies of the VMH and the LH in the context of the dual mechanism model. Some of these studies, especially those in which the limitations of lesion and stimulation

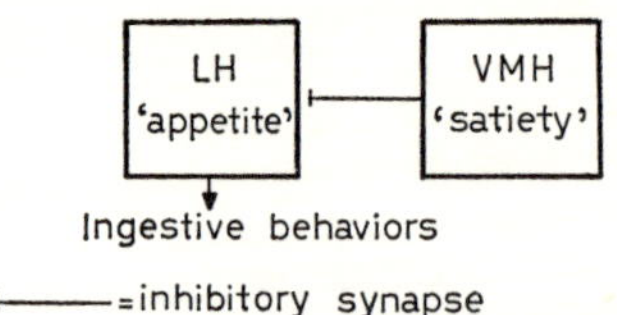

Fig. 1. The role of the hypothalamus in ingestive behaviors. According to the dual mechanism model the lateral hypothalamus *(LH)* 'appetite system' is inhibited by the ventromedial hypothalamus (VMH) 'satiety system'. After integrating hunger, thirst and satiety signals and modulating signals from limbic forebrain structures, the hypothalamus initiates ingestive responses by sending command signals to structures lower in the brainstem [see 77, fig. 8].

techniques are recognized and in which newer techniques and approaches have demonstrated the importance of extrahypothalamic structures, have led to a reassessment of the role of the hypothalamus in the initiation of ingestive behaviors. These developments are considered in the second section. In the final section an attempt is made to clarify some of the issues and to synthetize recent experimental evidence in relation to pathways that have been demonstrated with histological and histofluorescence techniques.

I. The Textbook View of the Hypothalamus and Ingestive Behavior Updated

A. The Role of the VMH

The first clues about the role of the hypothalamus in the control of food intake and energy balance came from lesion studies. When the VMH was lesioned in experimental animals, they increased their food intake and became obese [19], consistent with earlier clinical observations that hyperphagia and obesity were associated with the pathology of this region of the hypothalamus [92]. The marked increase in food intake during the first 2 or 3 weeks following bilateral lesions has been designated the dynamic phase of hypothalamic hyperphagia and is characterized by rapid weight gain and increased deposition of adipose tissue [119]. Eventually, food intake is reduced to prelesion levels and body weight plateaus; this is designated the static phase. The dynamic phase of hyperphagia can be restored by restriction of food for a few days. When food is again freely available the animals soon return to the same levels of adiposity and body weight [55, 118] suggesting that the set-point for the regulation of energy stores has been altered by the VMH lesions.

Recently, MROSOVSKY [85] has pointed out the similarities between animals with VMH lesions and the prehibernatory hyperphagia and obesity of hibernators. He proposes that the hyperphagia, finickiness and other behavioral changes of the prehibernation period could be due to a 'functional depression of the ventromedial nucleus' [85, p. 93] and suggests that, like VMH-lesioned animals and obese-prone people, they are less responsive to internal signals and more responsive to external food-related signals [see also SCHACHTER, 107; STUNKARD, 121].

Rats with VMH lesions are hypoactive as well as hyperphagic [37] and the resulting reduction in energy expenditure also contributes to the adiposity and weight gain. This appears to be the reason that VMH-lesioned rats gain more when they are given the same amount of food as unoperated controls [48].

The increased food intake following VMH lesions is frequently attributed to the destruction of satiety receptors or integrative systems which process satiety signals. Thus, although animals with VMH lesions increase their food intake when food is readily available they are not hyperphagic when they must work to obtain food [74], suggesting that the mechanisms that control the cessation of feeding are disrupted, not those that control appetite. MAYER [70] has proposed that glucoreceptors in the region of the VMH monitor the rate of glucose utilization and serve as satiety receptors for the short-term regulation of food intake and energy balance. Destruction of these postulated receptors by lesions should result in a reduced satiety effect of increased blood glucose levels after feeding. This suggestion is supported by the observation that rats with VMH lesions have much higher blood glucose levels when feeding terminates [114]. On the other hand, LE MAGNEN [64] has proposed that the release of insulin which results from the lipogenesis following food intake is monitored by the VMH. It is possible, however, that both glucose and insulin levels influence satiety since increased levels of insulin, initiated by feeding [114, 115], could enhance the uptake of glucose by the satiety (gluco) receptors, assuming of course that insulin crosses the blood-brain barrier in the hypothalamus and that the uptake of glucose by the central glucoreceptors is insulin-dependent [71].

The existence of glucoreceptors is supported by evidence from electrophysiological studies [3, 91]. Many of the glucosensitive neurons also respond to a variety of external stimuli [21] which may be the basis for the disruption of feeding by novelty, intense sounds and other emotion eliciting stimuli.

The evidence for central glucoreceptors is inconclusive and consideration has been given to the possibility of peripheral glucoreceptors, perhaps in the liver [106]. According to RUSSEK [106], the evidence for central glucoreceptors is inconclusive and he has postulated that innervated hepatocytes monitor both glucose and amino acids. According to his 'potentiostatic' hypothesis, the membrane potential of the hepatic receptors depends on hepatic glucose outflow so that these receptors would discharge more rapidly when this outflow is greater and the glycogen and protein reserves reduced. In support of this hypothesis, RUSSEK has observed that the average hepatocyte membrane potential is higher in *ad libitum* fed rats than in rats deprived for 24 h.

Feedback from lipid depots may also be involved in the central control of food intake, perhaps for long-term regulation of energy balance and body weight. According to KENNEDY [60] the nervous system is less sensitive, following VMH lesions, to the signals utilized in monitoring fat stores and, as a result, the set-point for the regulation of total fat stores is raised. The static phase of hyperphagia, characterized by reduced food intake and plateauing of body weight, apparently begins when the signals from fat stores become sufficiently high to influence the less sensitive central control systems for food intake and energy balance. The mechanism for the monitoring or feedback of fat stores is not known, although interesting suggestions have been made. HERVEY [52] proposed that a signal is provided by the tracer dilution of steroid hormones, known to accumulate in adipose tissue and depending on the concentration of free fatty acids [10]. BAILE *et al.* [7] have suggested that prostaglandins (PG), whose production in adipose tissue is related to the level of metabolism in such tissue [109], may signal the degree of adiposity. When PG were administered directly to the medial and perifornical region of the hypothalamus prostaglandin E_1 (PGE_1) produced a marked decrease in food intake whereas PGE_2 had little effect [7]. It has been suggested that depleted or overloaded adipose cells might produce different PG, which differentially alter the excitability of 'feeding centers' such as the VMH [106]. Another possibility is that food intake and body weight regulation depend on the ratio of sodium to calcium at some brain site since central infusion of calcium ions in the VMH and sodium ions in the *LH* induces vigorous feeding [86].

The VMH appears to be involved in mediating other factors which influence food intake. Gastric distention is known to reduce ingestive behaviors and since the activity of neurons in the VMH is increased [5, 18] it may result in the inhibition of the *LH* feeding system. The VMH may also be involved in mediating the effects of certain hormones on food intake; food intake and growth are reduced in female rats following puberty as compared to males and it has been proposed that the variations in food intake during the estrous cycle, pregnancy and pseudopregnancy are due to the influence of varying amounts of estrogens on the VMH [130]. WADE and ZUCKER have suggested that 'estrogens directly excite neurons in the VMH, which in turn, inhibit the *LH*' [130, p. 335], thereby reducing food intake. Furthermore, they suggest that estrogens fail to reduce feeding before puberty, not because the VMH is insensitive to estrogen at that time but because its effects are inhibited by growth hormone. 'The decline in growth hormone secretion which appears to occur at about 40 to 50 days of age [35] coincides with the ability of estrogens to modulate food consumption' [129]. Alternatively, the female may gain less because estrogens cause increased activity.

It should be clear from the foregoing that our knowledge of receptors and mechanisms for monitoring energy deficits and the role of the VMH in regulating energy stores is limited. BROBECK [17] has commented on this state of affairs as follows: 'Although these experiments clearly reveal that animals tend to preserve certain relationships between food intake, energy expenditure, and energy reserves, no one knows by what mechanism the body measures the size of the depots and the energy they contain. No kind of sensory cell is known to respond to body weight, body size, or body composition *per se*. Consequently, one may still consider the possibility that the apparent constancy of body composition is not maintained by any physiologic mechanism having this specific function but rather is an incidental result of the operation of mechanism serving other roles in bodily economy' [17, p. 509].

B. The Role of the LH

According to the dual mechanism model the *LH* is the focus of drinking and feeding systems: this region integrates 'hunger' and 'thirst' signals and sends 'command' signals to the midbrain or to other structures that control the motor sequences required for ingestive behaviors [see MOGENSON and HUANG, 77, fig. 8]. Following lesions of the *LH*, animals do not respond to osmotic, hypovolemic or glucoprivic stimuli [25, 27, 120]. STEVENSON [119] has suggested that *LH* lesions which cause adipsia and aphagia might be interrupting the pathways that transmit inputs from rostrally located receptor regions to more caudal integrative systems. There have been reports that receptors which signal water deficits are in the preoptic region [14, 26, 96]. Perhaps the drinking and feeding by electrical stimulation of the *LH* result from the activation of afferent fibers from osmotic, volume, glucose and other receptors [76, 78].

It has been reported that rats [125] as well as cats and monkeys [4] with lesions of the *LH* actively reject food for several days when it is placed in their mouths. This does not occur in animals in which the hypothalamus has been surgically isolated, although they are also aphagic and adipsic [23]. ELLISON has suggested that food rejection following lesions confined to the *LH* may be 'due to the unbalanced inhibiting signals emanating from the medial satiety centre' [23, p. 225]. On the other hand, the rejection of food could result from a reduction in saliva [47] and difficulty in swallowing; rats with *LH* lesions will accept liquid diet, especially if it is highly palatable [122].

Rats do recover ingestive responses following lesions of the lateral hypothalamus [79, 125], paralleling the development of ingestive behaviors at weaning [124], but the food and water intakes are apparently no longer initiated by deficit signals. Drinking is food related (prandial); dry mouth is a signal for drinking in these animals although the deficits to osmotic and hypovolemic signals persist [25]. Feeding, as well as drinking, depends on the sensory characteristics of the food or the liquid and on other external stimuli [57, 59, 97, 135], the recovered animals are considered to be finicky since the intakes are much greater when the food or liquid is palatable and the intakes drop when the sensory qualities of the food or liquid are unpleasant [44].

II. Reassessment of the Role of the Hypothalamus in the Control of Ingestive Behaviors

In recent years several features of the 'textbook view' of the hypothalamus in ingestive behaviors have been reassessed. The role of the lateral hypothalamus is being reconsidered, particularly the hypothesis that it is an integrative region for drinking and feeding and that it makes an essential contribution to ingestive responses [e.g., 76, 131]. The role of the VMH in obesity [88] and in processing depletion and repletion signals [93] is being reconsidered, as well as the widely held view that VMH-lesioned animals have reduced hunger motivation [61]. The relationship of the *LH* and the VMH has become a topic of keen interest and doubts have been expressed about the view that the VMH 'satiety center' inhibits the *LH* [99]. The specificity of the appetite (hunger and thirst) mechanism and of the satiety mechanism has also been questioned [46, 81, 127]. Increasing emphasis has been placed on the variety of conditions that initiate drinking and feeding with the possibility of separate neural mechanisms to subserve them. Nonhomeostatic signals appear to have a more prominent role in animals with more complex brains in which the limbic system and cerebral cortex participate to a greater degree in the control of ingestive and other motivated and regulatory behaviors [77]. Whether or not the hypothalamus is directly involved in the systems which subserve these signals is being investigated. These issues are considered in this section and the next.

It is not surprising in view of the limitations of the lesion and stimulation techniques that controversies have developed about the role of the hypothalamus in ingestive behaviors. The hypothalamus is a small structure concerned with many functions; lesions do not selectively destroy cells or fibers concerned with only feeding and drinking. There are reports that animals with *LH* lesions are hypoactive [37], apathetic and depressed [34] and exhibit motor deficits and disorders of locomotion [83]. MORRISON [83] has stressed that motor deficits are responsible for the changes in ingestive behaviors; lesions produce an 'apraxia of feeding' rather than a motivational deficit. In fact, it has even been reported that *LH* lesions cause increased food motivation [22].

Following lesions of the *LH,* rats are adipsic and aphagic [2], and, although as indicated earlier, they partially recover the control of feeding and drinking [79, 125], they do not respond to osmotic stimuli, hypovolemia or hypoglycemia [25]. This is taken as evidence that the *LH* is an integrative site for 'thirst' and 'hunger' signals [25] and that feeding [73] and drinking [78] elicited by electrical stimulation result from activation of neurons of the integrative mechanisms. It should be recognized, however, that lesion and stimulation techniques, which have been valuable in suggesting that the *LH* has a role in the initiation of feeding and drinking, do not provide proof that it has an integrative function. BROBECK [17] has made this clear by commenting that 'it is not even certain now that mechanisms for feeding originate in the regions where they can be destroyed or stimulated' [17, p. 510]. Lesions could disrupt and stimulation could activate fibers that pass through the *LH* caudally to the midbrain or rostrally to the amygdala or other limbic forebrain structures [for further discussion of this view see 54, 76, 77]. Neither do experiments in which aphagia and adipsia have been observed following surgical isolation of the hypothalamus [23] prove that the hypothalamus is the integrative site for 'hunger' and 'thirst' signals. The deficits could be due to the interruption of essential fiber connections between midbrain and limbic forebrain or basal ganglia.

The reciprocally related discharge rates recorded from neurons in the

LH and the VMH [90, 91] do not mean that these regions interact directly. Since definitive evidence of connections from VMH to *LH* is lacking [99] it is appropriate to consider alternatives to the classical view that the VMH 'satiety center' inhibits the *LH* 'appetite center'. One possibility is that inhibitory signals are projected caudalward from the VMH to the midbrain [43, 105] where they interact with facilitatory signals from the *LH* [77].

Lesions of the VMH also do not selectively disrupt the 'satiety system' since there are changes in emotional behavior [132] and gonadal function [129]. GROSSMAN [46] has proposed that VMH lesions produce a general disinhibition so that animals over-respond to a variety of stimuli (emotion provoking as well as food-related) and that the hyperphagia is not due to the destruction of a satiety system. Animals with VMH lesions, like obese humans [103] may show increased responsiveness to various external signals. Electrical stimulation of the *LH* and VMH also does not selectively activate neurons concerned with feeding and drinking or with satiety [in fact, it has been reported recently that low intensity stimulation of the VMH may elicit feeding, 94]. It has been suggested that feeding and drinking elicited by hypothalamic stimulation is due to a general activation of the animals, the behavior depending on the presence of food, water or other goal objects [127]. Furthermore, the cessation of ingestive behaviors produced by stimulation of the VMH has been attributed to the aversiveness of the stimulation [62], activation of a generalized inhibitory system [82] and activation of searching behavior which is incompatible with ingestive responses [8].

The multiple effects of lesions and stimulation of the hypothalamus appear to be due to the relatively crude nature of these techniques. However, this is not crucial evidence against the view that specific neural systems subserve feeding and drinking. With more discrete stimulation feeding only and drinking only have been elicited by central stimulation [56]. With discrete lesions adipsia has been observed in the absence of aphagia [79]. Furthermore, RODGERS *et al.* [102] have shown that there is a motivational deficit which persists after the initial motor deficit produced by *LH* lesions. The report that rats recover much faster from *LH* lesions when they are previously starved to lower their body weight [98] is also inconsistent with the view that the deficit in ingestive behaviors is due to a motor deficit. For further discussions of the specificity issue see MOGENSON [76], MOGENSON and HUANG [77], ROBERTS [100] and VALENSTEIN *et al.* [127].

Another recent development is the recognition of the importance of the forebrain and midbrain limbic systems, and of the cerebral cortex in the control of food and water intake [75, 77]. Many of the effects of lesions and stimulation on ingestive behaviors have been small and variable but in some cases dramatic changes in food and water intakes have been produced. For example, lesions of the septum cause hyperdipsia [49, 133], lesions of the amygdala produce both aphagia and hyperphagia [34], lesions of the midbrain tegmentum cause aphagia [66, 95, 113] and electrical stimulation of this region has been reported to elicit drinking [101] and feeding [134].

Limbic forebrain structures are interconnected with one another, with the hypothalamus and midbrain [87]. Because of the complexity of connections and relationships it has been very difficult to identify the neural systems and their normal role in the control of drinking and feeding. The usual view has been that limbic forebrain structures have modulatory influences on the control mechanisms represented in the hypothalamus and in the lower brainstem. The first evidence of modulatory influences came from studies of autonomic and endocrine responses. Stimulation of the amygdala, septum and other limbic forebrain structures both facilitated and inhibited heart rate, blood pressure, gastrointestinal motility, etc. [58]. Since the effects often varied with the state of the animal and since lesions of these structures typically had little or no effect on these responses it was suggested that they had a modulatory rather than an essential role [40]. Ingestive behaviors are influenced by fear and anxiety [16] as well as by the sexual state of the animal [20] and the taste and odor of food or fluid [57, 64]. Limbic structures have a role in the processing of these signals and they exert modulatory influences on hypothalamic sites concerned with ingestive behaviors [72, 111]. The hypothalamus, in turn, is thought to send 'command signals' to the lower brainstem [39].

Limbic forebrain structures and cerebral cortex appear to have a more important role and the hypothalamus a less critical one in animals in which these structures are more highly developed; there has been encephalization of neural systems controlling ingestive behaviors [123]. Thus, whereas surgical isolation of the hypothalamus causes adipsia and aphagia in rats, drinking and feeding were observed following such procedures in cats, suggesting that 'while the hypothalamus may be essential for the precise regulation of food intake, it should not be considered as an essential centre for appetitive behavior, at least in the cat' [24, p. 18].

Furthermore, drinking and feeding are more readily elicited by electrical stimulation of limbic forebrain structures of monkeys than of rats and cats [101]. SHARPE and MYERS [108] have suggested that 'the neural control system for intake in the monkey may be principally in the telencephalon rather than in the diencephalon' [108, p. 306]. In man, cognitive factors are even more important [107] and ingestive behaviors are less dependent on deficit signals. Several authors [31, 77] have emphasized that neural systems have evolved in higher species, presumably involving the limbic system and cerebral cortex, which enable the animal to make 'anticipatory' ingestive responses prior to the occurrence of water and energy deficits.

III. Some Current Views about the Role of the Hypothalamus in the Control of Ingestive Behaviors

Many of the issues and problems discussed in the previous section are under active investigation. It is not possible, therefore, to provide a definitive statement and model about the role of the hypothalamus in ingestive behaviors. Rather an attempt will be made in this final section to bring together some of the pertinent evidence about drinking and feeding in relation to some of the relevant anatomical pathways that have been demonstrated recently with histological, histochemical and histofluorescence techniques.

A. Chemical and Anatomical Evidence

The sites of stimulation or lesions of the hypothalamus which initiate or disrupt ingestive behaviors appear to overlap cholinergic [110] and catecholaminergic [36] pathways which have been demonstrated during the last few years using histochemical and histofluorescence techniques. Recently attempts have been made to relate the changes in food and water intakes from stimulation and lesioning of the hypothalamus, and other brain sites, to the activation or destruction of these pathways. Such studies have emphasized the importance of pathways that connect midbrain and forebrain limbic structures and have provided further reasons to reassess the role of the hypothalamus in the initiation of ingestive behavior.

Several lines of evidence suggest that the noradrenergic fibers which project from the midbrain along the medial forebrain bundle (MFB) to the preoptic region, amygdala, hippocampus and other forebrain regions [pathways designated A5 and A7 by FUXE *et al.,* 36] are associated with feeding. Since the endogeneous concentration [50] and uptake [136] of norepinephrine (NE) in the telencephalon are reduced after *LH*-MFB lesions the deficit in ingestive behavior that results may be due to the degeneration of noradrenergic fibers. STEIN and WISE [117] have demonstrated that stimulation of the MFB noradrenergic pathway causes the release of NE in the amygdala and has implicated it both in the reward associated with self-stimulation as well as in feeding. BERGER *et al.* [9] have reported that the infusion of NE into the ventricles of rats following lesions of the *LH* reversed the aphagia presumably by acting on the intact postsynaptic receptors of the partially destroyed noradrenergic fibers. The recovery of feeding typically observed following *LH* lesions, they suggest, is due to the supersensitivity of these partially denervated adrenergic fibers. Subsequently, in support of this proposal, GLICK *et al.* [38] have facilitated recovery of aphagia from *LH* lesions by pretreating rats with α-methyl-*p*-tyrosine. This compound, a potent inhibitor of tyrosine hydroxylase, blocks the synthesis of NE and thereby, according to these authors, causes a pharmacologically produced denervation supersensitivity of noradrenergic neurons. It has also been reported that rats recover more quickly following *LH* lesions if they are starved prior to lesioning; in some cases the animals were hyperphagic, not aphagic [98]. These authors suggested that the *LH* lesions had lowered the set-point for body weight regulation. An alternative interpretation is that starvation caused supersensitivity of noradrenergic neurons some of which were destroyed by the lesions.

Several years ago, GROSSMAN [45] reported that the local application of NE to the lateral hypothalamus elicits feeding. Subsequently, it was reported [e.g., 15] that low doses of NE elicited feeding and high doses suppressed feeding. According to LEIBOWITZ [63] the α-adrenergic 'hunger' system is activated at low doses and the β-adrenergic 'satiety' system at high doses. MARGULES [67] made an alternative proposal; the α-adrenergic system is concerned with satiety which follows feeding to homeostatic deficits whereas the β-adrenergic system is the satiety system which responds to unpleasant taste and mediates the finicky behavior of animals. Subsequently, MARGULES *et al.* [68] reported that NE administered intracerebrally during the light period of the day elicits feeding whereas during the dark period it suppresses feeding. They relate these findings to the circadian changes of NE in the hypothalamus and suggest that NE only elicits feeding when its levels are low during the light period.

Additional evidence implicating the noradrenergic pathway passing rostrally through the *LH* in the control of feeding has come from studies utilizing 6-hydroxydopamine (6-OH-DA), a compound that destroys catecholaminergic fibers. The application of 6-OH-DA via chronically placed cannulas causes aphagia, presumably because of destruction of the adrenergic fibers [126, experimental group 2]. On the other hand, by using small doses of 6-OH-DA it is possible initially to elicit feeding because of the release of NE to act at receptor sites [33].

Some investigators have noted that lesions which produce aphagia and adipsia are further lateral than the classical *LH* site of ANAND and BROBECK [2], bordering on or including the internal capsule and subthalamus [41, 79, 80]. Recent evidence suggests that the critical site is the nigrostriatal pathway which contains dopaminergic fibers projecting to the corpus striatum [pathways A8, A9, A10 of FUXE *et al.*, 36]. UNGERSTEDT [126] and FIBIGER *et al.* [28] reported aphagia and adipsia following destruction of this pathway by the local application of 6-OH-DA. OLTMANS and HARVEY [89] compared the effects of electrolytic lesions of the nigrostriatal pathway with lesions of the MFB-*LH* and observed that the former caused more severe aphagia and adipsia. They also reported that lesions of the nigrostriatal pathway caused a greater reduction of catecholamines in the telencephalon.

It appears, however, that the deficits from lesions of the nigrostriatal pathway are not the same as those following lesions of the MFB-*LH*. UNGERSTEDT [126] suggested that there may be a general motor deficit from nigrostriatal lesions and was reluctant to conclude that the lesions caused specific feeding and drinking deficits. Earlier, MORGANE [80] had observed striking differences in the deficits from MFB-*LH* lesions and those placed more laterally impinging on the internal capsule and presumably destroying fibers of the nigrostriatal pathway (or perhaps a descending pallidofugal pathway as suggested by MORGANE). Recovery of food and water intakes from MFB-*LH* lesions occurred much more quickly and MORGANE concluded that the deficit was in the motivation to feed and drink. There was a much more drastic aphagia and adipsia and weight loss with the far-lateral lesions which MORGANE called ‘metabolic decay’.

FISHER and COURY [29] were the first investigators to suggest that drinking was subserved by cholinergic pathways in the hypothalamus and limbic system. Subsequently, SHUTE and LEWIS [110] mapped with histochemical techniques cholinergic pathways which interconnect some of the structures implicated by FISHER and COURY. However, it is uncertain whether these pathways are involved in homeostatic drinking (see the next section).

B. An Attempted Synthesis of Recent Evidence

Some of the structures and pathways involved in feeding are shown in figure 2. As indicated above an ascending noradrenergic pathway which projects through the hypothalamus to the preoptic area, amygdala and other limbic forebrain structures has been implicated in feeding behavior. These ascending noradrenergic fibers are believed to inhibit limbic forebrain neurons which project, and provide an excitatory input, to the VMH [51, 116]. It has been suggested that lesions of the *LH* which destroy the ascending noradrenergic fibers cause aphagia because the VMH receives a stronger excitatory input and, in turn, exerts a greater inhibitory effect on feeding. On the other hand, stimulation of the *LH* activates these ascending noradrenergic fibers so that the excitatory input to the VMH is reduced and feeding occurs because of the reduced inhibitory effect of VMH.

It is not known whether the noradrenergic fibers which project rostrally through the *LH* transmit signals from receptors that monitor the depletion or repletion of blood glucose, lipids or amino acids. Rats recovered from *LH* lesions do not feed in response to hypoglycemia [25] but this is not necessarily due to the destruction of these noradrenergic fibers. An alternative suggestion has been made that the ascending noradrenergic fibers subserve feeding controlled by incentive or reward stimuli rather than in response to homeostatic deficit signals [54, 116]. Such a proposal does not imply that the ascending noradrenergic pathway is of little importance; there has been increasing emphasis on neural systems that initiate ingestive responses which anticipate homeostatic deficits [33, 64, 77].

The VMH may also receive direct inputs initiated by signals reflecting the depletion and repletion of energy. Lesions of the VMH cause hyperphagia and this has been attributed to the disruption of the response

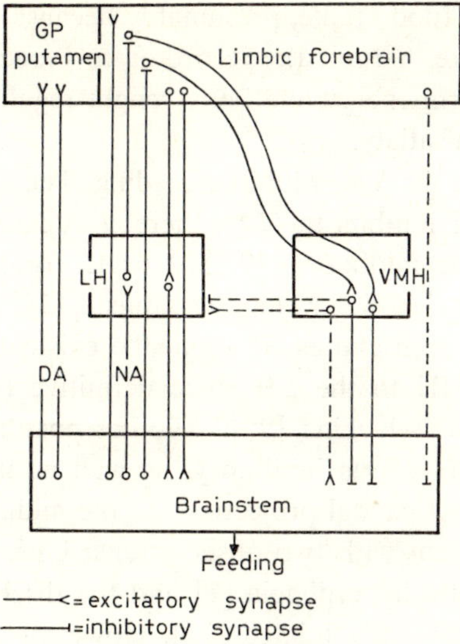

Fig. 2. The role of the hypothalamus in ingestive behaviors. Noradrenergic fibers (NA) identified with histofluorescence techniques, project from the midbrain and lower brainstem to hypothalamic and limbic forebrain structures. Ingestive responses are disrupted when these fibers are lesioned and feeding results from electrical and chemical stimulation of these fibers. Neurons inhibited by the NA fibers project to the VMH so that neurons of the VMH are indirectly inhibited by the NA fibers passing through the *LH*. When the VMH neurons are inhibited, their own inhibitory effects on feeding (either via the *LH* or more likely by caudal projections to the midbrain) are reduced and the animal eats. There may also be excitatory fibers from the VMH [93] to the *LH* or midbrain and excitatory projections from the *LH* to the midbrain and possibly both excitatory and inhibitory fibers from limbic forebrain to midbrain (the possible excitatory projections are shown by broken lines). Lesions of dopaminergic fibers (DA) of the nigrostriatal pathway also cause aphagia and adipsia but since stimulation of this pathway does not elicit feeding and drinking it is assumed that it is not involved in the initiation of ingestive behavior but rather with motor control. GP = globus pallidus.

to repletion and postingestional signals which are normally monitored by the VMH to inhibit feeding (see section IA). For example, the VMH is activated by gastric distention, an important postingestional signal which contributes to the cessation of drinking and feeding [5]. Panksepp [93] has suggested recently that the VMH also contains neurons (shown in fig. 2) that exert excitatory effects on feeding, raising the possibility that depletion signals are also mediated via the VMH (this explains why VMH lesions sometimes have no effect on food intake, presumably because both excitatory and inhibitory neurons are disrupted). According to PANKSEPP the VMH is concerned with long-term energy and body weight regulation whereas the *LH* mediates short-term satiety.

It is not clear by what pathways the VMH inhibits feeding. The traditional view has been that the VMH inhibits the *LH,* thereby attenuating command signals for feeding to the midbrain [39, 56] or to forebrain structures such as the amygdala [104] or hippocampus [128]. As indicated earlier, reservations have been expressed about the existence of inhibitory projections from the VMH to the *LH* since definitive histological evidence for such a pathway is lacking [99]. Another possibility, shown in figure 2, is that the VMH exerts inhibitory, as well as facilitatory, effects on feeding by means of caudal projections to the midbrain. This proposal seems to be inconsistent with two observations: knife cuts between the VMH and the *LH* cause hyperphagia [1] and the discharge of VMH and *LH* neurons are reciprocally related [91]. However, knife cuts between VMH and *LH* do not necessarily produce hyperphagia [69]. Also, since the effective knife cuts are at the rostral level of the VMH [42], the hyperphagia could be the result of disrupting the facilitatory input from the amygdala or preoptic region to the VMH rather than to destruction of fibers projecting from VMH to *LH.* Furthermore, the reciprocal discharge of VMH and *LH* neurons could occur because they interact with a third structure such as the midbrain or amygdala.

Lesions of the nigrostriatal pathway (dopaminergic fibers in fig. 2) also cause aphagia but it appears that the deficit is not the same as that produced by *LH* lesions which destroy the noradrenergic fibers of the MFB [80, 89]. It was suggested earlier that lesions which destroy the dopaminergic fibers of the nigrostriatal pathway do not produce a motivational deficit specific to feeding and drinking but rather a more general motor deficit [126]. The nigrostriatal pathway may be important in the integration of the various reflexes utilized in ingestive behaviors after the 'command signals' have been initiated by the hypothalamus and limbic system. In other words, the noradrenergic pathway through the *LH* may be concerned with the initiation of feeding whereas the nigrostriatal pathway may be concerned with the motor control of the ingestive responses.

It is more difficult to relate drinking systems to the new 'chemical anatomy' of the brain, although a few suggestions may be made. The classical observation that drinking is elicited by local cholinergic stimulation (e.g., carbachol) of the hypothalamus [45] is consistent, of course, with the demonstration of cholinergic pathways through the hypothalamus [110]. The regions to which this 'ascending cholinergic reticular system' [110] project, the preoptic region, diagonal band, septum, amygdala, cingulate region and anterior thalamus, are also sites where carbachol elicits drinking [30]. It has been assumed that carbachol acts at cholinergic synapses but it may also activate cholinergic fibers directly [65]. Furthermore, carbachol may reach the ventricles and be carried by the cerebrospinal fluid to a distant site, such as the subfornical organ [112].

The drinking subserved by cholinergic systems may not be the same as that induced by water deprivation or the 'natural' deficit signals for thirst. This suggestion is based on the observation that the administration of atropine sulphate, either systemically or directly to the preoptic region or *LH,* blocks carbachol-induced drinking but not drinking induced by water deprivation, systemic or intracranial dehydration, isomotic intravascular depletion or angiotensin [13]. Perhaps the drinking subserved by cholinergic pathways is initiated not by water deficits but by some nonhomeostatic signal.

There is some indication that catecholaminergic neurons are also involved in the initiation of drinking. This was suggested by the observation that the application of 6-OH-DA, a compound which destroys catecholamine neurons, to the preoptic region blocks drinking induced by administering angiotensin to the same brain site, while the response to carbachol remains [33]. Since the angiotensin-induced drinking is blocked by haloperidol, a catecholaminergic antagonist, but not by α- and β-adrenergic antagonists [33] it appears that dopaminergic fibers subserve drinking initiated by angiotensin. Also drinking is elicited by the injection of dopamine into the ventricles [32] and preoptic region [Wei and Mogenson, unpublished observations] but the dose levels are high.

Black *et al.* [12] have recorded from single neurons in the lateral hypothalamus of the urethane-anesthetized rat while administering angiotensin to the preoptic region, previously shown in chronic tests to induce drinking. The discharge rates of selected units were increased suggesting that the *LH* is involved in the processing of the neural signals initiated by angiotensin. When the *LH* was destroyed unilaterally by electrolytic lesions the angiotensin-induced drinking was greatly attenuated [11]. Whether the lesions destroy the integrative site for these 'thirst signals' or destroy the fibers projecting through the lateral hypothalamus to some other integrative site remains to be determined.

Summary

Views of the role of the hypothalamus in the control of ingestive behaviors have changed in recent years because of experimental results using newer techniques. Initially lesion and stimulation studies had suggested that the *LH* and VMH are involved in appetite and satiety, respectively. The notion that the *LH* is an integrative site for thirst and hunger signals has been a popular one supported by the observations that animals partially recovered from the adipsia and aphagia of *LH* lesions do not respond to hypovolemia, hyperosmolarity or hypoglycemia. However, lesion and stimulation techniques have the limitation that they do not permit definitive localization of ingestive behaviors to sites where they can be disrupted or initiated. Also the effects of these techniques are not confined to nerve cells or fibers specifically concerned with feeding and drinking.

Histochemical and histofluorescence techniques have been used to demonstrate cholinergic and catecholaminergic pathways which originate in the midbrain, pass through the hypothalamus to limbic forebrain structures, basal ganglia and cerebral cortex. Recent studies have indicated the aphagia and adipsia following hypothalamic lesions might be due to the destruction of these pathways. Drinking and feeding initiated by the intracerebral application of carbachol and noradrenalin, respectively, may be due to the activation of synapses in these pathways; the ingestive responses elicited by electrical stimulation might also be due to the activation of these pathways.

Pathways also project from limbic forebrain structures (e.g., amygdala, septum) to the hypothalamus and midbrain. Although they have not been characterized by histochemical techniques it seems likely that they play some role in the control of ingestive behaviors, since lesions and stimulation of limbic structures alter feeding and drinking, and electrophysiological studies have shown that these structures have modulatory influences on the hypothalamus and midbrain.

Clearly the hypothalamus can no longer be considered the center for hunger and thirst. On the other hand, it is probably not appropriate to propose that the hypothalamus merely contains fibers of passage interconnecting structures of the midbrain and limbic forebrain and basal ganglia which subserve ingestive responses. Feeding and drinking are only one aspect of energy and water balance; the other aspect is the utilization and dissipation of energy and the utilization and excretion of water and in these functions the hypothalamus is involved through its influences on the pituitary and endocrine system, and on the autonomic nervous system. The hypothalamus has a vital role in the functional coupling of autonomic, endocrine and behavioral responses for all aspects of homeostatic regulation including energy and water balance. The behavioral responses of feeding and drinking involve complex interactions with the external environment and the initiation of these responses requires interactions of hypothalamus with other regions of the brain including limbic forebrain and midbrain and cerebral cortex. For animals with greater encephalization of brain function, in which ingestive behaviors are less dependent on deficit signals, it appears that limbic forebrain structures and cerebral cortex play a relatively more important role in the initiation of feeding and drinking.

Note Added in Proof

Experiments in which 6-hydroxydopamine, a neurotoxin which can selectively destroy catecholaminergic neurons, has been administered into the ventricles or directly to the hypothalamus have clearly implicated catecholaminergic neurons in the control of feeding and drinking behaviors [126, 113a]. The behavioral deficits are similar to those following the classical electrolytic lesions of the lateral hypothalamus [28, 68a, 137] and the importance of the nigrostriatal dopaminergic pathway has been stressed. There is some evidence that catecholaminergic neurons are involved in the recovery from aphagia and adipsia following lesions produced electrolytically or with 6-hydroxydopamine [38, 137] but whether this is due to supersensitivity of postsynaptic receptors [126], to collateral sprouting of catecholaminergic axons [59a] or some other mechanism is not known. Furthermore, it is not certain that the lesions which cause aphagia and adipsia have been confined only to dopaminergic neurons and the relative contributions of the dorsal noradrenergic, ventral noradrenergic, nigrostriatal dopaminergic and mesolimbic dopaminergic pathways remain to be determined. Recently it was observed that lesions of the ventral noradrenergic pathway cause hyperphagia [6a] and that electrical stimulation of the dorsal noradrenergic pathway elicit feeding [Cioé and Mogenson, unpublished observations]. It is possible that the different catecholaminergic pathways make distinctive contributions to the control of food and water intake and energy and water balance.

Acknowledgements

The author wishes to acknowledge his appreciation to S. Black, B. Box, J. Cioe, M. Evered, A. Faiers, J. Kucharczyk, A. Mok, A. G. Phillips and R. Weick for constructive criticisms of earlier versions of this article, to Miss Anne Baxter who typed the several drafts of the manuscript and to Miss B. Box and Miss B. Woodside who assisted with the references and the illustration. The author's research is supported by grants from the Medical Research Council and the National Research Council.

References

1 Albert, D. J. and Storlien, L. H.: Hyperphagia in rats with cuts between the ventromedial and lateral hypothalamus. Science *165:* 599–600 (1969).

2 Anand, B. K. and Brobeck, J. R.: Hypothalamic control of food intake in rats and cats. Yale J. Biol. Med. *24:* 123–140 (1951).

3 Anand, B. K.; Chhina, G. S.; Sharma, K. N.; Dua, S., and Singh, B.: Activity of single neurons in the hypothalamic feeding center. Effect of glucose. Amer. J. Physiol. *207:* 1146–1154 (1964).

4 ANAND, B. K.; DUA, S., and SCHOENBERG, K.: Hypothalamic control of food intake in cats and monkeys. J. Physiol., Lond. *127:* 143–152 (1955).

5 ANAND, B. K. and PILLAI, R. V.: Activity of single neurons in the hypothalamic feeding centre. Effect of gastric distension. J. Physiol., Lond. *192:* 63–77 (1967).

6 ANDERSSON, B. and WYRWICKA, W.: The elicitation of a drinking motor conditioned reaction by electrical stimulation of the 'drinking area' in the goat. Acta physiol. scand. *41:* 194–198 (1957).

6a AHLSKOG, J. E. and HOEBEL, B. G.: Overeating and obesity from damage to a noradrenergic system in the brain. Science *182:* 166–169 (1973).

7 BAILE, C. A.; BEAN, S. M.; SIMPSON, C. W., and JACOBS, H. L.: Feeding effects of hypothalamic injections of prostaglandins. Fed. Proc. *30:* 375 (1971).

8 BALL, G. G.: Self-stimulation in the ventromedial hypothalamus. Science *178:* 72–73 (1972).

9 BERGER, B. C.; WISE, C. D., and STEIN, L.: Norepinephrine. Reversal of anorexia in rats with lateral hypothalamic damage. Science *172:* 281–284 (1971).

10 BIZZI, A.; TACCOMI, A. M., and GARATTINI, S.: Relationship between lipolysis and storage of corticosterone in adipose tissue. Biochem. Pharmacol. *21:* 999–1008 (1972).

11 BLACK, S. L.; KUCHARCZYK, J., and MOGENSON, G. J.: Lesions of the lateral hypothalamus disrupt drinking induced by the preoptic injection of angiotensin. Paper presented at Canadian Psychological Association, Victoria 1973.

12 BLACK, S. L.; MOK, A.; COPE, D., and MOGENSON, G. J.: Activation of lateral hypothalamic neurons by the injection of angiotensin into the preoptic area of the rat. Abstract 930. Fed. Proc. *32:* (1973).

13 BLASS, E. M. and CHAPMAN, H. W.: An evaluation of the contribution of cholinergic mechanism to thirst. Physiol. Behav. *7:* 679–686 (1971).

14 BLASS, E. M. and EPSTEIN, A. N.: A lateral preoptic osmosensitive zone for thirst in the rat. J. comp. physiol. Psychol. *76:* 378–394 (1971).

15 BOOTH, D. A.: Mechanism of action of norepinephrine in eliciting an eating response on injection into the rat hypothalamus. J. Pharmacol. exp. Ther. *160:* 336–348 (1968).

16 BRADY, J. V.: Emotional behavior and the nervous system. Trans. N. Y. Acad. Sci. *18:* 601–612 (1956).

17 BROBECK, J. R.: in MOUNTCASTLE Medical physiology, vol. 1, pp. 498–519 (Mosby, St. Louis 1968).

18 BROBECK, J. R.; LARSSON, S., and REYES, E.: A study of the electrical activity of the hypothalamic feeding mechanism. J. Physiol., Lond. *132:* 358–364 (1956).

19 BROBECK, J. R.; TEPPERMAN, J., and LONG, C. N. H.: Experimental hypothalamic hyperphagia in the albino rat. Yale J. Biol. Med. *15:* 831–853 (1943).

20 BROOKS, C. MCC.: The relative importance of changes in activity in the development of experimentally produced obesity in the rat. Amer. J. Physiol. *147:* 708–716 (1946).

21 CAMPBELL, J. F.; BINDRA, D.; KREBS, H., and FERENCHAK, R. P.: Responses of single units of the hypothalamic ventromedial nucleus to environmental stimuli. Physiol. Behav. *4:* 183–187 (1969).

22 DEVENPORT, L. D. and BALAGURA, S.: Lateral hypothalamus. Reevaluation of function in motivated feeding behavior. Science *172:* 744–746 (1971).

23 ELLISON, G. D.: Appetite behavior in rats after circumsection of the hypothalamus. Physiol. Behav. *3:* 221–226 (1968).

24 ELLISON, G. D. and FLYNN, J. P.: Organized aggressive behavior in cats after surgical isolation of the hypothalamus. Arch. Ital. Biol. *106:* 1–20 (1968).

25 EPSTEIN, A. N.: in STELLAR and SPRAGUE Progress in physiological psychology, vol. 4, pp. 263–317 (Academic Press, New York 1971).

26 EPSTEIN, A. N.; FITZSIMONS, J. T., and ROLLS, B. J.: Drinking induced by injection of angiotensin into the brain of the rat. J. Physiol., Lond. *210:* 457–474 (1970).

27 EPSTEIN, A. N. and TEITELBAUM, P.: in WAYNER Thirst, pp. 395–406 (Pergamon Press, London 1964).

28 FIBIGER, H. C.; ZIS, A., and MCGEER, E. G.: Feeding and drinking deficits after 6-hydroxydopamine administration in the rat. Similarities to the lateral hypothalamic syndrome. Brain Res. *55:* 135–148 (1973).

29 FISHER, A. E. and COURY, J. N.: Cholinergic tracing of a central neural circuit underlying the thirst drive. Science *138:* 691–693 (1962).

30 FISHER, A. E. and COURY, J. N.: in WAYNER Thirst, pp. 515–531 (Pergamon Press, New York 1964).

31 FITZSIMONS, J. T.: Thirst. Physiol. Rev. *52:* 468–561 (1972).

32 FITZSIMONS, J. T.: in MOGENSON and CALARESU Central control of physiological regulations and behavior (University of Toronto Press, Toronto 1973).

33 FITZSIMONS, J. T. and SETLER, P.: Catecholaminergic mechanisms in angiotensin-induced drinking. J. Physiol., Lond. *218:* 43–44 (1971).

34 FONBERG, E.: The role of the hypothalamus and amygdala in food intake, alimentary motivation and emotional reactions. Act. biol. exp., Warszawa *29:* 335–358 (1969).

35 FROHMAN, L. A. and BERNARDIS, L. L.: Growth hormone and insulin levels in weanling rats with ventromedial hypothalamic lesions. Endocrinology *82:* 1125–1132 (1968).

36 FUXE, K.; HOKFELT, T., and UNGERSTEDT, U.: Morphological and functional aspects of central monoamine neurons. Int. Rev. Neurobiol. *13:* 93–126 (1970).

37 GLADFELTER, W. E. and BROBECK, J. R.: Decreased spontaneous locomotor activity in the rat induced by hypothalamic lesions. Amer. J. Physiol. *203:* 811–817 (1962).

38 GLICK, S. D.; GREENSTEIN, S., and ZIMMERBERG, B.: Facilitation of recovery by α-methyl-*p*-tyrosine after lateral hypothalamic damage. Science *177:* 534–535 (1972).

39 GLICKMAN, S. E. and SCHIFF, B. B.: A biological theory of reinforcement. Psychol. Rev. *74:* 81–109 (1967).

40 GLOOR, P.: in FIELD, MAGOUN and HALL Handbook of physiology, vol. II (1), pp. 1395–1420 (Williams & Wilkins, Baltimore 1960).

41 GOLD, R. M.: Aphagia and adipsia following unilateral and bilaterally asymmetrical lesions in rats. Physiol. Behav. *2:* 211–220 (1967).

42 GOLD, R. M.: Hypothalamic hyperphagia produced by parasagittal knife cuts. Physiol. Behav. *5:* 23–25 (1970).

43 GOLD, R. M.; QUACKENBUSH, P. M., and KAPATOS, G.: Obesity following combination of rostrolateral to VMH cut and contralateral mamillary area lesion. J. comp. physiol. Psychol. *79:* 210–218 (1972).

44 GRAFF, H. and STELLAR, E.: Hyperphagia, obesity and finickiness. J. comp. physiol. Psychol. *55:* 418–424 (1962).

45 GROSSMAN, S. P.: Direct adrenergic and cholinergic stimulation of hypothalamic mechanisms. Amer. J. Physiol. *202:* 872–882 (1962).

46 GROSSMAN, S. P.: The VMH. A center for affective reactions, satiety, or both? Physiol. Behav. *1:* 1–10 (1966).

47 HAINSWORTH, F. R. and EPSTEIN, A. N.: Severe impairment of heat induced saliva-spreading in rats recovered from lateral hypothalamic lesions. Science *153:* 1255–1257 (1966).

48 HAN, P. W.: Energy metabolism of tube-fed hypophysectionized rats bearing hypothalamic lesions. Amer. J. Physiol. *215:* 1343–1350 (1968).

49 HARVEY, J. A. and HUNT, H. F.: Effect of septal lesions on thirst in the rat as indicated by water consumption and operant responding for water and reward. J. comp. physiol. Psychol. *59:* 49–56 (1965).

50 HELLER, A. and MOORE, R. Y.: Effect of central nervous system lesions on brain monamines in the rat. J. Pharmacol. exp. Ther. *150:* 1–9 (1965).

51 HERBERG, L. J. and FRANKLIN, K. B. J.: Adrenergic feeding. Its blockade or reversal by posterior VMH lesions; and a new hypothesis. Physiol. Behav. *8:* 1029–1034 (1972).

52 HERVEY, G. R.: Regulation of energy balance. Nature, Lond. *223:* 629–631 (1969).

53 HOEBEL, B. G.: Feeding and self-stimulation. Ann. N. Y. Acad. Sci. *157:* 758–778 (1969).

54 HOEBEL, B. G.: Feeding. Neural control of intake. Annu. Rev. Physiol. *33:* 533–568 (1971).

55 HOEBEL, B. G. and TEITELBAUM, P.: Weight regulation in normal and hypothalamic hyperphagic rats. J. comp. physiol. Psychol. *61:* 189–193 (1966).

56 HUANG, Y. H. and MOGENSON, G. J.: Neural pathways mediating drinking and feeding in rats. Exp. Neurol. *37:* 269–286 (1972).

57 JACOBS, H. L. and SHARMA, K. N.: Taste versus calories. Sensory and metabolic signals in the control of food intake. Ann. N. Y. Acad. Sci. *157:* 1084–1125 (1969).

58 KAADA, B. R.: Somatomotor, autonomic and electrocorticographic responses to electrical stimulation of 'rhinencephalic' and other structures in primates, cat and dog. Acta physiol. scand. *24:* suppl. 83, pp. 1–285 (1951).

59 KARE, M. R.: in KARE and HALPERN Comparative aspects of the sense of taste. The physiological and behavioral aspects of taste, pp. 6–15 (University of Chicago Press, Chicago 1961).

59a Katzman, R.; Bjorklund, A.; Owman, C.; Stenevi, U., and West, K.: Evidence for regenerative axon sprouting of central catecholamine neurons in the rat mesencephalon following electrolytic lesions. Brain Res. *25:* 579–596 (1971).

60 Kennedy, G. C.: The role of depot fat in the hypothalamic control of food intake in the rat. Proc. roy. Soc. B *140:* 578–592 (1953).

61 Kent, M. A. and Peters, R. H.: Effects of ventromedial hypothalamic lesions on hunger-motivated behavior in rats. J. comp. physiol. Psychol. *83:* 92–97 (1973).

62 Krasne, F. B.: General disruption resulting from electrical stimulation of ventromedial hypothalamus. Science *138:* 822–823 (1962).

63 Leibowitz, S.: A hypothalamic beta-adrenergic 'satiety' system antagonizes an alpha-adrenergic 'hunger' system in the rat. Nature, Lond. *226:* 963–964 (1970).

64 Le Magnen, J.: in Stellar and Sprague Progress in physiological psychology, vol. 4, pp. 203–261 (Academic Press, New York 1971).

65 Levitt, R. A. and O'Hearn, J. Y.: Drinking elicited by cholinergic stimulation of CNS fibers. Physiol. Behav. *8:* 641–644 (1972).

66 Lyon, M.; Halpern, N. M., and Mintz, E. Y.: The significance of the mesencephalon for coordinated feeding behavior. Acta neurol. scand. *24:* 323–346 (1968).

67 Margules, D. L.: Alpha- and beta-adrenergic receptors in perifornical hypothalamus for the suppression of feeding behavior by satiety and taste. Fed. Proc. *29:* 485 (1970).

68 Margules, D. L.; Lewis, M. J.; Dragovich, J. A., and Margules, A. S.: Hypothalamic norepinephrine. Circadian rhythms and the control of feeding behavior. Science *178:* 640–643 (1972).

68a Marshall, J. F. and Teitelbaum, P.: A comparison of the eating in response to hypothermic and glucoprivic challenges after nigral 6-hydroxydopamine and lateral hypothalamic electrolytic lesions in rats. Brain Res. *55:* 229–233 (1973).

69 Maul, G.: Ventromedial hypothalamus may not directly inhibit the lateral hypothalamus. Paper presented at Eastern Psychological Association, Washington 1973.

70 Mayer, J.: Regulation of energy intake and body weight. The glucostatic theory and the lipostatic hypothesis. Ann. N. Y. Acad. Sci. *63:* 15–43 (1955).

71 Mayer, J. and Thomas, D. W.: Regulation of food intake and obesity. Science *156:* 328–337 (1967).

72 Miller, J. J. and Mogenson, G. J.: Effect of septal stimulation on lateral hypothalamic unit activity in the rat. Brain Res. *32:* 125–142 (1971).

73 Miller, N. E.: Motivational effects of brain stimulation and drugs. Fed. Proc. *19:* 846–854 (1960).

74 Miller, N. E.; Bailey, C. J., and Stevenson, J. A. F.: Decreased 'hunger' but increased food intake resulting from hypothalamic lesions. Science *112:* 256–259 (1950).

75 MILNER, P. M.: Physiological psychology (Holt, Rinehart & Winston, New York 1970).

76 MOGENSON, G. J.: in STELLAR, EPSTEIN and KISSILEFF The neuropsychology of thirst, pp. 119–142 (Winston, New York 1973).

77 MOGENSON, G. J. and HUANG, Y. H.: in KERKUT and PHILLIS Progress in neurobiology, vol. 1 (1), pp. 53–83 (Pergamon Press, London 1973).

78 MOGENSON, G. J. and STEVENSON, J. A. F.: Drinking and self-stimulation with electrical stimulation of the lateral hypothalamus. Physiol. Behav. *1:* 251–254 (1966).

79 MONTEMURRO, D. G. and STEVENSON, J. A. F.: Adipsia produced by hypothalamic lesions in the rat. Canad. J. Biochem. *35:* 31–37 (1957).

80 MORGANE, P. J.: Medial forebrain bundle and 'feeding centers' of the hypothalamus. J. comp. Neurol. *117:* 1–26 (1961).

81 MORGANE, P. J.: The function of the limbic and rhinic forebrain-limbic midbrain systems and reticular formation in the regulation of food and water intake. Ann. N. Y. Acad. Sci. *157:* 806–838 (1969).

82 MORGANE, P. J. and JACOBS, H. L.: in BOURNE World review of nutrition and dietetics, vol. 10, pp. 100–123 (Karger, Basel 1969).

83 MORRISON, S. D.: The relationship of energy expenditure and spontaneous activity to the aphagia of rats with lesions in the lateral hypothalamus. J. Physiol., Lond. *197:* 325–343 (1968).

84 MORRISON, S. D.; BARRNETT, R. D., and MAYER, J.: Localization of lesions in the lateral hypothalamus of rats with induced adipsia and aphagia. Amer. J. Physiol. *193:* 230–234 (1958).

85 MROSOVSKY, N.: Hibernation and the hypothalamus (Appleton-Century-Crofts, New York 1971).

86 MYERS, R. D. and VEALE, W. L.: Spontaneous feeding in the satiated cat evoked by sodium or calcium ions perfused within the hypothalamus. Physiol. Behav. *6:* 507–512 (1971).

87 NAUTA, W. J. H.: Hippocampal projections and related neural pathways to the midbrain in the cat. Brain *81:* 319–340 (1958).

88 NISBETT, R. E.: Hunger, obesity and the ventromedial hypothalamus. Psychol. Rev. *79:* 433–453 (1972).

89 OLTMANS, G. A. and HARVEY, J. A.: LH syndrome and brain catecholamine levels after lesions of the nigrostriatal bundle. Physiol. Behav. *8:* 69–78 (1972).

90 OOMURA, Y.; KIMURA, K.; OOYAMA, H.; MAENO, T.; IKI, M., and KUNIYOSHI, M.: Reciprocal activities of the ventromedial and lateral hypothalamic area of cats. Science *143:* 484–485 (1964).

91 OOMURA, Y.; OOYAMA, H.; YAMAMOTO, T.; NAKA, F.; KOBAYASHI, N., and ONO, T.: in ADEY and TOKIZANE Structure and function of the limbic system. Progr. Brain Res., vol. 27, pp. 1–33 (Elsevier, Amsterdam 1967).

92 PAGET, S.: On cases of voracious hunger and thirst from injury or disease of the brain. Trans. clin. Soc. *30:* 113–119 (1897).

93 PANKSEPP, J.: A re-examination of the role of the ventromedial hypothalamus in feeding behavior. Physiol. Behav. *7:* 385–394 (1971).

94 PANKSEPP, J.: Reanalysis of feeding patterns in the rat. J. comp. physiol. Psychol. *82:* 78–94 (1973).

95 PARKER, W. and FELDMAN, S. M.: Effect of mesencephalic lesions on feeding behavior in rats. Exp. Neurol. *17:* 313–326 (1967).

96 PECK, J. W. and NOVIN, D.: Evidence that osmoreceptors mediating drinking in rabbits are in the lateral preoptic area. J. comp. physiol. Psychol. *74:* 134–147 (1971).

97 PFAFFMAN, C.: The pleasures of sensation. Psychol. Rev. *67:* 253–268 (1960).

98 POWLEY, T. L. and KEESEY, R. E.: Relationship of body weight to the lateral hypothalamic feeding syndrome. J. comp. physiol. Psychol. *70:* 25–36 (1970).

99 RABIN, B. M.: Ventromedial hypothalamic control of food intake and satiety. A reappraisal. Brain Res. *43:* 317–342 (1972).

100 ROBERTS, W. W.: in WHALEN The neural control of behavior, pp. 175–207 (Academic Press, New York 1970).

101 ROBINSON, B. W. and MISHKIN, N.: Alimentary responses to forebrain stimulation in monkeys. Exp. Brain Res. *4:* 330–366 (1968).

102 RODGERS, W. L.; EPSTEIN, A. N., and TEITELBAUM, P.: Lateral hypothalamic aphagia. Motor or motivational deficit? Amer. J. Physiol. *208:* 334–342 (1965).

103 RODIN, J.: Effects of distraction on performance of obese and normal subjects. J. comp. physiol. Psychol. *83:* 68–75 (1973).

104 ROLLS, E. T.: Activation of amygdaloid neurones in reward, eating and drinking elicited by electrical stimulation of the brain. Brain Res. *45:* 365–381 (1972).

105 RUCH, T. C.; PATTON, H. D., and BROBECK, J. R.: Hyperphagia and adiposity in relation to disturbances of taste. Fed. Proc. *1:* 76–77 (1942).

106 RUSSEK, M.: in MOGENSON and CALARESU Central control of physiological regulations and behavior (University of Toronto Press, Toronto 1973).

107 SCHACHTER, S.: Obesity and eating. Science *161:* 25–36 (1968).

108 SHARPE, L. G. and MYERS, R. D.: Feeding and drinking following stimulation of the diencephalon of the monkey with amines and other substances. Brain Res. *8:* 295–310 (1969).

109 SHAW, J. E. and RAMWELL, P. W.: Release of prostaglandin from rat epididymal fat pad on nervous and hormonal stimulation. J. biol. Chem. *243:* 1498–1503 (1968).

110 SHUTE, C. C. D. and LEWIS, P. R.: The ascending cholinergic reticular system. Neocortical, olfactory and subcortical projections. Brain *90:* 497–520 (1967).

111 SIBOLE, W.; MILLER, J. J., and MOGENSON, G. J.: Effects of septal stimulation on drinking elicited by electrical stimulation of the lateral hypothalamus. Exp. Neurol. *32:* 466–477 (1971).

112 SIMPSON, J. B. and ROUTTENBERG, A.: The subfornical organ and carbachol-induced drinking. Brain Res. *45:* 135–152 (1972).

113 SKULTETY, F. M.: Stimulation of the periaqueductal gray and hypothalamus. Arch. Neurol., Chicago *8:* 608–620 (1963).

113a SMITH, G. P.; STROHMAYER, A. J., and REIS, D. J.: Effect of lateral hypothalamic injections of 6-hydroxydopamine on food and water intake in rats. Nature New Biol. *235:* 27–29 (1972).

114 STEFFENS, A. B.: The influence of insulin injections and infusions on eating and blood blucose level in the rat. Physiol. Behav. *4:* 823–828 (1969).

115 STEFFENS, A. B.; MOGENSON, G. J., and STEVENSON, J. A. F.: Blood, glucose, insulin and free fatty acids after stimulation and lesions of the hypothalamus. Amer. J. Physiol. *222:* 1446–1452 (1972).

116 STEIN, L.: in TAPP Reinforcement and behavior, pp. 329–352 (Academic Press, New York 1969).

117 STEIN, L. and WISE, C. D.: Release of norepinephrine from hypothalamus and amygdala by rewarding medial forebrain bundle stimulation and amphetamine. J. comp. physiol. Psychol. *67:* 189–198 (1969).

118 STEVENSON, J. A. F.: Effects of hypothalamic lesions on water and energy metabolism in the rat. Recent Progr. Hormone Res. *4:* 363–394 (1949).

119 STEVENSON, J. A. F.: in HAYMAKER, ANDERSON and NAUTA The hypothalamus, pp. 524–621 (Thomas, Springfield 1969).

120 STRICKER, E. M. and WOLF, G.: The effects of hypovolemia on drinking in rats with lateral hypothalamic damage. Proc. Soc. exp. Biol. Med. *124:* 816–820 (1967).

121 STUNKARD, A. J.: Environment and obesity. Recent advances in our understanding of regulation of food intake in man. Fed. Proc. *27:* 1367–1373 (1968).

122 TEITELBAUM, P.: Sensory control of hypothalamic hyperphagia. J. comp. physiol. Psychol. *48:* 156–163 (1955).

123 TEITELBAUM, P.: in STELLAR and SPRAGUE Progress in physiological psychology, vol. 4, pp. 319–350 (Academic Press, New York 1971).

124 TEITELBAUM, P.; CHENG, M. F., and ROZIN, P.: Development of feeding parallels its recovery after hypothalamic damage. J. comp. physiol. Psychol. *67:* 430–441 (1969).

125 TEITELBAUM, P. and EPSTEIN, A. N.: The lateral hypothalamic syndrome. Recovery of feeding and drinking after lateral hypothalamic lesions. Psychol. Rev. *69:* 74–90 (1962).

126 UNGERSTEDT, U.: Adipsia and aphagia after 6-hydroxydopamine induced degeneration of the nigrostriatal dopamine system. Acta physiol. scand. *367:* suppl., pp. 95–122 (1971).

127 VALENSTEIN, E. S.; COX, V. C., and KAKOLEWSKI, J. W.: Modification of motivated behavior elicited by electrical stimulation of the hypothalamus. Science *159:* 1119–1121 (1968).

128 VANDERWOLF, C. H.; BLAND, B. H., and WHISHAW, I. Q.: in MASER Efferent organization and the integration of behavior, pp. 229–262 (Academic Press, New York 1973).

129 WADE, G. N. and ZUCKER, I.: Development of hormonal control over food intake and body weight in female rats. J. comp. physiol. Psychol. *70:* 213–220 (1970).

130 WADE, G. N. and ZUCKER, I.: Modulation of food intake and locomotor activity in female rats by diencephalic hormone implants. J. comp. physiol. Psychol. *72:* 328–336 (1970).

131 WAYNER, M. J. and CAREY, R. J.: Basic drives. Annu. Rev. Psychol. *24:* 53–80 (1973).

132 Wheatley, M. D.: The hypothalamus and affective behavior in cats. Arch. Neurol., Chicago *52:* 296–316 (1944).

133 Wishart, T. B. and Mogenson, G. J.: Effects of food deprivation on water intake in rats with septal lesions. Physiol. Behav. *5:* 1481–1486 (1970).

134 Wyrwicka, W. and Doty, R. W.: Feeding induced in cats by electrical stimulation of the brain stem. Exp. Brain Res. *1:* 152–160 (1966).

135 Young, P. T.: Appetite, palatability, and feeding habit. Psychol. Bull. *45:* 289–320 (1948).

136 Zigmond, M. J.; Chalmers, J. P.; Simpson, J. R., and Wurtman, R. J.: Effect of lateral hypothalamic lesions on uptake of norepinephrine by brain homogenates. J. Pharmacol. exp. Ther. *179:* 20–28 (1971).

137 Zigmund, M. J. and Stricker, E. M.: Recovery of feeding and drinking by rats after intraventricular 6-hydroxydopamine or lateral hypothalamic lesions. Science *182:* 717–720 (1973).

Author's address: Dr. G. J. Mogenson, Departments of Physiology and Psychology, University of Western Ontario, *London, Ontario* (Canada)

Recent Studies of Hypothalamic Function
Int. Symp. Calgary 1973, pp. 294–305 (Karger, Basel 1974)

Range of Control of Cardiovascular Variables by the Hypothalamus[1]

O. A. SMITH, R. B. STEPHENSON and D. C. RANDALL

Regional Primate Research Center, and Department of Physiology and Biophysics, University of Washington, Seattle, Wash.

The influence of the hypothalamus on the cardiovascular system has been recognized since the turn of the century, and has been the object of continuous study since then. Until recently the studies usually involved electrical stimulation in the vicinity of the hypothalamus and measurement of the resulting changes in heart rate or blood pressure. Because of the great variability in these responses, little more could be said than that the hypothalamus has some influence on the cardiovascular system.

One would like, however, to answer more significant questions. How widespread is this control? Is the control absolute or merely modulating? Is the control concerned only with total blood pressure effects or can it influence the regulation of a single organ or vascular bed? Does the hypothalamus control the varying patterns of cardiovascular adjustments required by various behavioral and environmental stresses? Or, in more general terms, what is the functional significance of this control, i.e. how does it help to maintain the homeostasis of the organism?

In recent times, the dependent variable has been given more status with the advent of chronically implantable flow meters, assorted strain gauges, radioactive tracers, etc. In parallel with these new devices have come innovations in behavior control and the use of nonhuman primates, whose brain and cardiovascular system are similar to man's. These technological changes have yielded new understanding about how the cardiovascular system can be controlled or influenced by the hypothalamus.

1 Supported by NIH Grant RR00166 to the Regional Primate Research Center at the University of Washington, and grants PHS HL04741, PHS GM00260 and NASA NGR48–022–131.

This paper presents two examples of how the new methodologies enable us to answer the more functional questions posed above. The two examples selected are particularly interesting because intrinsic regulatory mechanisms in the organs involved could conceivably carry on basic functions in the absence of hypothalamic influence. The first example, regulation of myocardial contractility, could possibly be understood entirely within the context of Starling's law of the heart, i.e. the force of contraction of the heart is regulated by the imposed loads on the heart via the degree of stretch of the myocardium. However, during exercise, when the contractile force is tremendously increased, the size (and therefore the amount of stretch) of the heart is decreased [6]. This means that some other regulating factor must be playing a predominent role. The demonstration that stimulation of the cardiac sympathetic nerves profoundly influences the force of contraction [5] provides a hypothesis for explaining the regulation. This neural control has been further analyzed by measuring the contractile state of the myocardium during a series of normal behavioral situations [4]. In this case monkeys were maintained in restraint chairs after undergoing surgical placement of strain gauge arches on the ventricles and of pressure measuring cannulae in the left ventricle and brachial artery. The monkeys were then trained to exercise, to show an emotional response and to manipulate a lever. Their responses to these and to normal behaviors such as eating, sleeping, being startled, etc., were recorded. An important point to emphasize is that all of these behaviors were brought under experimental control so that the monkey exercised, ate or became afraid as desired by the investigator. This control proved to be critical in establishing exact time-relationships between the behavior and the resulting cardiovascular responses. These studies demonstrated large changes in the force of myocardial contraction coincident with several behaviors and smaller changes with others. It is possible to state, with some certainty, that these contractile changes were due to the action of the sympathetic nerves upon the heart rather than to changes in the amount of stretch (preload), arterial pressure (afterload) or interval between activations (heart rate) [4]. Figure 1 illustrates the changes in right and left ventricular contractility, arterial pressure and heart rate during a response to a signal (conditional stimulus, CS) which always preceded a painful shock (unconditional stimulus, UCS) by 1 min. Subsequently the monkey was anesthetized and his hypothalamus was explored with stimulating electrodes until cardiovascular changes approximating this response were elicited. Bilateral lesions made in this location

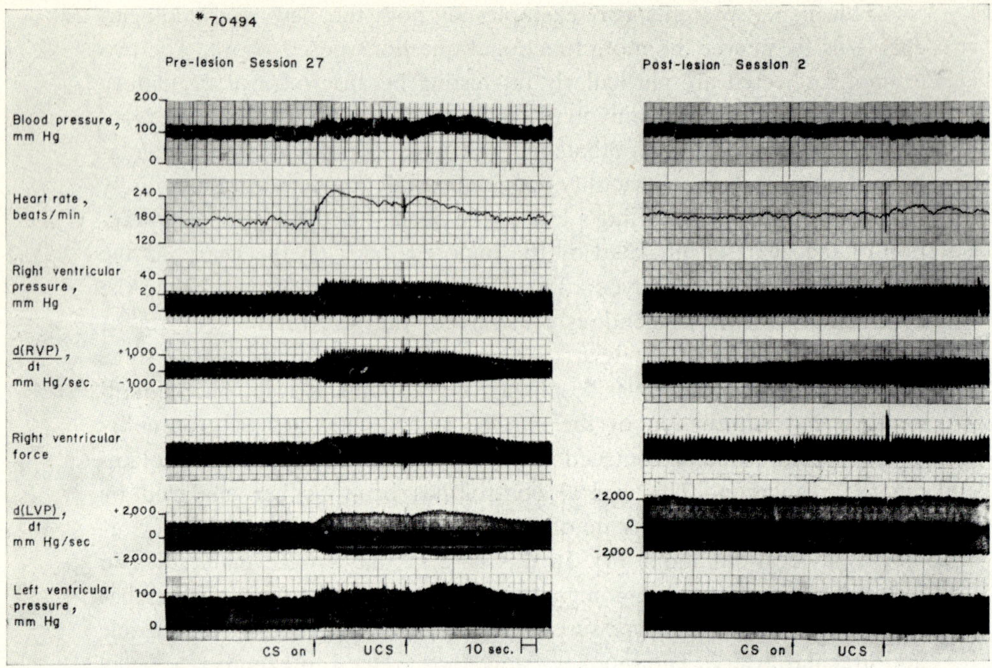

Fig. 1. Classical conditioning. Comparison of cardiovascular responses in the rhesus monkey before and after hypothalamic lesions to light signal (conditional stimulus, CS) which precedes a painful shock (unconditional stimulus, UCS) by 1 min. RVP = right ventricular pressure; LVP = left ventricular pressure.

are shown in figure 2, and the effects of these lesions on the response after full recovery of the monkey are shown in figure 1. Adjunct studies have shown that the disappearance of the response is not due to the animal 'forgetting' the significance of the stimulus (CS), but that the animal is simply incapable of making the response. This was demonstrated by training monkeys to press a lever to obtain a food reward. After stable behavior was obtained, the CS was presented during lever pressing. The monkey characteristically stopped lever-pressing until after the shock had been delivered. This cessation of lever-pressing has been called the conditioned emotional response (CER) and has been considered a defining characteristic for emotional behavior. Monkeys with hypothalamic lesions that eliminate the cardiovascular response to such situations still show

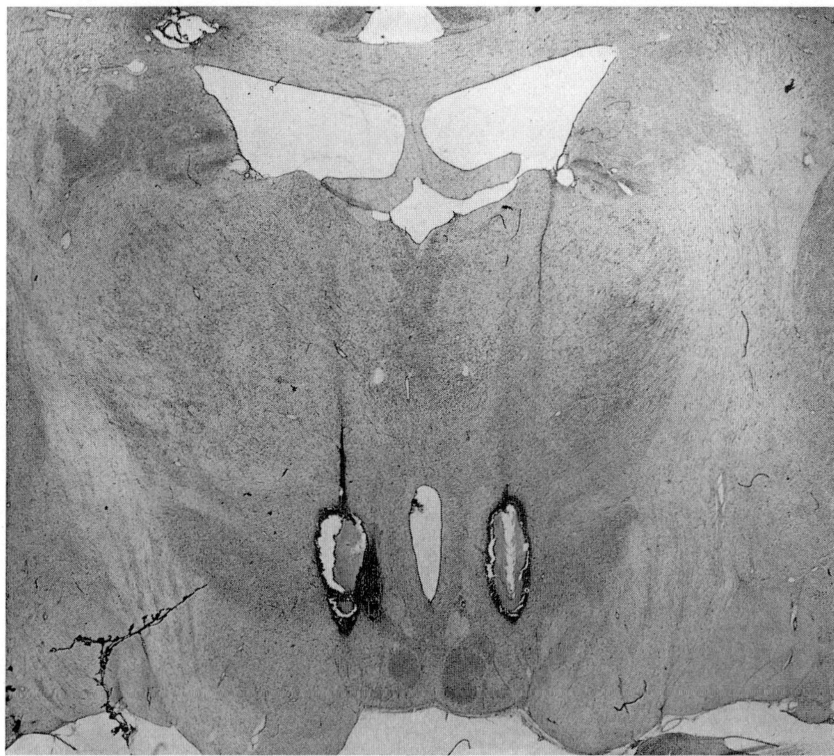

Fig. 2. Sites of hypothalamic lesions responsible for the change in cardiovascular response shown in figure 1 (posterior hypothalamus, just dorsal and lateral to mammillary bodies and encroaching on the fields of Forel).

the cessation of lever pressing, indicating an appropriate behavioral (emotional) response to the signal, but without the cardiovascular component. We can now say that a very specific part of the hypothalamus regulates the changes in myocardial contractility that accompany emotion. We do not yet have enough substantial data to draw conclusions about the influence this lesion exerts on other behaviors.

The second example concerns hypothalamic control of renal blood flow. Where the heart had Starling's law to explain its regulatory behavior, 'autoregulation' could serve a similar function for the kidney. In autoregulation, renal blood flow remains nearly constant over a wide range of arterial pressure, which implies that its critical functions are maintained despite large fluctuations in other systems [7]. However, the advent of

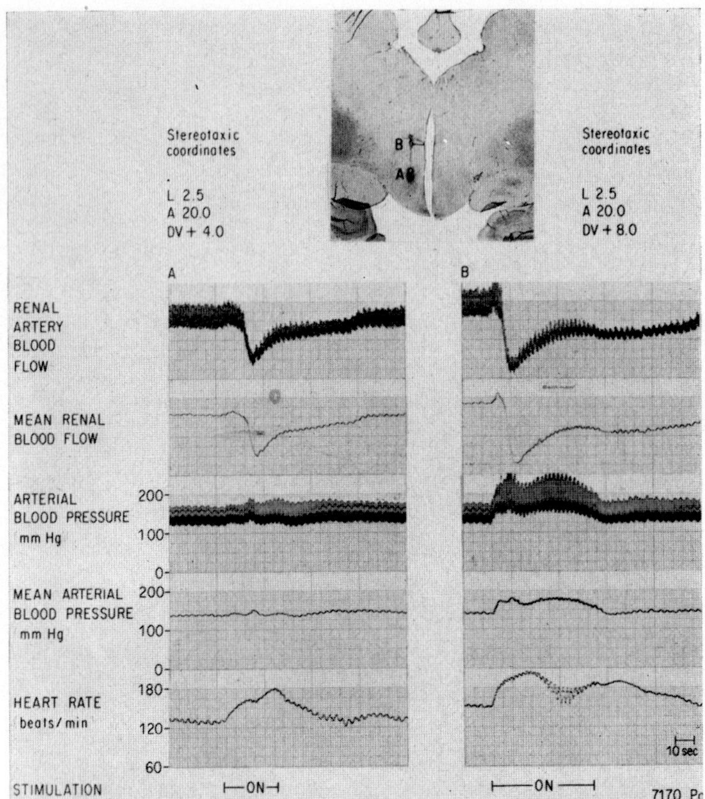

Fig. 3. Cardiovascular response to hypothalamic stimulation in anesthetized baboon. Decreased renal flow delayed relative to heart rate and blood pressure changes. Point A, between posterior portion of paraventricular nucleus and descending columns of fornix; point B, medial to lateral hypothalamic area and dorsal to posterior portion of ventromedial nucleus. 1.0 mA stimulation.

renal clearance techniques made it clear that renal blood flow could be modified by emotion or other behavioral stresses [8]. The heavy sympathetic innervation to the kidney can hardly be ignored in this regard.

Hypothalamic influence on renal blood flow has been demonstrated repeatedly by stimulation studies in acute preparations [2, 3, 9]. Another look at this relationship has begun in our laboratory using monkeys as experimental subjects. This study was undertaken because of the possibility that neurally mediated decreases in renal blood flow might be

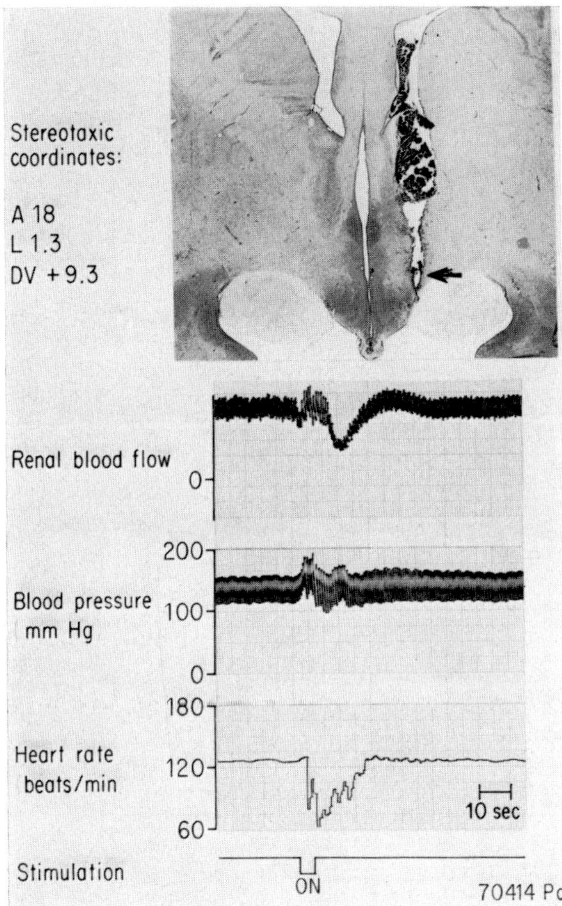

Fig. 4. Persistence of delayed decrease in renal flow after short stimulus. Initial rise in renal flow considered to be passive. 1.0 mA stimulation in lateral hypothalamus area (medial forebrain bundle) of anesthetized baboon.

involved in essential hypertension. In the initial experiments, monkeys were prepared with an electromagnetic blood flow section on the renal artery and, occasionally, one on the femoral artery as well. Blood pressure and heart rate were recorded and mean pressure and flows were computed. Brain stimulation was produced by a constant current stimulator through either a monopolar or coaxial electrode. Two instances of a response commonly produced by hypothalamic stimulation are demonstrated in figure 3. In both cases renal flow was reduced. In one case the reduction

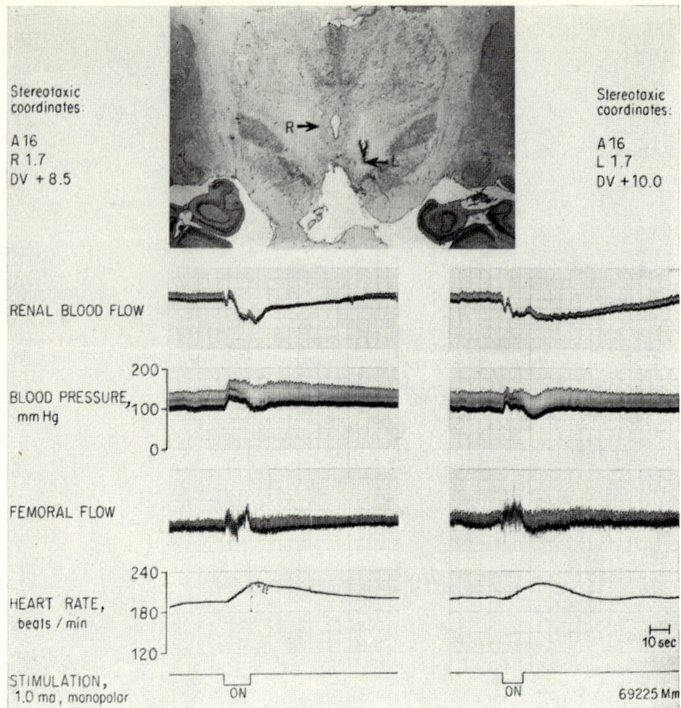

Fig. 5. Biphasic decreases in renal blood flow resulting from subthalamic stimulation in anesthetized baboon. Point R, dorsomedial portion of H-field of Forel; point L, ventromedial portion of H-field of Forel.

occurred in spite of a large pressure increase. Also, the flow changes were delayed by about 7 sec after the start of stimulation while heart rate and blood pressure changes occurred much earlier. These time differences point strongly to a humoral or indirect effect on the kidney rather than a direct neural effect. Even when stimulation is very brief and ends well before onset of the response, the delayed response is still produced (fig. 4), supporting the idea that this response is humorally mediated. The hypothesis that the delayed response resulted from activation of the hypothalamic-hypophysial system was investigated by collecting all jugular blood during stimulation and then reinfusing it. Collecting the blood failed to prevent the delayed response, and reinfusing it failed to mimic the

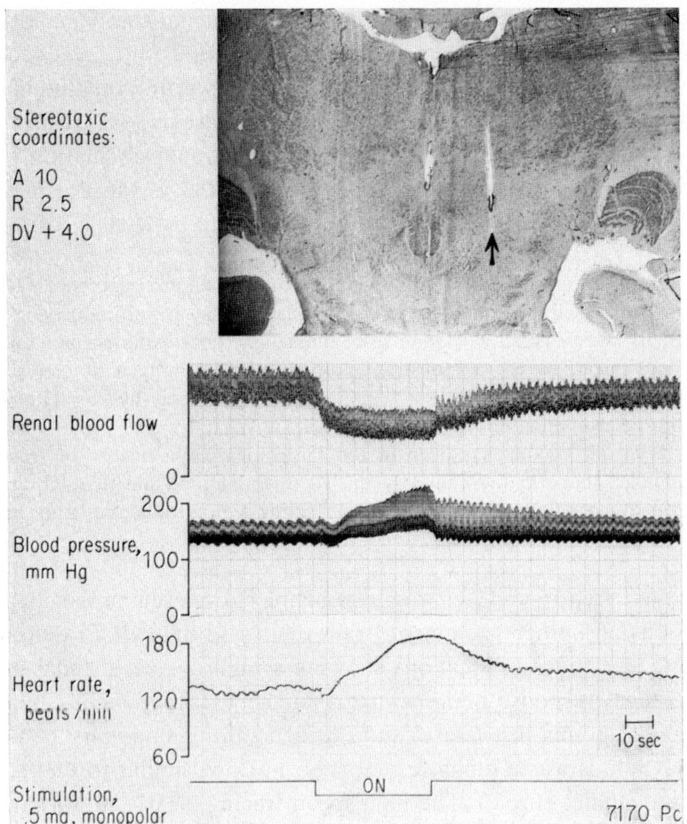

Fig. 6. Short-latency decrease in renal blood flow with stimulation of medial midbrain reticular formation in anesthetized baboon.

response, leading us to reject this idea. We suggest instead that the response is mediated through the hypothalamic-adrenal axis via the sympathetic nervous system [3].

Not all renal responses to stimulation are delayed. Stimulation in the Forel fields of the subthalamus produced an immediate decrease in renal blood flow followed by the delayed, long-acting second response (fig. 5). This biphasic response leads to the belief that the hypothalamus has available at least two different routes of control over renal blood flow, the direct neural route and a humoral route. This rapid direct effect can also be produced by brainstem stimulation, independent of the delayed, long-acting effect (fig. 6). One cannot help but be impressed with the potential utility of such control to an organism facing the demands of a wide variety of external or internal stresses.

These acute studies demonstrated the very powerful potential control of the hypothalamus over renal circulation. However, we realize that definitive statements about the neural control of renal circulation in the awake, intact animal cannot be inferred from studies of the anesthetized, traumatized animal. This was eloquently pointed out by HOMER SMITH [8]:

> 'If emotional disturbance, even of an extreme nature, can so markedly influence the renal circulation in man, what must we say of procedures in laboratory animals which entail profound and almost maximal sympathetic excitation by alarm and fright, by the administration of anesthetic, by excitation of sensory nerves, by surgical incisions, by the handling of the viscera, by circulatory inadequacy, by the manipulation of the kidney itself?... What I would plead is that the history of renal physiology has been in too large measure a history of traumatic procedures which have in the end only misled investigation; and that far more is to be gained by spending several years in perfecting a non-traumatic, truly physiological method than one year in applying a traumatic one.'

Our next step, therefore, was to prepare chronic animals with renal flow sections and other equipment, train them and implant chronic brain stimulating electrodes in locations showing a high degree of renal reactivity. The renal response of an awake monkey to simultaneous bilateral stimulation of points just lateral and dorsal to the mammillary bodies in the hypothalamus was a biphasic reduction in flow, similar to that shown in the acute studies (fig. 7). The high-speed tracing clearly shows that the reduction approaches a complete shutdown of flow.

In figure 8 the cardiovascular response to a 1-min stimulation of the hypothalamus is compared with the response to the type of emotion producing situation described earlier. While the two responses in figure 8 are very similar, the similarity does not prove that the structures activated by the stimulation are the same ones responsible for the cardiovascular change accompanying emotion. Unfortunately, a lesion was not made in this animal to see if the cardiovascular change could be eliminated. However, close inspection of the histology shows that the tips of the stimulating electrodes are located just dorsal and lateral to the mammillary bodies – precisely the same location that was shown to be critical for the myocardial contractility changes accompanying emotion (fig. 2). In a classic paper ABRAHAMS *et al.* [1] found that stimulation in this same region produced cholinergic muscle vasodilation, vasoconstriction in skin and intestine and increased heart rate and blood pressure in the anesthe-

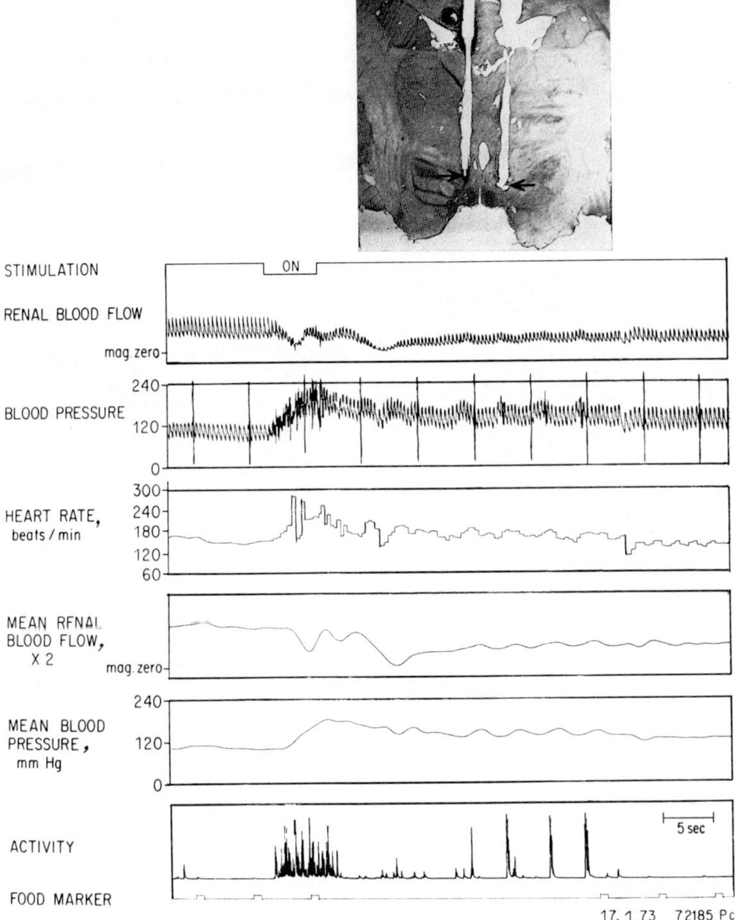

Fig. 7. Biphasic decrease in renal blood flow with simultaneous, bilateral stimulation of posterior hypothalamus (dorsal and lateral to mammillary bodies) in awake baboon. Stimulation parameters: 0.4 mA, 0.3 msec, 100 Hz. Renal flow zero set equal to magnet (mag.) (electric) zero at time of implant. Mean renal flow twice the gain of renal flow. Activity recorded as output of accelerometer attached to restraint chair. Deflection of food marker indicates baboon licked feeder tube for food reinforcement during preceding 6 sec.

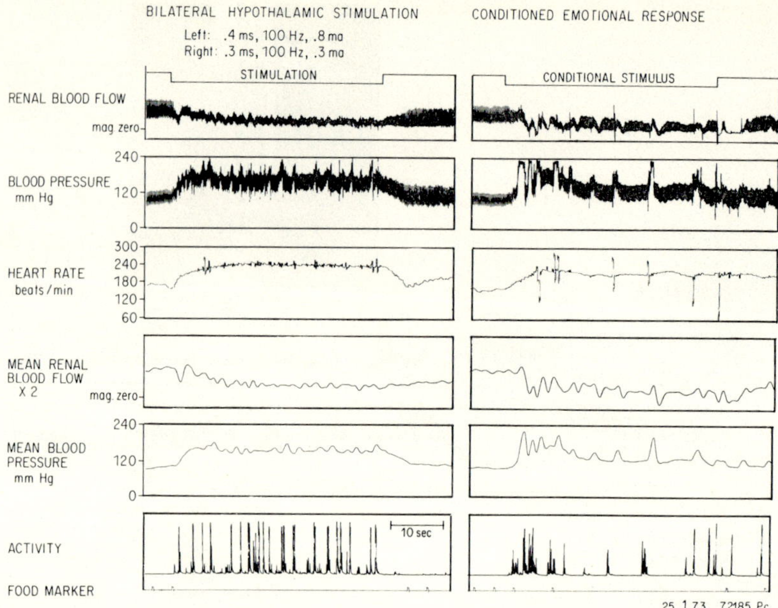

Fig. 8. Comparison of responses to hypothalamic stimulation and emotional situation in awake baboon. Site of simultaneous bilateral stimulation shown in figure 7. Conditional stimulus is white noise usually followed by shock (no shock in this record). Peaks of blood pressure correlated with spontaneous isometric contractions. Note suppression of eating. Additional details as in figure 7. Mag. = magnet.

tized cat; it produced defense-like behavior in the conscious cat. Collectively, these data point to this region as a critical autonomic outflow pathway which is involved in the regulation of the integrated cardiovascular response to highly emotional situations. This response includes changes in myocardial contractility, renal constriction, muscle vasodilation, skin and intestinal constriction, and tachycardia.

Studies of the neural-cardiovascular interactions accompanying other behaviors are being conducted, with the objective of elucidating how the central nervous system is organized to produce a given cardiovascular response associated with a definable behavior pattern. The critical elements in these kinds of studies are the use of the most accurate and stable transducers and electronic equipment, precise behavior control and the scientific management and care of experimental animals. In addition to increasing our understanding of normal regulation, this information may also contribute to our understanding of cardiovascular pathology.

References

1 ABRAHAMS, V. C.; HILTON, S. M., and ZBROŽYNA, A.: Active muscle vasodilation produced by stimulation of the brainstem. Its significance in the defence reaction. J. Physiol., Lond. *154:* 491–513 (1960).

2 FEIGL, E. O.: Vasoconstriction resulting from diencephalic stimulation. Acta physiol. scand. *60:* 372–380 (1964).

3 JURF, A. N. and BLAKE, W. D.: Renal response to electrical stimulation of the septum and diencephalon of rabbits. Circulat. Res. *30:* 322–331 (1972).

4 RANDALL, D.: Right and left ventricular myocardial contractility during exercise and emotional conditioning in the non-human primate; Ph. D. thesis, Seattle (1971).

5 RANDALL, D. C.; ARMOUR, J. A., and RANDALL, W. C.: Dynamic responses to cardiac nerve stimulation in the baboon. Amer. J. Physiol. *220* (2)*:* 526–533 (1971).

6 RUSHMER, R. F. and SMITH, O. A., jr.: Cardiac control. Physiol. Rev. *39*(1): 41–68 (1959).

7 SHIPLEY, R. E. and STUDY, R. S.: Changes in renal blood flow, extraction of inulin, glomerular filtration rate, tissue pressure, and urine flow with acute alterations of renal artery blood pressure. Amer. J. Physiol. *167:* 676–688 (1951).

8 SMITH, H. W.: Physiology of the renal circulation. Harvey Lect. *35:* 166–222 (1940).

9 TAKEUCHI, J.; SHIGERU, Y.; TAKEDA, T.; UCHIDA, E.; INOUE, G., and UEDA, H.: Experimental studies on the nervous control of the renal circulation. Changes in renal circulation induced by electrical stimulation of the diencephalon and effects of several drugs on the circulatory responses. Jap. Heart J. *3:* 57–72 (1962).

Authors' addresses: Dr. O. A. SMITH and Dr. R. B. STEPHENSON, Regional Primate Research Center, and Department of Physiology and Biophysics, University of Washington, *Seattle, WA 98195;* Dr. D. C. RANDALL, Division of Behavioral Biology, Johns Hopkins University School of Medicine, 720 Rutland Avenue, *Baltimore, MD 21205* (USA)

Recent Studies of Hypothalamic Function
Int. Symp. Calgary 1973, pp. 306–314 (Karger, Basel 1974)

The Role of the Hypothalamus in the Organisation of Patterns of Cardiovascular Response

S. M. HILTON

Department of Physiology, Medical School, Birmingham

The idea dies hard that the most significant areas of the central nervous system for basic control of the heart and circulation lie in the medulla. Not only in general text-books, but also in those dealing especially with the circulation, higher parts of the nervous system are at most allowed a modulatory role, as in the initiation of the autonomic accompaniments of fear and rage reactions – temperature regulation just happens to be an inconvenient exception, for tradition has firmly placed this function in the hypothalamus, which thus manages in this particular case to operate independently of the medulla, which as we have been told for about 100 years, is the site of the vasomotor centre.

The old idea of a special centre in the medulla for cardiac and vasomotor control is so embedded in the physiological tradition that, to point out that there never was any substantial evidence for it, is to be met with incomprehension or disbelief. The situation at present is so unsatisfactory as to constitute a positive embarrassment to anyone who has worked on this problem and who has the responsibility of teaching it to undergraduates. And when it comes to marking examination papers, how can one penalise the student who has carefully read and understood his text-book? If this seems to be labouring the point, it is only because earlier attempts to make it plain [19, 33] appear to have fallen on deaf ears; so I hope you will bear with me if I marshall the evidence once again.

The suggestion of a medullary vasomotor centre came originally from experiments of OWSJANNIKOW [32] and DITTMAR [12]. They sectioned the brain-stem of anaesthetised animals, at intervals proceeding caudally, and finally reached a level at which the blood pressure fell profoundly. The conclusion reached was that a tonically active ‘centre’, which normally maintains vasomotor tone, is situated just above the last section, which was made 4 mm rostral to the calamus scriptorius in the medulla. Although no one would interpret a finding of this kind in this way to-day, it has somehow stuck with us. PEISS [33] felt that theoretical reasoning alone (based on our familiarity with spinal shock) would not be sufficient to overcome the traditional view, and so he painstakingly repeated the old experiments, to find that the arterial blood pressure was reduced by sections that excluded the hypothalamus and that 50 % of the total fall seen in the spinal animal was obtained *before* the 'classical' level in the medulla has been reached.

The world of physiology remained unmoved, perhaps because the established view had received a particular impetus from the experiments of RANSON, who with BILLINGSLEY, in 1916 [34], electrically stimulated points of the dorsal medulla, in the floor of the IV ventricle, while recording arterial blood pressure and found that the pressure rose when they stimulated in some positions and fell when they stimulated at others. This led BAYLISS to postulate, in his monograph entitled 'The vaso-motor system' published in 1923 [7], that there is a medullary vasomotor centre which consists of two parts, one excitatory and one inhibitory, each with its own output. This is still the conventional view of the day. The most recent and complete study exemplifying it is that of ALEXANDER [4] whose carefully prepared maps of the pressor and depressor regions now appear in specialised books as well as the standard texts. Matters have been taken even further than this when, for example, investigators have looked for all those parts of the brain-stem from which a slowing of the heart-rate can be produced on electrical stimulation, and have then put forward the thesis they have been mapping a pathway for bradycardia [3]. I have parodied this approach already elsewhere [20] by suggesting that, if I were interested as an experimental physiologist in what I may call the 'come-hither' response (a simple, but significant, gesture of beckoning), I should therefore study its central nervous organisation by attempting to identify all those parts of the central nervous system from which, by electrical stimulation, I could excite contractions of the flexor digitorum longus muscle.

The central nervous system produces integrated patterns of response and it is almost too obvious to state that each variable, although it can be described or measured independently, is actually a component of several such patterns. But in all experimental work we can only get answers to the questions that we ask and, in so much work of this kind to-date, the over-riding approach has been to regard each variable as having its own brain-stem centre, all as separate entities, both functionally and topographically. This has led to ever-increasing artificiality, as assumption has been built upon assumption, the final result being not only that we are left with the virtually groundless concept of a special vasomotor centre located in the medulla, but also that this centre is held to be tonically active; to contain effectively the final common paths for the vasoconstrictor outputs to all regions of the body, with separate neurone pools for each part of each vascular bed; and to interact in some mutually inhibitory manner with neurones in the vasodepressor part of the centre. And when questions are posed, such as how does the baroreceptor afferent input act, the answer glibly follows that it excites the neurones of the inhibitory region or directly inhibits the neurone pools in the excitatory region (or possibly both).

In case it be thought that this summary is wildly exaggerated, some illustrative quotations may be appropriate from a recent paper from a laboratory long-connected with first-class investigations of vascular regulations. This study was of the vasoconstrictor fibre discharge to different vascular beds (muscle, skin, kidney and intestine) at different levels of baroreceptor afferent activity. In the Discussion it is stated that, since the vasoconstrictor nerves were intact during the experiments, 'a certain tonic, spontaneous nerve activity may have been present...' 'The present study thus demonstrates that the increased vasoconstrictor fibre discharge' is 'induced when the bulbar vasomotor centre is released from baroreceptor inhibition'. Finally, 'the background of the differentiated vasoconstrictor fibre outflow to the various parallel-coupled systemic circuits when baroreceptor restraint is gradually reduced is not fully known. The observations that there was a considerably lower discharge rate in renal vasoconstrictor fibres than in those running to e.g. the skeletal muscle also when the vasomotor centre was *completely* released from baroreceptor inhibitory influence and thus allowed to display its 'inherent' activity, suggest that the differentiation is primarily related to differences in the level of spontaneous activity in the separate neuron pools in the vasomotor centre'.

These quotations are from KENDRICK *et al.* [29]. It is not just, however, to stigmatise a particular paper or group of workers, so prevalent is the view expressed in the lines above: it just happened that the reprint was lying on my desk, recently from the press of Acta physiol. scand., as I was writing. Indeed, the experimental work forming the background to the Discussion from which I quoted was impeccable, but the theoretical treatment was largely myth. Recent work in our department in Birmingham by COOTE and MACLEOD [10] has gone far to show that the baroreceptor afferent input produces most of its effects by exciting a *descending* inhibitory pathway which is first clearly located in the ventrolateral region of the medulla, far from the traditionally postulated 'vasomotor centre', and that the effects of this pathway are directly exerted on neurones of the lateral horn region of the spinal cord. These neurones doubtless constitute the real final common paths.

The descending, inhibitory, spinal pathway probably has an important functional connection with the hypothalamus (and even higher parts of the brain); for it was recently shown by HILTON and SPYER [21, 22] that the depressor area in the anterior hypothalamus which has been known to exist since the work of KABAT *et al.* [27] is neither a parasympathetic centre, as was suggested by GELLHORN [16], nor a sympatho-inhibitory area, as was proposed by FOLKOW *et al.* [15], but that it elicits a pattern of cardiovascular response involving parasympathetic activation and sympathetic inhibition, accompanied by respiratory inhibition, all of which resembles the pattern of response elicited by the baroreceptor afferent endings of the carotid sinus. This area was located ventral and caudal to the anterior commissure, from which it extends caudally in the dorsal hypothalamus, dorsal to the fornix. Lesions destroying this area bilaterally reduced the response to baroreceptor afferent stimulation. Lesions in the medullary depressor area which spared a large part of the nucleus of the tractus solitarius also reduced the reflex response. Lesions had to be made in both regions to abolish the reflex. Thus, in every part of the brain-stem from the hypothalamus, through the mid-brain, to the medulla there are specific areas that appear to act together as a functional unit which integrates the response to baroreceptor afferent stimulation. Moreover, the region near the tractus solitarius, particularly its more caudal extension, which is part of the long-known medullary depressor area, seems most likely to represent a section of the *afferent* pathway for the reflex, as indicated by a large number of recent studies [8, 24, 30, 31, 35, 36]. The most recent work of my colleagues, MCALLEN and SPYER, in Birmingham is providing detailed information on the sinus nerve afferent connections in this region.

Thus, the question may be asked whether the division of the brain-stem into a hypothalamus, mid-brain, pons and medulla may not, in fact, be largely an anatomical abstraction with little relevance to an understanding of physiological function. This may be emphasised briefly in a slightly different context by reference to the defence reaction, as it was called by HESS and BRUGGER [23], or the fear and rage reaction, to use CANNON's terminology. The early work of these two pioneers in the field of the neurophysiological basis of emotional expression and response certainly needs no re-emphasis at a symposium of this kind. Both of them concluded that hypothalamic structures were mainly responsible for the integration of the somatic and visceral components of the defence response, and my interest in it was stimulated by the original observation of ELIASSON *et al.* [14] that electrical stimulation in parts of the hypothalamus could activate the sympathetic vasodilator nerve fibres which run to the skeletal muscles in the cat. We showed that the regions in the hypothalamus and mid-brain which integrate the somatic movements characteristic of the defence reaction play a similar part with regard to the cardiovascular (and other visceral) components of the response [1]. The pattern of cardiovascular response includes a mobilisation of venous reserves, an increase in cardiac output and redistribution of blood flow such that this increase is diverted chiefly to the skeletal muscles [18]. We concluded that these regions of the brain-stem could be regarded as a functional unit acting virtually as a reflex centre for the reaction [2].

Nevertheless, we knew at that time that the overall integrative region could include more caudal parts of the brain-stem; for KELLER [28] had obtained what he regarded as typical rage reactions in response to mild stimuli in cats chronically decerebrated just rostral to the pons, and BARD and MACHT [5] had seen many features of the reaction when the decerebration had removed even the rostral part of the pons, though in this case stronger noxious stimulation had to be applied. COOTE, ZBROZYNA and I looked into this more recently [9], and we found that electrical stimulation within a narrow and sharply localised strip 2.5 mm on each side of the mid-line, in the caudal mesencephalon, pons and medulla could elicit a response with the behavioural, cardiovascular and other visceral components of the defence reaction. This strip lies in the dorsal part of the well-known pressor area, ending caudally in the medulla, close to the floor of the IVth ventricle. I would emphasise two points about this work in relation to my present thesis: firstly, that within the so-called pressor area of the medulla there are regions capable of eliciting highly organised patterns of response which include somatic (or behavioural) components; and, secondly, that the muscle vasodilatation was found, in this case, to arise entirely from a selective inhibition of on-going activity in the vasoconstrictor nerve supply to skeletal muscle – in other words, there is within this region a form of nervous organisation which is capable of producing highly differentiated responses in different parts of the cardiovascular system.

Thus, when considering the overall functional, neurophysiological organisation of the defence reaction we have, once again, to deal with regions in the hypothalamus, mid-brain, pons and medulla which must act together in the normal, conscious animal to produce the appropriate, integrated response. A final point worth making is that these two basic reactions of the organism appear to be mutually antagonistic; for there is a good evidence that they inhibit each other. WHEATLEY [37] first found that, after chronic lesions had been placed in the anterior hypothalamus, cats exhibited apparently unprovoked outbursts of rage behaviour, which led KAADA [26] to point out that the areas that had been destroyed included those from which vasodepressor responses had been known to be elicited. In the high decerebrate cat, which also shows such unprovoked rage behaviour, the outbursts can be inhibited by baroreceptor afferent stimulation and potentiated by bilateral occlusion of the carotid arteries below the carotid sinuses [6]. The fact that this is mutual was demonstrated by the finding that electrical stimulation of the brainstem areas integrating the defence reaction, which produces a pattern of cardiovascular response appropriate to a preparation for exercise [2], in which arterial blood pressure and heart-rate rise together, leads to strong inhibition of all components of the baroreceptor reflex [17, 18]. This has been questioned by FOLKOW and his colleagues [13] and by HUMPHREYS *et al.* [25], but recent work by COOTE and PEREZ-GONZALEZ [11] has established the point incontrovertibly.

These responses are in constant interplay in the daily lives of animals and men. It would be fascinating to explore the eventualities that such findings may possibly open up for an understanding of alerting and sleep, and maybe even for the pathogenesis of hypertension, but this would be to go far beyond the proper confines of this talk. However, we can certainly conclude that there is a longitudinal form of organisation, for cardiovascular control at least, extending from the hypothalamus, through the mid-brain, to the medulla which functions in such a way as to produce patterns of response. It has a complexity of organisation far greater, no doubt, than we can yet comprehend but which, even as we see it now, allows for a variety of possibilities for activating both systems that I have dealt with today, by the visceral and somatic afferent inputs more caudally and by the limbic system and cerebral cortex more rostrally. This appears to be a fruitful way of thinking about the central nervous control of the circulation.

References

1 ABRAHAMS, V. C.; HILTON, S. M., and ZBROZYNA, A.: Active muscle vasodilatation produced by stimulation of the brain stem. Its significance in the defence reaction. J. Physiol., Lond. *154:* 491–513 (1960).

2 ABRAHAMS, V. C.; HILTON, S. M., and ZBROZYNA, A. W.: The role of active muscle vasodilatation in the alerting stage of the defence reaction. J. Physiol., Lond. *171:* 189–202 (1964).

3 ACHARI, N. K.; DOWNMAN, C. B. B., and WEBER, W. V.: A cardio-inhibitory pathway in the brain stem of the cat. J. Physiol., Lond. *197:* 35P (1968).

4 ALEXANDER, R. S.: Tonic and reflex functions of medullary sympathetic cardiovascular centers. J. Neurophysiol. *9:* 205–217 (1946).

5 BARD, P. and MACHT, M. B.: The behaviour of chronically decerebrate cats; in WOLSTENHOLM and O'CONNOR Neurological basis of behaviour, pp. 57–71 (Churchill, London 1958).

6 BARTORELLI, C.; BIZZI, E.; LIBRETTI, A., and ZANCHETTI, A.: Inhibitory control of sinocarotid pressoceptive afferents on hypothalamic autonomic activity and sham-rage behaviour. Arch. ital. Biol. *98:* 308–326 (1960).

7 BAYLISS, W. M.: The vaso-motor system (Longmans, Green, London 1923).

8 BISCOE, T. J. and SAMPSON, S. R.: Responses of cells in the brain stem of the cat to stimulation of the sinus, glossopharyngeal, aortic and superior laryngeal nerves. J. Physiol., Lond. *209:* 359–374 (1970).

9 COOTE, J. H.; HILTON, S. M., and ZBROZYNA, A. W.: The ponto-medullary area integrating the defence reaction in the cat and its influence on muscle blood flow. J. Physiol., Lond. *229:* 257–274 (1973).

10 COOTE, J. H. and MACLEOD, V. H.: The possibility that noradrenaline is a sympatho-inhibitory transmitter in the spinal cord. J. Physiol., Lond. *225:* 44–46P (1972).

11 COOTE, J. H. and PEREZ-GONZALEZ, J. F.: The baroreceptor reflex during stimulation of the hypothalamic defence region. J. Physiol., Lond. *224:* 74–75P (1972).

12 DITTMAR, C.: Ueber die Lage des sogenannten Gefässcentrums der Medulla Oblongata. K. sächs. Ges. Wiss. math.-physiol. Kl. *25:* 449 (1873).

13 DJOJOSUGITO, A. M.; FOLKOW, B.; KYLSTRA, P. H.; LISANDER, B., and TUTTLE, R. S.: Differentiated interaction between the hypothalamic defence reaction and baroreceptor reflexes. 1. Effects on heart rate and regional flow resistance. Acta physiol. scand. *78:* 376–385 (1970).

14 ELIASSON, S.; FOLKOW, B.; LINDGREN, P., and UVNAS, B.: Activation of sympathetic vasodilator nerves to the skeletal muscles in the cat by hypothalamic stimulation. Acta physiol. scand. *23:* 333–351 (1951).

15 Folkow, B.; Johanssen, B., and Oberg, B.: A hypothalamic structure with a marked inhibitory effect on tonic sympathetic activity. Acta physiol. scand. *47:* 262–270 (1959).

16 Gellhorn, E.: Autonomic imbalance and the hypothalamus (University of Minnesota Press, Minneapolis 1957).

17 Hilton, S. M.: Inhibition of baroreceptor reflexes on hypothalamic stimulation. J. Physiol., Lond. *165:* 56–57P (1963).

18 Hilton, S. M.: Hypothalamic control of the cardiovascular responses in fear and rage. In: The scientific basis of medicine annual reviews, pp. 217–238 (Athlone Press, London 1965).

19 Hilton, S. M.: Hypothalamic regulation of the cardiovascular system. Brit. med. Bull. *22:* 243–248 (1966).

20 Hilton, S. M.: A critique of current ideas of the nervous system control of circulation; in Bartorelli and Zanchetti Cardiovascular regulation in health and disease, pp. 57–62 (Cardiovascular Research Institute, Milan 1970).

21 Hilton, S. M. and Spyer, K. M.: The hypothalamic depressor area and the baroreceptor reflex. J. Physiol., Lond. *200:* 107–108P (1969).

22 Hilton, S. M. and Spyer, K. M.: Participation of the anterior hypothalamus in the baroreceptor reflex. J. Physiol., Lond. *218:* 271–293 (1971).

23 Hess, W. R. und Brugger, M.: Das subkortical Zentrum der affektiven Abwehrreaktion. Helv. physiol. Acta *1:* 33–52 (1943).

24 Humphrey, D. R.: Neuronal activity in medulla oblongata of the cat evoked by stimulation of the carotid sinus nerve, in Kezdi Baroreceptors and hypertension, pp. 131–167 (Pergamon Press, New York 1967).

25 Humphreys, P. W.; Joels, N., and McAllen, R. M.: Modification of the reflex response to stimulation of carotid sinus baroreceptors during and following stimulation of the hypothalamic defence area in the cat. J. Physiol., Lond. *216:* 461–482 (1971).

26 Kaada, B. R.: Brain mechanisms related to agressive behaviour; in Clemente and Lindsley Aggression and defence. Brain function, vol. V, pp. 25–134 (University of California Press, Berkeley 1967).

27 Kabat, H.; Magoun, H. W., and Ranson, S. W.: Electrical stimulation of points in the forebrain and midbrain. Resultant alterations in blood pressure. Arch. Neurol., Chicago *34:* 931–955 (1935).

28 Keller, A. D.: Autonomic discharges elicited by physiological stimuli in midbrain preparations. Amer. J. Physiol. *100:* 576–586 (1932).

29 Kendrick, E.; Oberg, B., and Wennergren, G.: Vasoconstrictor fibre discharge to skeletal muscle, kidney, intestine and skin at varying levels of arterial baroreceptor activity in the cat. Acta. physiol. scand. *85:* 464–476 (1972).

30 McAllen, R. M. and Spyer, K. M.: 'Baroreceptor' neurones in the medulla of the cat. J. Physiol., Lond. *222:* 68–69P (1972).

31 Miura, M. and Reis, D. J.: Termination and secondary projections of carotid sinus nerve in the cat brainstem. Amer. J. Physiol. *217:* 142–153 (1969).

32 Owsjannikow, P.: Die tonischen und reflextorischen Centren der Gefassnerven. K. sächs. Ges. Wiss. math.-phys. Kl. *23:* 135–147 (1871).

33 PEISS, C. N.: Concepts of cardiovascular regulation. Past, present and future; in RANDALL Nervous control of the heart, pp. 154–197 (Williams & Wilkins, Baltimore 1965).

34 RANSON, S. W. and BILLINGSLEY, P. R.: Vasomotor reactions from stimulation of the floor of the IVth ventricle. III. Studies in vasomotor reflex arcs. Amer. J. Physiol. *41:* 85–90 (1916).

35 SELLER, H. and ILLERT, M.: The localisation of the first synapse in the carotid sinus baroreceptor reflex pathway and its alteration of the afferent input. Pflügers Arch. ges. Physiol. *306:* 1–19 (1969).

36 SPYER, K. M. and WOLSTENCROFT, J. H.: Problems of the afferent input to the paramedian reticular nucleus, and the central connections of the sinus nerve. Brain Res. *26:* 411–414 (1971).

37 WHEATLEY, M. D.: The hypothalamus and affective behaviour in cats. A study of the effects of experimental lesions, with anatomic correlations. Arch. Neurol., Chicago *52:* 296–316 (1944).

Author's address: Dr. S. M. HILTON, Department of Physiology, Medical School, Vincent Drive, *Birmingham B15 2TJ* (England)

Recent Studies of Hypothalamic Function
Int. Symp. Calgary 1973, pp. 315–327 (Karger, Basel 1974)

Neuronal Models of Hypothalamic Temperature Regulation

J. Bligh

Agricultural Research Council, Institute of Animal Physiology, Babraham, Cambridge

The central nervous control of body temperature is clearly a function of central neurones which lie between the pathways from temperature sensors and those to the thermoregulatory effectors, but little is known of how an orderly relation between thermal disturbance and thermoregulatory response is achieved. Concepts of these events can be best expressed in the analogous terminology of control systems engineering. By comparing the disturbance-response patterns of physiological and physical thermoregulation, Hardy [14] showed that the biological thermostat behaves as if there is a set-point against which a signal representative of body temperature is being compared, and as if the intensity of the signals to the thermoregulatory effector processes is proportional to the error between set-point and the controlled body temperature (fig. 1).

A further stage in the elucidation of the mechanisms of the central nervous regulation of body temperature is to try to understand the relations between the afferent signals from temperature sensors and the efferent signals to thermoregulatory effectors in neuronal terms.

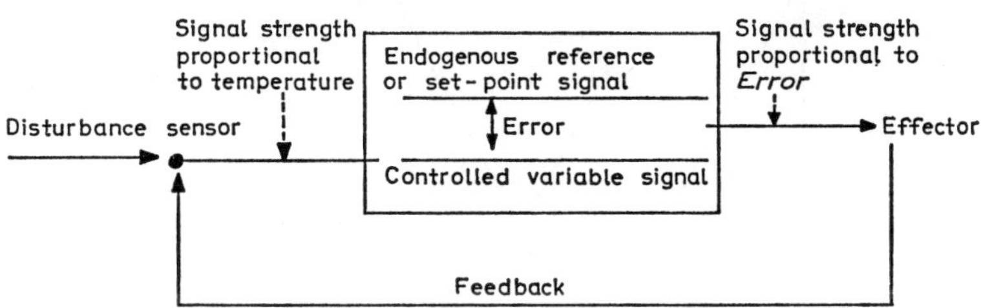

Fig. 1. A simple representation of the proportional control of temperature in a physical system (see text).

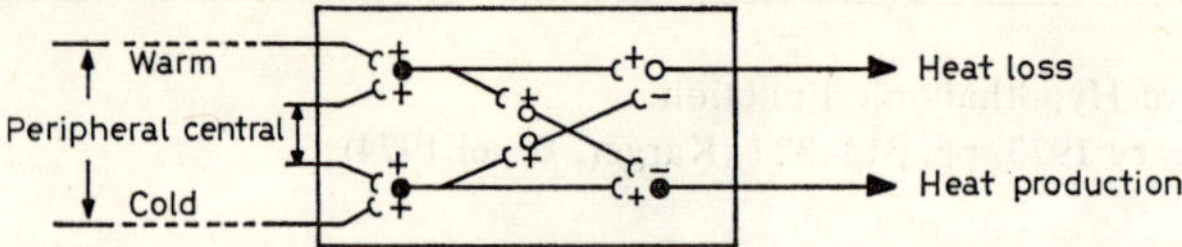

Fig. 2. A neuronal representation of the apparent relation between thermal disturbances and thermoregulatory responses, constructed from a line drawing by WYNDHAM and ATKINS [24] of these apparent relations in man. In this, and subsequent, figures + = excitatory influence; – = inhibitory influence.

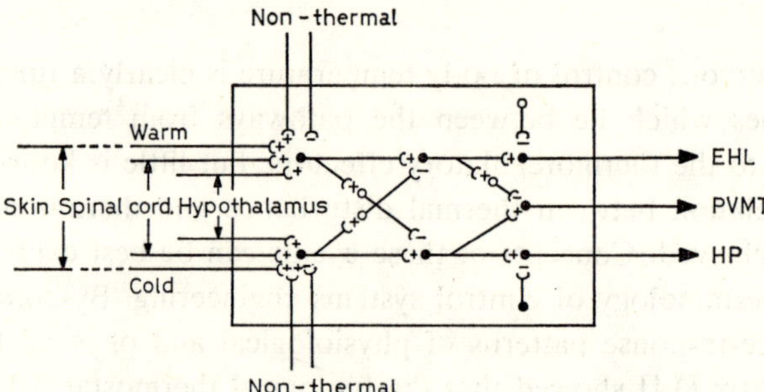

Fig. 3. A neuronal model based on a more detailed discussion (see text) of the relation between thermal and non-thermal disturbances and thermoregulatory responses. In this, and subsequent, figures EHL = evaporative heat loss; PVMT = peripheral vasomotor tone; HP = heat production.

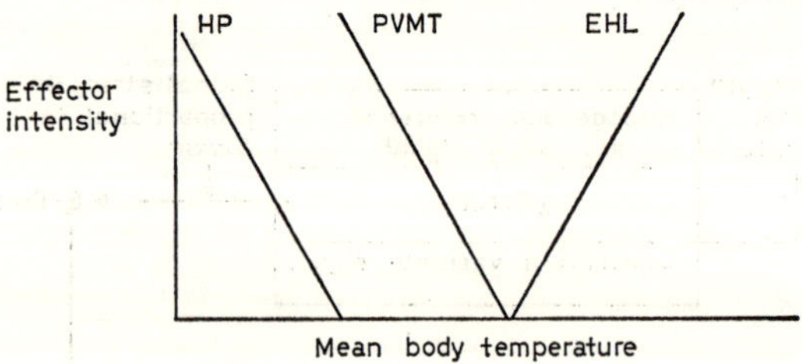

Fig. 4. A non-specific representation of the relations in mammals between the intensity of thermoregulatory effector functions and a body temperature considered (see text) to be best represented by mean body temperature.

A Neuronal Model Based on Thermal Disturbance-Thermoregulatory Response Relations

A construction in neuronal format (fig. 2) of a simple physiological scheme for the control in man of sweating and heat conductance by the thermal receptors in the hypothalamus and the skin [24] has 3 principal features.

1. The convergence of the pathways from the skin and core temperature sensors.

2. Two main pathways – one from 'warm' sensors to heat loss effectors, and the other from 'cold' sensors to heat conservation or heat production (HP) effectors.

3. Crossing inhibitory connexions between these two sensor-to-effector pathways.

In addition to the well established thermosensitivity of hypothalamic and cutaneous structures, there is now much evidence of extra-hypothalamic deep-body thermosensitivity, especially in the spinal cord [23]. There is also some evidence that the onset and intensity of thermoregulatory effector functions may correlate best with mean body temperature or, at least, with some temperature which is intermediate between that of the skin and the core [6, 22]. This is what would be expected if the afferent pathways from the temperature sensors converge and the activity in the pathways to the effectors is determined by the summed synaptic influences of temperature sensors wherever located. These relations are represented in figure 3.

There are 3 principal autonomic thermoregulatory functions: evaporative heat loss (EHL), peripheral vasomotor tone (PVMT) and HP. Figure 4 indicates the general relations between these thermoregulatory effector functions and body temperature in mammals. Although these thermoregulatory effector functions are more usually plotted against a core temperature, they are influenced by both peripheral and core temperatures and may correlate best with mean body temperature ($\overline{T}_b$) or something close to it. The general relations expressed in figure 4 are, therefore, as follows.

1. As $\overline{T}_b$ falls, there is no increase in HP until some lower (HP) threshold is reached then, as $\overline{T}_b$ falls below this threshold, HP rises proportional to the extent that $\overline{T}_b$ is below the HP threshold.

2. As $\overline{T}_b$ rises, there is no activation of EHL (sweating or panting) until some upper (EHL) threshold is reached then, as $\overline{T}_b$ rises above this

threshold, evaporative heat loss increases in proportion to the extent that $\overline{T}_b$ is above the EHL threshold.

3. At temperatures between these thresholds for HP and EHL, PVMT decreases as $\overline{T}_b$ increases.

Clearly a neuronal model of temperature regulation must offer some explanation for these thresholds, and a hypothetical explanation can be borrowed from MURAKAMI *et al.* [20]: activity in the pathways excited by signals from warm and cold sensors might be inhibited by the synaptic influences of temperature-insensitive neurones. Such inhibitory influences would impose threshold levels of excitatory activity along the pre-synaptic pathways from the sensors, that must be exceeded before signals would be passed on towards the EHL and HP effectors. Beyond these thresholds the post-synaptic activities would be proportional to the signals derived from the integrated influences of the warm or the cold sensors (fig. 3). The thermoregulatory control of PVMT operates in the thermal zone which lies between the HP and EHL thresholds. Thus, the inhibitory influence of warm sensors and the excitatory influence of cold sensors on PVMT are represented in figure 3 as being unaffected by the threshold mechanisms on the HP and EHL pathways.

The many non-thermal influences which act on the hypothalamic neurone pools and modify the temperature sensor-thermoregulatory effector relations [1, 13], are represented in the model (fig. 3) by non-thermal excitatory and inhibitory influences acting on the same synapses as those which represent the point of convergence of the pathways from hypothalamic and extra-hypothalamic temperature sensors. Such non-thermal excitatory or inhibitory influences could shift the balance of activities in the two principal sensor-effector pathways, and so cause a variation in the regulated temperature.

Neuronal Models Based on Hypothalamic Unit Activity Studies

The neuronal model discussed above is derived from theoretical interpretations of disturbance-response patterns in the whole organism and does not rely on any knowledge of events within the central nervous system (CNS). A detailed discussion of neuronal models based on records of the electrical activities of temperature-responsive and temperature-unresponsive hypothalamic neurones is outside the scope of this paper,

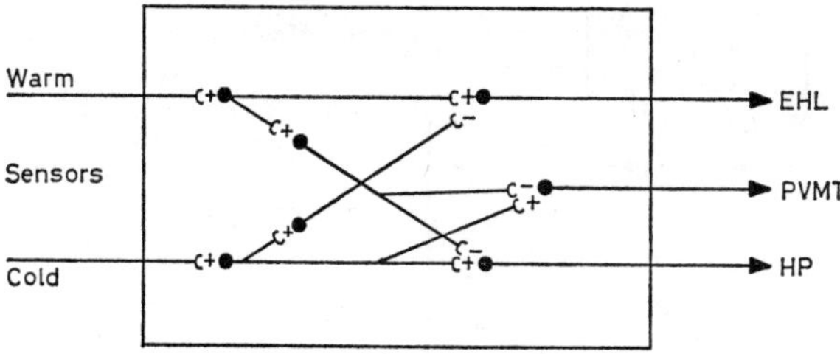

Fig. 5. A reconstruction of the neuronal relations between temperature sensors and thermoregulatory effectors proposed by HAMMEL [12] on the basis of hypothalamic unit activity studies.

but it is necessary to consider to what extent this evidence is consistent with the disturbance-response neuronal model of figure 3.

On the basis of some early unit activity studies, HAMMEL [12] proposed a neuronal model of mammalian thermoregulation (fig. 5), all the features of which are common to some of those of the disturbance-response model (fig. 3). Subsequent electrical activity studies have revealed many different unit activity-temperature relations, and HARDY and GUIEU [15] utilised all 22 patterns reported at that time from several different species, to construct an ingenious neuronal model. Because of its complexity, this model is not readily comparable with the disturbance-response model. While admitting the simplicity, and perhaps naïvety, of the disturbance-response model, it may be doubted whether that of HARDY and GUIEU is necessarily any nearer to physiological reality. It is assumed that all reported activity-temperature relations are fixed characteristics which can be related to fixed functions, but all units with biphasic activity-temperature curves are probably interneurones [7] affected by excitatory and inhibitory synaptic influences which will vary from moment to moment. Thus, it is doubtful whether activities of interneurones should be treated as fixed characteristics with specific functions in a neuronal model.

Theoretical activity patterns in the simple disturbance-response model cannot be expected to match all those actually recorded, especially since it is only an assumption that all recorded patterns relate to temperature regulation. However, theoretical activity-temperature characteristics consistent with the threshold-creating synaptic inhibition on the pathways to EHL and HP (fig. 6) are very similar to some recorded activity-temperature patterns [2]. Unit activity studies also indicate the convergence of

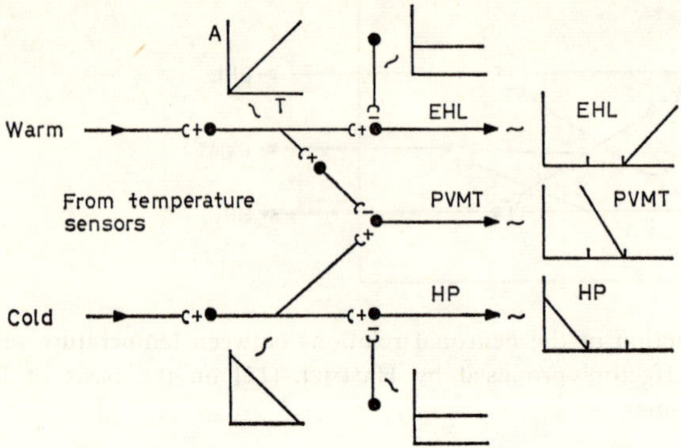

Fig. 6. An expression of the electrical activity (A) – temperature (T) relations which might be recorded if the hypothalamus really contained neurones with the characteristics and inter-relations suggested in figure 3.

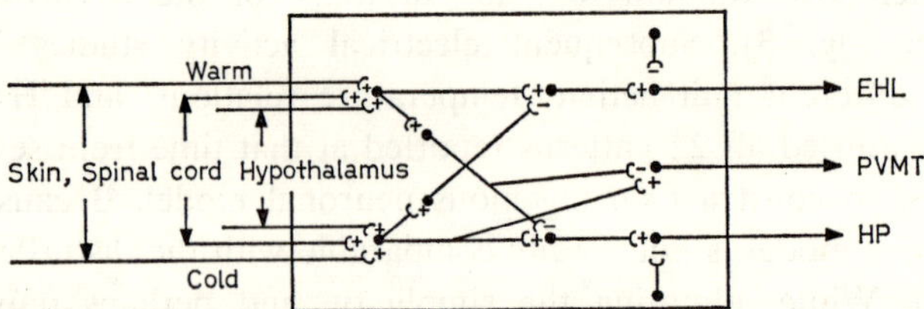

Fig. 7. A neuronal model developed from figure 5 by interpreting further evidence of the electrical activities of temperature-sensitive and -insensitive units in the hypothalamus (see text).

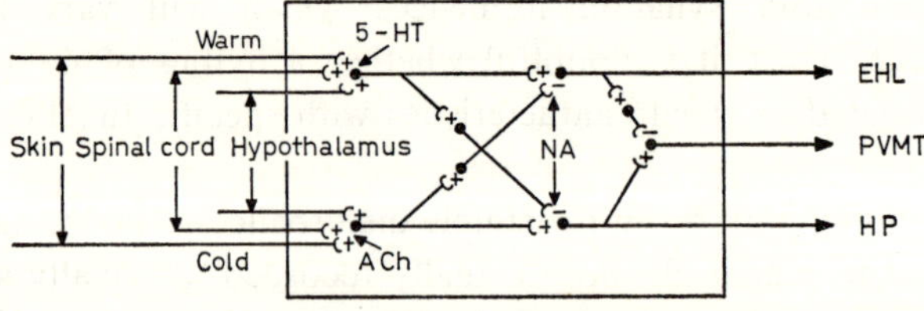

Fig. 8. A neuronal model based entirely on the evidence of synaptic interference studies on sheep. During peripheral, spinal or hypothalamic heating or cooling, 5-hydroxytryptamine (5-HT), noradrenaline (NA) and cholinomimetic substances (ACh) injected into a lateral cerebral ventricle have thermoregulatory effects which are as if these substances are acting at the sites indicated in the model.

the pathways from extra-hypothalamic and hypothalamic temperature sensors [7, 16]. By incorporating interpretations of unit activity studies, the model of HAMMEL [12] can be extended (fig. 7). This unit activity model is virtually the same as the disturbance-response model (fig. 3).

Neuronal Models Based on Pharmacological Interference with Hypothalamic Synaptic Functions

Attempts to interfere with synaptic chemistry as a means of learning something about the organisation in the neurone pools concerned in thermoregulation do not commend themselves at the theoretical level. There is no *a priori* reason why the same excitatory and inhibitory transmitter substances should not be acting in many different pathways, including pathways connected to opposing effector functions. Thus, the swamping of even a small region of the brain with synaptically active substances might be expected to produce such bizarre and unco-ordinated outflows of signals to effector organs as to render interpretation impossible. Contrary to this prediction, FELDBERG and MYERS [8] obtained clear and reproducible effects on body temperatures of cats when they injected the monoamines adrenaline, noradrenaline (NA) and 5-hydroxytryptamine (5-HT) into the lateral cerebral ventricles. Since then, many studies using many means to interfere with synaptic events have yielded almost as many different and often conflicting response patterns, some of which indicate species differences in the involvement of putative transmitter substances in the central regulation of body temperature.

The synaptic interference studies made in my laboratory on sheep have resulted in consistent and orderly patterns of thermoregulatory responses to the monoamines NA and 5-HT, and to cholinomimetic substances, when introduced into the lateral cerebral ventricles during (1) exposure to different levels of ambient temperature [3]; (2) local heating or cooling of the pre-optic-anterior hypothalamic region of the brain [17], and (3) during local heating and cooling of the spinal cord [BLIGH and MASKREY, unpublished information]. All 3 results can be expressed in terms of a simple neuronal model (fig. 8) which, although conceived quite independently, bears a remarkable similarity to the theoretical disturbance-response model (fig. 3). In all our experiments, 5-HT behaved as if it were an excitatory transmitter substance on the pathway between warm sensors and heat loss effectors, acting beyond the convergence of the pathways from the central and extra-central sensors and before the crossing inhibitory influence on the cold sensor to HP pathway. Cholinomimetic substances behaved as if acetylcholine (ACh) were an excitatory transmitter substance on the pathway between the cold sensors and the HP effectors, subsequent to the convergence of the pathways from the central and extra-central sensors and before the crossing inhibitory influence on the warm sensor to heat loss pathway. NA exerted an inhibitory influence on whichever set of effectors was active. When warm sensors were being stimulated, HL was inhibited by NA. When cold sensors were being stimulated, HP was inhibited by NA. Within the thermoneutral range, NA had little or no effect on body temperature, but a small rise in ear-skin temperature was indicative of an inhibition of active peripheral vasoconstriction. These effects of NA were consistent with the inhibitory influence of the pathways crossing from each main sensor-effector pathway to the other.

5-HT induced a rise in ear-skin temperature only when ambient temperature was within the range of thermoneutrality when the ear blood vessels were neither fully dilated nor fully constricted. Similarly, cholinomimetic substances induced a fall in ear-skin temperature only when ambient temperature was within the thermoneutral range. This evidence of the influence of putative hypothalamic transmitter substances on PVMT is consistent with the representation of its control in figure 8.

Further Experiments to Test the Validity of the Model Based on Synaptic Interference Studies

Whether this model is anything more than a useful description of a limited range of observations may depend on the extent to which it is capable of accommodating further experimental observations, and can be used to predict the outcome of further studies. The 4 experiments briefly summarised below indicate that, so far, the model has survived such tests.

A. The Thermoregulatory Effect of a Sympathomimetic Drug with Alleged Central Nervous Activity

MASKREY *et al.* [18] found that clonidine, when injected into the cerebral ventricles of sheep at different ambient temperatures, had thermoregulatory effects which were very similar to those of NA: at high ambient temperature panting was inhibited and core temperature rose, while at low ambient temperature shivering was inhibited and core temperature fell. This suggests that the thermoregulatory effects of NA in the sheep are not due to mere side-effects of this particular substance, but relate to a more general effect of sympathomimetic substances.

B. The Effect of Intraventricular Atropine on Intraventricular Injections of Cholinomimetic Substances

MASKREY and BLIGH [unpublished information] reasoned that if ACh acted only on the pathway from cold sensors to HP effectors, then atropine should block the HP resulting from a cold stimulus, but should have no effect on the EHL effects resulting from a warm stimulus. This is exactly what they found.

That atropine yielded the predicted effect was surprising, as we never supposed that this model represented exclusive roles of transmitter substances at one particular location. The same transmitters might be acting at different synapses along both HP and EHL pathways. Indeed, MYERS and YAKSH [21] have indicated that in the monkey, ACh acts in both pathways in the posterior hypothalamus, and HALL [11] has shown that in cats ACh exerts a nicotinic excitatory action on the EHL pathway and a muscarinic action on the HP pathway. Our clear result of atropine block on only the HP pathway could be due to its blocking only the muscarinic cholinergic sites, or because the effect of the intraventricular atropine was limited to cholinergic synapses in or near to the anterior hypothalamus.

C. The Interactions between the Pyrogenic Effects of TAB Vaccine and Intraventricularly Injected 5-HT and NA

Fever in sheep induced by TAB vaccine given intravenously is characterised by the inhibition of panting, the onset of shivering and a concomitant rise in core temperature. If the pyrogen exerts its direct or indirect effect on the afferent side of the points where the putative transmitter substances seem to be acting, then an intraventricular injection of NA during the rising phase of fever might be expected to inhibit the shivering, but not to activate panting. An intraventricular injection of

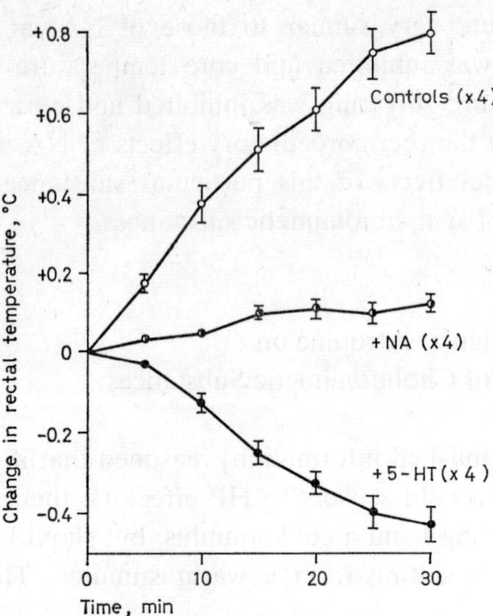

Fig. 9. The effect on the rectal temperature of sheep of an intravenous injection of a pyrogen (TAB vaccine), and the modification to this temperature pattern when noradrenaline (NA) or 5-hydroxytryptamine (5-HT) is injected into a lateral cerebral ventricle during the fever. Zero time is that point during the ascending phase of fever when the putative transmitter substances were injected. Each curve represents the average of 4 experiments.

5-HT might be expected to inhibit shivering and to activate panting. This is precisely what BLIGH and MASKREY [4] observed: NA, by inhibiting shivering, halted the rise in core temperature but did not reverse it; 5-HT, by inhibiting shivering and activating panting, caused a sharp reversal in core temperature (fig. 9).

D. An Analysis of the Effect of Prostaglandin E_1 on the Thermoregulation of the Sheep

Prostaglandin E_1 (PGE_1) causes hyperthermia when injected into the ventricles of the cat, rat and rabbit [9, 10, 19]. Unlike the putative transmitter substances, NA, 5-HT and ACh, which have different thermoregulatory effects on these 3 species, PGE_1 has similar effects. PGE_1, might thus be expected to increase HP and/or decrease EHL when injected into the cerebral ventricles of sheep, depending on the ambient temperature and whether PGE_1 acts on the HP pathway before or after the point of departure of the crossing inhibitory influence.

When BLIGH and MILTON [5] infused PGE_1 into the lateral cerebral ventricle of sheep at different ambient temperatures, the effects were very similar indeed to those of intraventricular injections of cholinomimetic substances. At low temperatures there was an initiation or intensification of shivering and a rise in core temperature. At high ambient temperatures there was an inhibition of panting and a rise in core temperature. Thus, PGE_1 on sheep, like the cholinomimetic substances, seems to be acting somewhere in the pathway between cold sensors and HP effectors before the origin of a crossed inhibitory influence on the pathway between warm sensors and heat loss effectors.

Discussion

Despite these further tests of its adequacy, the neuronal model remains no more than a description of what seems to happen when natural synaptic events in the neuronal pools concerned with temperature regulation are disturbed.

At least two severe limitations render this model unsatisfactory. The first is that it relates only to the sheep, and many similar studies on other species have yielded results that cannot be represented in this neuronal format. The second is that the model is based on experiments that allow no anatomical discretion: substances injected into a lateral cerebral ventricle are not necessarily acting on the anterior hypothalamus, or on this area alone. On the other hand, there are two considerations which may strengthen its position as a serious neuronal model. The first is the extraordinary similarity between this model and those which can be constructed from whole animal disturbance-response studies, and hypothalamic unit activity studies. The second is that predictions based on the model have been fulfilled experimentally. However, despite its descriptive and predictive integrity at the present time, the model is a very simple representation of a much more complex pattern of interrelating neurones. Further studies will, no doubt, reveal the limitations and fallacies of these early attempts to transfer our thoughts from analogous engineering models to neuronal models of mammalian temperature regulation, and so bring us progressively nearer to reality.

References

1 BLIGH, J.: The thermosensitivity of the hypothalamus and thermoregulation in mammals. Biol. Rev. *41:* 317–367 (1966).

2 BLIGH, J.: Neuronal models of mammalian temperature regulation; in BLIGH and MOORE Essays on temperature regulation (North-Holland, Amsterdam 1972).

3 BLIGH, J.; COTTLE, W. H., and MASKREY, M.: Influence of ambient temperature on the thermoregulatory responses to 5-hydroxytryptamine, noradrenaline and acetyl choline injected into the lateral cerebral ventricles of sheep, goats and rabbits. J. Physiol., Lond. *212:* 377–392 (1971).

4 BLIGH, J. and MASKREY, M.: The interactions between the effects on thermoregulation of TAB vaccine injected intravenously and monoamines injected into a lateral cerebral ventricle of the Welsh Mountain sheep. J. Physiol., Lond. *213:* 60–62P (1971).

5 BLIGH, J. and MILTON, A. S.: The thermoregulatory effects of prostaglandin E_1 when infused into a lateral cerebral ventricle of the Welsh Mountain sheep at different ambient temperatures. J. Physiol., Lond. *229:* 30–31P (1973).

6 BRÜCK, K. und WÜNNENBERG, W.: Die Steuerung des Kaltezitterns beim Meerschweinchen. Pflügers Arch. ges. Physiol. *293:* 215–225 (1967).

7 EISENMAN, J. S.: Unit activity studies of thermoresponsive neurons; in BLIGH and MOORE Essays in temperature regulation (North-Holland, Amsterdam 1972).

8 FELDBERG, W. and MYERS, R. D.: Effects on temperature of amines injected into the cerebral ventricles. A new concept of temperature regulation. J. Physiol., Lond. *173:* 226–237 (1964).

9 FELDBERG, W. and SAXENA, P. N.: Fever produced by prostaglandin E_1. J. Physiol., Lond. *217:* 547–556 (1971).

10 FELDBERG, W. and SAXENA, P. N.: Further studies on prostaglandin E_1 fever in cats. J. Physiol., Lond. *219:* 739–745 (1971).

11 HALL, G. H.: Changes in body temperature produced by cholinomimetic substances injected into the cerebral ventricles of unanaesthetized cats. Brit. J. Pharmacol. *44:* 634–641 (1972).

12 HAMMEL, H. T.: Neurones and temperature regulation; in YAMAMOTO and BROBECK Physiological regulation and control (Saunders, Philadelphia 1965).

13 HAMMEL, H. T.; JACKSON, D. C.; STOLWIJK, J. A.; HARDY, J. D., and STRØMME, S. B.: Temperature regulation by hypothalamic proportional control with an adjustable set-point. J. appl. Physiol. *18:* 1146–1154 (1963).

14 HARDY, J. D.: Physiology of temperature regulation. Physiol. Rev. *41:* 521–566 (1961).

15 HARDY, J. D. and GUIEU, J. D.: Integrative activity of preoptic units. II. Hypothetical network. J. Physiol., Paris *63:* 264–267 (1971).

16 HELLON, R. F.: The stimulation of hypothalamic neurones by changes in ambient temperature. Pflügers arch. ges. Physiol. *321:* 56–66 (1970).

17 Maskrey, M. and Bligh, J.: Interactions between the thermoregulatory responses to injections into a lateral cerebral ventricle of the Welsh Mountain sheep of putative neurotransmitter substances, and of local changes in anterior hypothalamic temperature. Int. J. Biometeorol. *15* (2–4): 129–133 (1971).

18 Maskrey, M.; Vogt, M., and Bligh, J.: Central effects of clonidine (2-(2,6-dichlorophenylamino)-2-imidazolinehydrochloride, St 155) upon thermoregulation in the sheep and goat. Europ. J. Pharmacol. *12:* 297–302 (1970).

19 Milton, A. S. and Wendlandt, S.: A possible role for prostaglandin E_1 as a modulator for temperature regulation in the central nervous system of the cat. J. Physiol., Lond. *207:* 76–77P (1970).

20 Murakami, N.; Stolwijk, J. A. J., and Hardy, J. D.: Responses of preoptic neurons to anaesthetics and peripheral stimulation. Amer. J. Physiol. *213:* 1015–1024 (1967).

21 Myers, R. D. and Yaksh, T. L.: Control of body temperature in the unanaesthetized monkey by cholinergic and aminergic systems in the hypothalamus. J. Physiol., Lond. *202:* 483–500 (1969).

22 Snellen, J. W.: Mean body temperature and the control of thermal sweating. Acta physiol. pharmacol. neerl. *14:* 99–170 (1966).

23 Thauer, R. and Simon, E.: Spinal cord and temperature regulation; in Ito, Ogata and Yoshimura Advances in climatic physiology (Igaku Shoin, Tokyo 1972).

24 Wyndham, C. H. and Atkins, A. R.: A physiological scheme and mathematical model of temperature regulation in man. Pflügers Arch. ges. Physiol. *303:* 14–30 (1968).

Author's address: Dr. J. Bligh, Agricultural Research Council, Institute of Animal Physiology, *Babraham, Cambridge CB2 4AT* (England)

Recent Studies of Hypothalamic Function
Int. Symp. Calgary 1973, pp. 328–340 (Karger, Basel 1974)

Unit Studies of Brainstem Projections to the Preoptic Area and Hypothalamus[1]

J. S. EISENMAN
Department of Physiology, Mount Sinai School of Medicine,
City University of New York, New York, N.Y.

I. Introduction

Since the early studies of RANSON and MAGOUN [13], attention has been focussed on the hypothalamus as both integrative network and a detector system in the regulation of body temperature. It is clear that the integrative functions of the hypothalamus require inputs from a variety of sources, in addition to that from temperature receptors. Thermoregulatory activity will be elicited by changes in thermal load (both internal and environmental) and also by shifts in general activity, level of cerebral cortical arousal and hormonal balances. Thus, the hyperthermia of exercise, circadian fluctuations of body temperature and the postovulatory rise in temperature all represent imposed modifications of thermoregulation. This suggests that brainstem areas related to control of arousal should project to the hypothalamic thermoregulator. Since lesions of, or stimulation in, these brainstem areas produce a variety of physiological and behavioral responses, specific analysis of the thermoregulatory role of brainstem projections has not been possible. Several recent lines of investigation have provided data which should now enable us to begin this type of analysis.

One of these lines of research is that of FELDBERG [7] and coworkers, who studied the effects on body temperature of amines injected into the cerebral ventricles or directly into the hypothalamus. In cats, it was found that injections of catecholamines (norepinephrine [NE] or epinephrine)

1 This research was supported by grant No. NS 09998 from the National Institute of Neurological Disease and Stroke, USPHS.

lowered body temperature, while injection of the indoleamine, 5-hydroxytryptamine (5-HT) (serotonin) caused a prolonged hyperthermia (fig. 1). The suggestion was made that heat-loss effectors were activated by adrenergic mechanisms, while heat production and conservation were activated by serotonergic mechanisms. Although species differences have been seen [cf. 3], the general observation that thermoregulatory effects can be activated by intrahypothalamic amines or acetylcholine [14] has been confirmed repeatedly.

These observations were extended by analysis of response of single neurons in the preoptic area to locally applied amines and acetylcholine [2, 5]. By correlating thermosensitivity and drug sensitivity, we attempted to confirm the observations of FELDBERG and colleagues. It was found that NE inhibited activity in warm-sensitive neurons which, because of their thermal response patterns [6], were thought to be interneurons in the regulatory pathways. In contrast, NE excited the firing of cool-sensitive

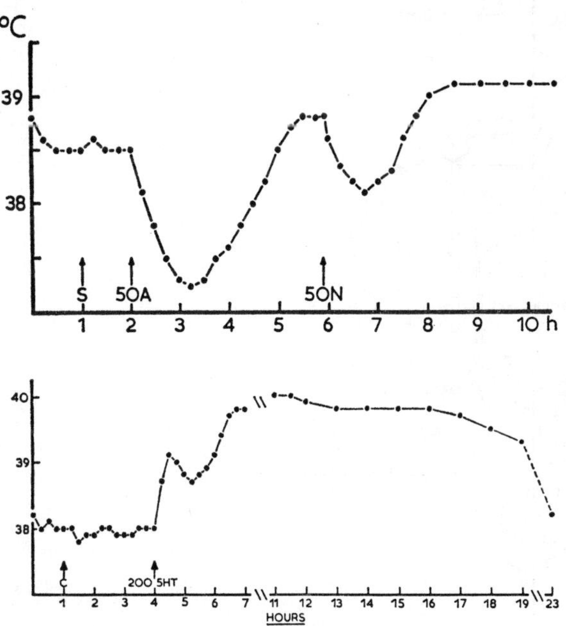

Fig. 1. Rectal temperature changes in unanesthetized cats in response to intracerebroventricular injection of saline, 50 μg adrenaline (A) and 50 μg noradrenaline (N) (upper figure); 100 μg creatinine sulphate and 200 μg 5-hydroxytryptamine (5-HT) (lower figure) [from HELLON, 11; after FELDBERG and MYERS, 8].

neurons. The high Q_{10} thermosensitive neurons, thought to be the central thermodetectors, were generally not affected by locally applied drugs, but did respond to nonspecific ionic currents. The responses to 5-HT and acetylcholine were not striking. These data indicated that neurons in the preoptic area with presumed thermoregulatory functions did, in fact, respond to catecholamine in a specific way. Further, it appeared that the thermodetector group was not sensitive to transmitter amines.

The final line of evidence that contributed to this analysis came from the localization studies of FUXE and his group [1]. Using a histochemical fluorescence technique, they determined that the catecholamines and indoleamines found in the preoptic area and hypothalamus are contained in axon terminals of neurons whose cell bodies lie in specific brainstem areas (fig. 2). Neurons in the hypothalamus itself do not contain NE or serotonin. Although most of this work was done on rats, some of these

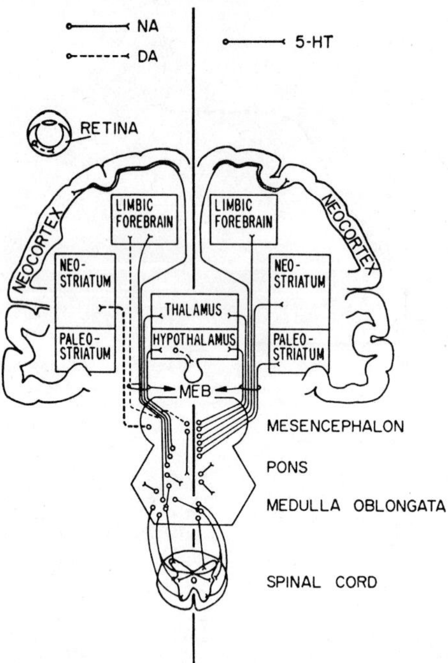

Fig. 2. Diagrammatic representation of the locations and projections of catecholamine containing neurons (on left) and serotonergic neurons (on right) in brainstem and forebrain. NA = noradrenaline; DA = dopamine; 5-HT = 5-hydroxytryptamine; MFB = medial forebrain bundle [from ANDEN *et al.*, 1].

observations have been confirmed in cats [12]. The implication of these findings is that the intracerebral microinjection studies and the iontophoretic single unit studies were actually examining responses of the hypothalamic neurons to extrahypothalamic inputs and not the activity of intrahypothalamic neural networks.

Thus, it seemed appropriate to study the responses of preoptic neurons to inputs from the brainstem areas containing the noradrenergic and serotonergic neurons and pathways by electrically stimulating these areas and recording activity changes in preoptic area neurons.

II. Methods

Urethane anesthetized cats were prepared with an array of 3 parasagittal thermodes (fig. 3). Another tube, open at both ends, was placed lateral to these to hold the thermistor for monitoring tissue temperature. The microelectrode was placed in the contralateral preoptic area or anterior hypothalamus. The relative positions of the thermistor, thermode array and microelectrode were adjusted so that the thermistor readings would reflect the tissue temperature at the microelectrode recording site.

For stimulation in the mesencephalic reticular formation, a transverse array of 5 monopolar electrodes was placed ipsilateral to the microelectrode at Horsley-Clarke planes A2.0–A4.0. For stimulation of the midline raphe nuclei, an array of 3 coaxial electrodes was placed in the midline, at an angle of 30° posterior to the vertical. Only one array was placed in a given cat. Conduction distances to the preoptic area (assuming a direct pathway) were 10–12 mm from the mesencephalon and 14–18 mm from the raphe nuclei.

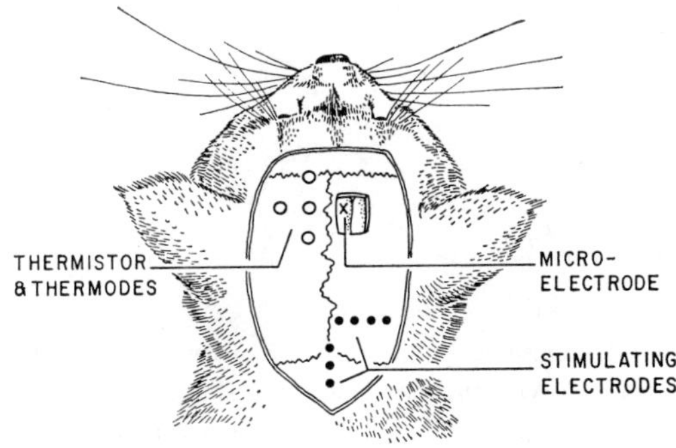

Fig. 3. Placements of thermodes and thermistor for varying tissue temperature, microelectrode for unit recording and electrode arrays for brainstem stimulation.

Stimuli were rectangular, constant current pulses, 0.3 msec in duration, with amplitudes up to 1 mA, given at a 1/sec rate. Responses in preoptic area (POA) neurons could generally be obtained with currents of 500–750 μA.

III. Results

Stimulation in the mesencephalic reticular formation influenced the firing of preoptic neurons when applied in two areas (fig. 4); a ventromedial region, just lateral to the interpeduncular nucleus, and a more dorsolateral area, above the medial lemniscus. Facilitation or inhibition of POA firing could be obtained from either area. Most commonly, a facilitatory response was followed by a period of decreased firing probability or inhibition, while initial inhibition was followed by a period of 'rebound', or enhanced firing probability.

Midline pontine stimulation was effective only when applied to the more rostral raphe nuclei (fig. 5). No units were found that responded to stimulation in the caudal pons or medulla. The response *patterns* to pontine stimulation were similar to those produced by mesencephalic stimuli, although the latencies and percentage responsiveness were different.

The latencies of responses to stimulation varied markedly for different neurons, and also varied somewhat for a given unit. Latencies of

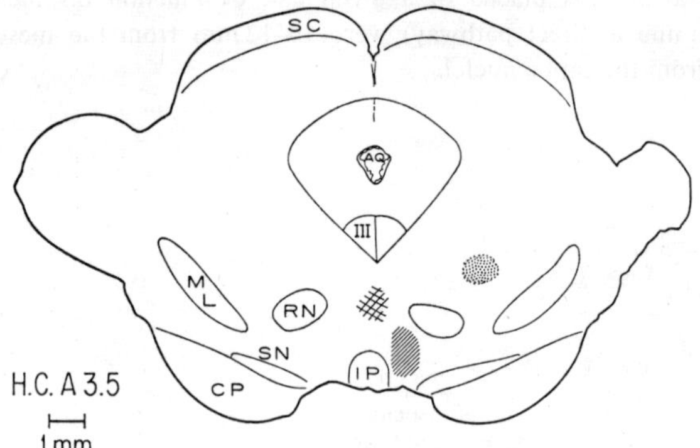

Fig. 4. Locations of areas in mesencephalic reticular formation, stimulation of which affected preoptic area (POA) neurons. Diagonal shading: ventromedial zone. Stippling: dorsolateral zone. Aq. = aqueduct; CP = cerebral peduncle; IP = interpeduncular nucleus; ML = medial lemniscus; RN = red nucleus; SC = superior colliculus; SN = substantia nigra; III = oculomotor complex. H.C.A3.5 = sterotaxic plane, A3.5.

facilitatory actions ranged from 3 to 50 msec for mesencephalic stimulation, and from 5 to 50 msec for pontine stimulation. In general, the strongest facilitatory actions had the shortest latencies and durations, and showed the least variation (fig. 6, 7). Short-latency responses had durations of 3–10 msec. Weaker facilitatory effects had longer latencies and durations of action. Durations of 30–50 msec were common in responses with latencies of about 50 msec; in a few instances, durations of 80–100 msec were seen.

Figure 6 illustrates a very short latency response (5 msec average), followed by an inhibition of firing for about 50 msec. Figure 7 shows a short latency facilitation with no later inhibitory effect. Not all strong facilitatory responses had short latencies, however. Figure 8 illustrates driving of a warm-sensitive POA neuron with a very constant latency of 34 msec. This was not followed by any apparent inhibition. One POA unit was found which was antidromically activated by mesencephalic stimulation, with a latency of 5 msec.

Inhibitory latencies and durations were generally longer than facilitatory ones. Inhibition of POA neuron firing by raphe nuclear stimulation occurred with latencies of 12–200 msec and durations of 40–350 msec. Reticular stimulation produced inhibition with latencies of 20–50 msec and durations up to 100 msec (fig. 9). Again, short latencies were associated with short durations.

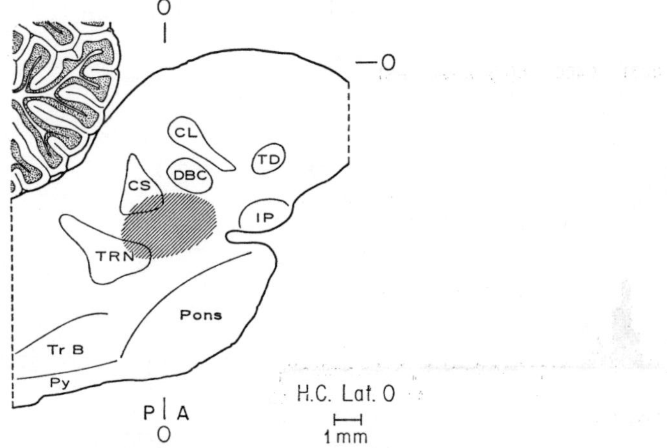

Fig. 5. Location of active area in pontine raphe nuclei. CL, CS = central linear and superior nuclei; DBC = decussation of the brachium conjunctivum; PY = pyramid; TRN = tegmental reticular nucleus; Tr B = trapezoid body; IP = interpeduncular nucleus; TD = decussation; H.C. lat.0 = lateral 0 plane.

Stimulation in the mesencephalic reticular formation affected about 50 % of the POA units tested, for both thermosensitive and thermo-insensitive cells. Overall, 56 cells of 107 tested (52 %) responded (including one $Q_{10}1$ cell which was fired antidromically) (table I). Of these, 29 of 62 $Q_{10}1$ cells (48 %) were affected by the stimulus, while

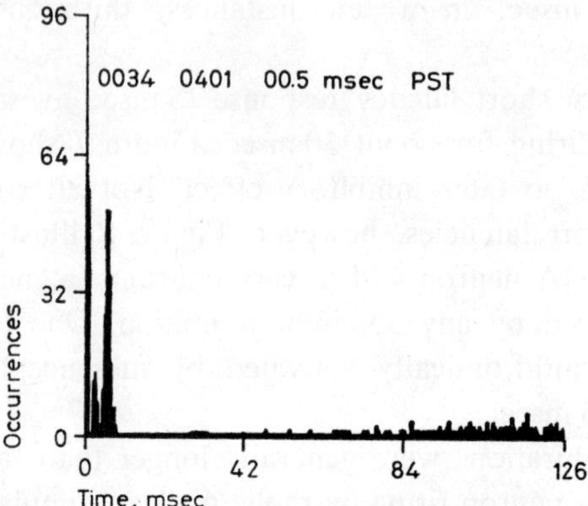

Fig. 6. Poststimulus-time (PST) histogram for a cool-sensitive interneuron. Short-latency driving of spike from dorsolateral mesencephalic reticular formation. Mean latency, 5 msec. (Early response is due to stimulus artifact.) Note marked inhibition of firing following driven spike. 257 sweeps at 1/sec.

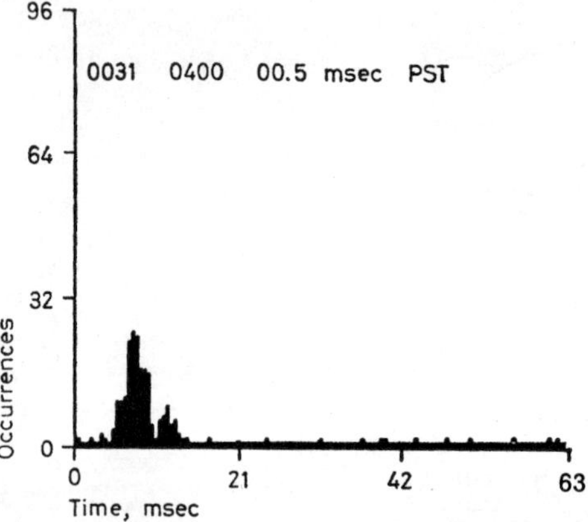

Fig. 7. Poststimulus-time histogram for a thermo-insensitive POA neuron. Driving of spike is seen, with a mean latency of 9 msec. Stimulating electrode in dorsolateral reticular formation. Note slight inhibition of firing following driven response. 256 sweeps at 1/sec.

26 of 42 thermosensitive cells (62 %) responded. In the latter group are included 14 cells that were classified as $Q_{10}>2$ thermodetector-type neurons, 7 of which (50 %) were inhibited by the mesencephalic stimulus.

With pontine raphe nuclear stimulation, 40 of 52 POA neurons tested (77 %) were affected (table II); 28 of 35 $Q_{10}1$ cells (80 %) and

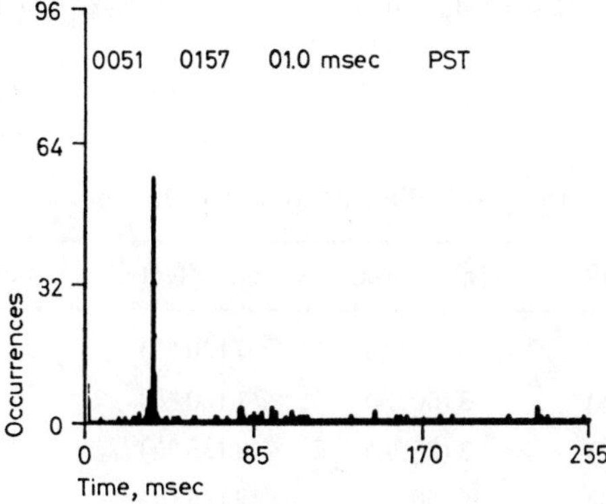

Fig. 8. Poststimulus-time histogram for warm-sensitive interneuron. Driving, with latency of 34 msec. Note lack of postdriving inhibition. 111 sweeps at 1/sec. Stimulating electrode in medial reticular formation.

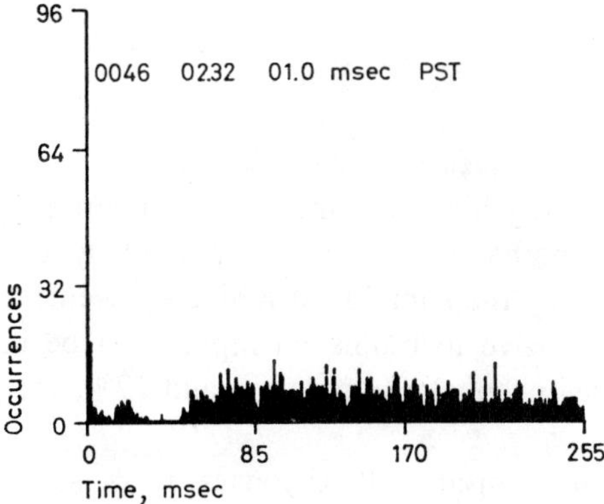

Fig. 9. Poststimulus-time histogram for thermodetector-type POA neuron. Short-duration inhibition of unit firing with latency of about 30 msec, evoked from medial reticular formation. Note postinhibitory 'rebound' of long duration. 154 sweeps at 1/sec.

Table I. Responses of preoptic area (POA) neurons to mesencephalic reticular stimulation

Unit type	Facilitation	Inhibition	Antidromic	No effect	Total
$Q_{10}1$	14 (23 %)	15 (24 %)	1 (2 %)	32 (51 %)	62
$Q_{10}>2$	0	7 (50 %)	0	7 (50 %)	14
Interneuronal	7 (25 %)	12 (43 %)	0	9 (32 %)	28
Not tested	0	0	0	3	3
Total	21 (20 %)	34 (31 %)	1 (1 %)	51 (48 %)	107

Table II. Responses of POA neurons to midline pontine stimulation

Unit type	Facilitation	Inhibition	No effect	Total
$Q_{10}1$	8 (23 %)	20 (57 %)	7 (20 %)	35
$Q_{10}>2$	1 (20 %)	3 (60 %)	1 (20 %)	5
Interneuronal	5 (42 %)	3 (25 %)	4 (33 %)	12
Total	14 (27 %)	26 (50 %)	12 (23 %)	52

12 of 17 thermosensitive cells (70 %). The thermosensitive cell group includes 5 $Q_{10}>2$ detector-type cells, 4 of which responded, 1 was facilitated and 3 inhibited.

IV. Discussion

These results demonstrate that mesencephalic and pontine brainstem areas influence the activity of a very high percentage of the neurons in the preoptic area and anterior hypothalamus. Since it is unlikely that all ascending fibers were activated by the stimulus at a given placement, the actual percentage of cells responsive to brainstem input must be higher than the observed 50 % for mesencephalic stimulation and 77 % for pontine stimulation.

The mesencephalic stimulus produced responses in a somewhat higher percentage of *thermosensitive* neurons (62 %) than it did in *thermally insensitive* cells (48 %). Further, while about equal numbers of $Q_{10}1$ cells were facilitated and inhibited, the predominant effect of the input on thermosensitive neurons was inhibitory; 19 of the 26 cells responding (73 %) were inhibited. All of the detector-type POA neurons ($Q_{10}>2$) that responded to reticular stimulation were inhibited.

With midline pontine stimulation, the response patterns were reversed. Equal numbers of thermosensitive cells were facilitated and inhibited, while 20 of the 28 $Q_{10}1$ cells responding (71 %) were inhibited. For the small group of detector-type cells studied, 3 of the 4 cells responding to pontine stimulation were inhibited and only 1 was facilitated.

Overall, for both the mesencephalic and pontine stimulations, about two-thirds of the responsive cells were inhibited. The thermosensitive cell group was more strongly inhibited by mesencephalic than by pontine stimulation.

The most surprising result to appear in these experiments was that the $Q_{10}>2$ detector-type neurons were affected by brainstem stimulation, since earlier findings [5, 6] had suggested that these neurons were primary receptor-type cells with an inherent or pacemaker activity, and no synaptic input. The present results do not contradict the hypothesis that the detector cells are inherently active, but now add the possibility of neural modification of this activity. In view of these findings, the low sensitivity of detector cells to iontophoretically applied transmitter agents [2] serves to emphasize that adequate stimulation of neural synapses may not always be possible with this mode of application. Alternately, the transmitter involved in this system may be one *not* tested in the iontophoresis series.

The wide range of latencies observed for responses to brainstem and pontine stimulation could be due to activation of many ascending fibers of different axon diameters (and, thus, conduction velocities) and to activation of pathways with different numbers of synapses contributing to the overall conduction time from stimulus site to recording site. Both of these factors undoubtedly operate. The long duration of action, seen in some instances, suggests activity in a complex multisynaptic network with some measure of re-excitation. The short-latency activation responses, which showed very little latency spread, could be due to a fairly direct input. If monosynaptic, such excitation would be carried by fibers with maximal conduction velocities of about 3–4 M/sec for mesencephalic input and 2–3 M/sec for the pontine afferents. On the other hand, the strong, precise driving of the unit shown in figure 8, if monosynaptic, would require an afferent conduction velocity of 0.35 M/sec.

For inhibitory effects, maximal conduction velocities (assuming a monosynaptic path) would be 0.6–0.8 M/sec for mesencephalic inputs and

0.9–1.0 M/sec for raphe nuclear afferents. It is probably more reasonable to assume that the inhibitory pathway is multisynaptic and complex, especially since very short durations of inhibition, comparable to the short excitatory responses, were not seen.

Following responses to the stimulus, most neurons showed a 'rebound' in firing; depressed firing following facilitation (fig. 6) or enhanced firing following inhibition (fig. 9). Depression of firing probability following activation need not represent an active inhibitory process, but may simply be due to a cycle of depressed excitability following firing of the cell. In this regard, depression of firing frequently lasted for a time-period comparable to the minimal interspike interval seen in the spontaneously firing cell. Rebound facilitation, following an inhibitory response, is more likely an active process since the cell should not cycle into a hyper-excitable state with removal of inhibition.

Part of the rationale behind these experiment was to provide a correlation between responses of POA neurons and activation of histochemically identified brainstem pathways as part of the analysis of organisation of the hypothalamic thermoregulatory system. Thus, the stimulation sites were chosen to allow activation of the major ascending noradrenergic and serotonergic systems. The ventromedial mesencephalic area is a region of passage of both NE- and 5-HT-containing axons, although some spatial separation exists [1, 16]. The dorsolateral area appears to be mainly noradrenergic. The midline raphe nuclei are purely serotonergic.

Since stimuli at any active site could produce primary facilitation of some neurons and inhibition of others, we are left with the possibility that the same transmitter may facilitate one neuron and inhibit another, or that one of the functional pathways contains an additional interneuron to convert facilitation to inhibition. The generally longer latencies associated with inhibition suggest that the latter alternative may be correct, although the former is theoretically possible. Direct correlation of amine sensitivity and responsiveness to brainstem stimulation should provide the solution to this problem.

There have been several studies, reported in the literature, describing effects of reticular formation activation on hypothalamic units. Most of these have dealt with responses in the posterior and tuberal hypothalamus in relation to neuro-endocrine function and limbic system mechanisms [cf. 10]. FELDMAN [9] has reported that stimulation in the dorsal mesencephalic reticular formation produced a long-lasting inhibition of the potentials evoked in the hypothalamus by stimulation of the sciatic nerve.

DAFNY and FELDMAN [4] found that dorsal mesencephalic reticular stimulation could facilitate or inhibit posterior hypothalamic neurons. Lesions in this same reticular area produced a drop in the spontaneous activity of posterior hypothalamic cells. TSUBOKOWA and SUTIN [15] reported that cells in the ventromedial nucleus of the hypothalamus could be facilitated or inhibited by high-frequency reticular stimulation. Single-shock stimuli facilitated only 4 % of the cells studied.

It is, thus, clear that reticular formation inputs influence the activity of neurons in a variety of hypothalamic loci and, presumably, with a variety of functions in addition to thermoregulation. The interpretation of responses to injected transmitter agents must take account of these factors.

Summary

Single-shock, electrical stimulation in the ventromedial mesencephalic reticular formation or in the midline pontine raphe nuclei can influence the firing of neurons in the preoptic area and anterior hypothalamus. Both inhibitory and facilitatory effects can be obtained from either active site. Mesencephalic stimulation affected 62 % of the thermosensitive POA cells studied, with 73 % of those responding being inhibited and 27 % being facilitated. This stimulus affected only 48 % of the thermally insensitive cells studied, with equal numbers being facilitated and inhibited.

Pontine stimulation affected 77 % of the cells studied, with equal numbers of thermosensitive cells being facilitated and inhibited.

High Q_{10}, thermodetector-type neurons in the POA were among those thermosensitive cells that were inhibited by mesencephalic and pontine stimulation.

Conduction latencies suggest that a monosynaptic or oligosynaptic pathway may mediate the facilitatory effects, but that a more complex pathway probably mediates the inhibitory responses.

The stimulus sites correspond to areas of origin and passage of ascending serotonergic and noradrenergic pathways. Since both facilitation and inhibition of different neurons could be obtained from each site, the same pathway seems able to mediate both actions, perhaps with the introduction of an additional interneuron to change excitation to inhibition. Alternatively, the same transmitter agent could act to excite some neurons and to inhibit others.

References

1 ANDEN, N. E.; DAHLSTRÖM, A.; FUXE, K.; LARSSON, K.; OLSON, L., and UNGERSTEDT, U.: Ascending monoamine neurons to the telencephalon and diencephalon. Acta physiol. scand. *67:* 313–326 (1966).

2 BECKMAN, A. L. and EISENMAN, J. S.: Microelectrophoresis of biogenic amines on hypothalamic thermosensitive cells. Science *170:* 334–336 (1970).

3 BLIGH, J.; COTTLE, W. H., and MASKREY, M.: Influence of ambient temperature on the thermoregulatory responses to 5-hydroxytryptamine, noradrenaline and acetylcholine injected into the lateral cerebral ventricles of sheep, goats and rabbits. J. Physiol., Lond. *212:* 377–392 (1971).

4 DAFNY, N. and FELDMAN, S. M.: Effects of stimulating reticular formation, hippocampus and septum on single cells in the posterior hypothalamus. Electroenceph. clin. Neurophysiol. *26:* 578–587 (1969).

5 EISENMAN, J. S.: Unit activity studies of thermosensitive neurons; in BLIGH and MOORE Essays on temperature regulation, pp. 55–69 (North Holland, Amsterdam 1972).

6 EISENMAN, J. S. and JACKSON, D. C.: Thermal response patterns of septal and preoptic neurons in cats. Exp. Neurol. *19:* 33–45 (1967).

7 FELDBERG, W. S.: The monoamines of the hypothalamus as mediators of temperature responses; in HARDY Physiological and behavioral temperature regulation, pp. 493–506 (Thomas, Springfield 1970).

8 FELDBERG, W. and MYERS, R. D.: Effects on temperature of amines injected into the cerebral ventricle. A new concept of temperature regulation. J. Physiol., Lond. *173:* 226–237 (1964).

9 FELDMAN, S. M.: Neurophysiological mechanisms modifying afferent hypothalamo-hippocampal conduction. Exp. Neurol. *5:* 269–291 (1962).

10 FELDMAN, S. M. and DAFNY, N.: Effects of extrahypothalamic structures on sensory projections to the hypothalamus; in MARTINI The hypothalamus, pp. 103–114 (Academic Press, New York 1970).

11 HELLON, R.: Central transmitters and thermoregulation; in BLIGH and MOORE Essays on temperature regulation, pp. 71–86 (North Holland, Amsterdam 1972).

12 PIN, C.; JONES, B. et JOUVET, M.: Topographie des neurones monoaminergiques du tronc cérébral du chat. Etude par histofluorescence. C.R. Soc. Biol. *162:* 2136–2141 (1968).

13 RANSON, S. W. and MAGOUN, H. W.: The hypothalamus. Ergeb. Physiol. *41:* 56–163 (1939).

14 RUDY, T. A. and WOLF, H. H.: Effect of intracerebral injection of carbamylcholine and acetylcholine on temperature regulation in the cat. Brain Res. *38:* 117–130 (1972).

15 TSUBOKOWA, T. and SUTIN, J.: Mesencephalic influences upon the hypothalamic ventromedial nucleus. Electroenceph. clin. Neurophysiol. *15:* 804–810 (1963).

16 UNGERSTEDT, U.: Stereotaxic mapping of the monoamine pathways of the rat brain. Acta physiol. scand. *82:* suppl. 367, pp. 1–39 (1971).

Author's address: Dr. J. S. EISENMAN, Department of Physiology, Mount Sinai School of Medicine, City University of New York, *New York, NY 10029* (USA)

Recent Studies of Hypothalamic Function
Int. Symp. Calgary 1973, pp. 341–358 (Karger, Basel 1974)

Hypothalamic Control of Thermoregulatory Behavior

Preoptic-Posterior Hypothalamic Interaction

ELEANOR R. ADAIR

John B. Pierce Foundation Laboratory, and Yale University, New Haven, Conn.

Introduction

The notion of two brainstem centers for the regulation of body temperature, originally formalized by HANS MEYER in 1913 [21], is still widely held among physiologists and physicians. Every medical student learns that there is a thermolytic region in the anterior hypothalamus that regulates the dissipation of body heat, whereas an opposing thermogenic region, located more caudally, regulates the conservation or generation of body heat. On the basis of abundant neurophysiological, physiological and behavioral evidence, there can be little doubt today that one of these regions, the preoptic-anterior hypothalamus, serves both thermolytic and thermogenic control functions. Indeed, the implication is strong that this particular brainstem area is the major central thermosensitive site in endotherms.

The role of the posterior hypothalamus in temperature regulation is less well understood but is clearly more than thermogenic control. Neurophysiological studies show the presence of thermosensitive units in this region that respond with changes in firing rate to both local heating and cooling [7, 8, 28] and also to temperature changes in the preoptic area, the spinal cord and the skin [11, 17, 23–25]. On the other hand, autonomic responses to heating and cooling this area are reported to be weak [12]. KELLER and HARE [18] showed that surgical removal of the posterior hypothalamus in dogs resulted in loss of thermoregulation in the cold and impaired thermoregulation in the heat. Subsequent observations by MURGATROYD *et al.* [22] on KELLER's dogs suggested that behavioral thermoregulation might be very difficult for these animals. HARDY, after surveying much of the experimental evidence related to thermoregulatory functions of the posterior hypothalamus, ventured that '...the caudal posterior hypothalamus is the area in which the reticular integration is completed and the central (thermoregulatory) effector signal developed' [15].

Thermal stimulation of the preoptic-anterior hypothalamus results in clear-cut changes in any behavior that alters the body heat content. For example, preoptic warming stimulates animals to move to a cooler part of a thermal gradient [6], take cool showers [10] or lower the temperature of the test chamber in which they are confined [2]. However, sometimes prolonged preoptic cooling is not as effective in altering thermoregulatory behavior as it was when first imposed, indicating the possible shift of control to some other thermosensitive site [2]. Furthermore, lesions of the preoptic area, while impairing autonomic thermoregulatory responses, appear to leave thermoregulatory behavior intact [4, 13, 19, 26]. While areas such as the spinal cord [17] and the medulla [16, 20] must not be ruled out, it is not unreasonable to ask if the posterior hypothalamus may play a role in the behavioral regulation of body temperature.

The experiments to be reported here examine the interplay between the preoptic and posterior hypothalamic thermosensitive areas in the behavioral control of body temperature. They present clear evidence that thermal stimulation of the posterior hypothalamus can alter thermoregulatory behavior nearly as effectively as can comparable thermal stimulation of the preoptic area. When the temperatures of both preoptic and posterior hypothalamic areas are simultaneously displaced, the resulting behaviorally produced body temperature changes in the steady state reflect additivity of action. However, the posterior hypothalamus may exert control over behavioral regulation alone; no change in autonomic responses of heat production or heat loss could be detected when this area alone was thermally stimulated.

Subjects and Thermode Preparation

Four adult male squirrel monkeys, *Saimiri sciureus* (Leticia, Columbia) served as subjects. Their estimated age range was 24–36 months, and they weighed 700–1000 g at the time of testing.

Each animal was unilaterally implanted with thermode and thermocouple re-entrant tubes in both the medial preoptic area and the posterior hypothalamic area. The stereotaxic coordinates for the preoptic thermode were A 13.0, L 0.5 and D +1.0 and for the posterior hypothalamic thermode, A 7.5, L 1.5 and D +1.5, according to the atlas of EMMERS and AKERT [9]. The re-entrant tube placements were verified by X-rays in the coronal and sagittal planes [for details of the surgical procedures, see 2]. Water perfusion of the thermodes (at a rate of 12 ml/min) adjusted preoptic and/or posterior hypothalamic temperature to discrete levels between 33 and 43 °C. The local temperatures thus produced were measured by thermocouples inserted in re-entrant tubes centered 2 mm lateral to each thermode.

Measurement of Autonomic Responses

Each monkey was first tested for autonomic responses to local heating and cooling of the preoptic and posterior hypothalamic areas. The animal was restrained in a neutral environment, 32–33 °C, and the following temperatures were monitored: preoptic, posterior hypothalamic, rectal, ambient and 4 representative skin temperatures (abdomen, tail, leg and foot) from which a weighted mean skin temperature was computed. In some experiments, oxygen consumption and respiratory evaporative heat loss were also monitored but these parameters showed little measurable change as a result of local hypothalamic heating and cooling.

Figure 1 shows representative results for one monkey of 1-hour displacements of preoptic temperature, first to 35 and then to 42 °C, holding posterior hypothalamic temperature constant. Cooling the preoptic area results in heat storage, as evidenced by the abrupt rise in rectal temperature. This occurs as a result of some peripheral vasoconstriction and a slight increase in metabolic heat production. Conversely, when the preoptic area is warmed, rectal temperature falls and evidence of peripheral vasodilation is indicated by the sudden rise in foot skin temperature.

A comparable experiment, on the same monkey, showing the effects of local cooling and heating of the posterior hypothalamus is presented in figure 2. During 1-hour displacements of posterior hypothalamic temperature, keeping preoptic temperature constant, there is no change in rectal temperature, and the changes in skin temperature that occur seem uncor-

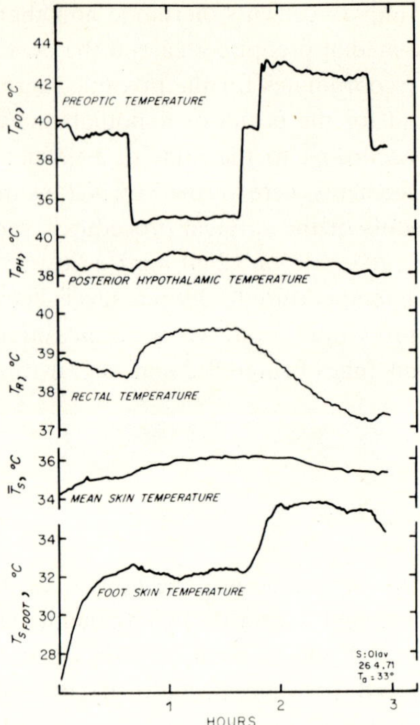

Fig. 1. Autonomic responses of one monkey to cooling and warming the preoptic area. Preoptic cooling stimulates heat storage and preoptic warming stimulates heat dissipation. Ambient temperature = 33 °C.

related with thermal stimulation of the posterior hypothalamus. The fluctuations in foot skin temperature seen here have been studied in the squirrel monkey by STITT and HARDY [27]. These fluctuations reflect changes in peripheral vasomotor state when the ambient temperature is between 30 and 35 °C. Thus, this monkey, representative of the 4 monkeys studied, shows normal thermoregulatory responses to preoptic thermal stimulation, but apparently no such responses to comparable stimulation of the posterior hypothalamus.

Measurement of Behavioral Responses

Behavioral thermoregulation. Prior to surgery, each monkey received extensive training to control the temperature of his environment. The

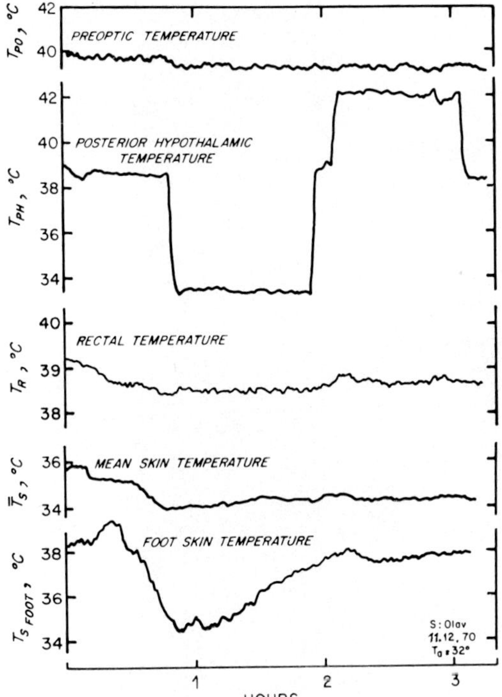

Fig. 2. Lack of autonomic responses to cooling and warming the posterior hypothalamic area of one monkey. Fluctuations in skin temperature reflect changes in peripheral vasomotor state in a neutral environment (32 °C).

monkey was restrained, as shown in figure 3, in a small chamber that was heated and cooled by forced convection. A valve system, activated by the animal's response chain, allowed air from one of two thermostatically controlled air sources (10 and 50 °C) to circulate through the chamber. [A detailed description of the apparatus and training procedures may be found in 2.] Under normal conditions, squirrel monkeys choose an average air temperature of about 35 °C by alternating between the two available source temperatures.

The behavioral thermoregulatory responses of the monkeys were measured under conditions of both brief and prolonged thermal stimulation of the preoptic and posterior hypothalamic areas. In some experiments, only one local temperature was changed, with the temperature of the other area held as closely as possible to the neutral level. In other

Fig. 3. View of restrained monkey in the test chamber. Air enters at lower right and exits at upper left.

experiments, both local temperatures were changed simultaneously to the same or different levels. During all experiments, preoptic, posterior hypothalamic, rectal, ambient and the 4 skin temperatures were monitored.

Effects of brief thermal displacements. Each monkey was initially tested for effects on thermoregulatory behavior of brief thermal displacements of the two brain areas singly. Figure 4 shows portions of two such experiments on one monkey. In the experiment on the left, local preoptic temperature was raised and lowered every 10 min by different

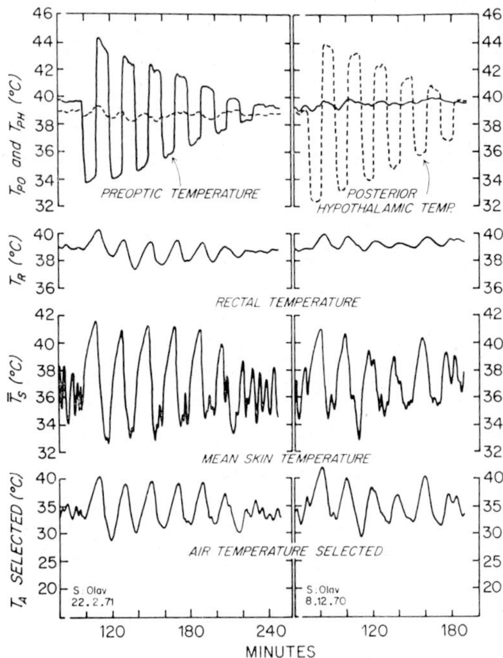

Fig. 4. Portions of two experiments showing changes in air temperature selected and resulting skin and rectal temperature changes when preoptic (left panel) or posterior hypothalamic (right panel) temperature is displaced every 10 min.

amounts relative to the neutral level, while posterior hypothalamic temperature was held as constant as possible. A comparable experiment in which posterior hypothalamic temperature was displaced every 10 min is shown on the right. The results of both experiments appear similar. When either preoptic or posterior hypothalamic temperature is lowered, the monkey turns on the warm air, driving skin and core temperature up. Conversely, when either brain area is warmed, the monkey chooses a cooler air temperature, driving skin and core temperature down. The figure also shows that progressively smaller thermal displacements of either area around the neutral level produce body and air temperature changes in the appropriate direction but of progressively smaller magnitude.

A quantitative representation of this relationship is given in figure 5, where results from several experiments on this monkey are combined. The percentage of time the animal chose 50 °C air (and the average air temperature produced by this behavior) is plotted as a function of the average pre-

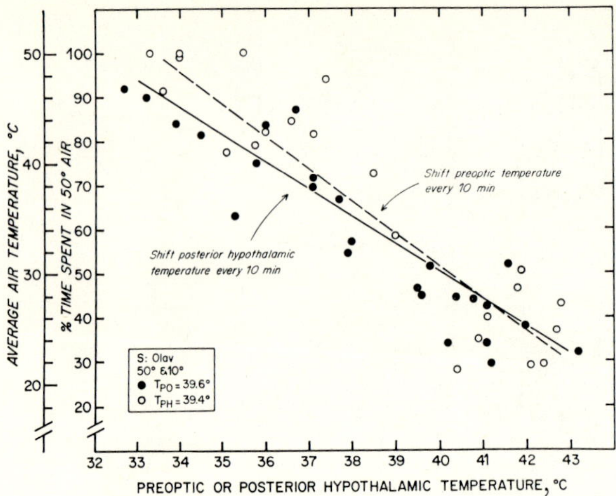

Fig. 5. Behavioral analysis of several experiments like those shown in figure 4. Percentage of each 10-min interval 50 °C air was chosen (and resulting average air temperature) is plotted as a function of preoptic (open circles, dashed line) or posterior hypothalamic (closed circles, solid line) temperature.

optic (open circles) or posterior hypothalamic (closed circles) temperature during each 10-min interval. Both relations are linear over the range explored and the slopes represent, to a first approximation, the degree of control exerted on thermoregulatory behavior by signals from the preoptic and posterior hypothalamic areas. The regression lines, calculated by the method of least squares, have slopes of –7.3 for preoptic thermal stimulation (r = –0.937) and –6.2 for posterior hypothalamic thermal stimulation (r = –0.925). For this monkey, preoptic cooling produces a greater change in behavior per degree displacement than does comparable posterior hypothalamic cooling, while the sensitivity of the two areas to warming is about equal. In terms of the proportion of time spent in 50 °C air per degree displacement of the central temperature, the overall difference is about 1 % per degree.

A second monkey showed intermediate sensitivity; the slope difference between the calculated regression lines was 3.3 % per degree. The other two monkeys showed no behavioral changes when posterior hypothalamic temperature was changed. Clearly the anatomical locus of effective stimulation must be small and has not yet been identified histo-

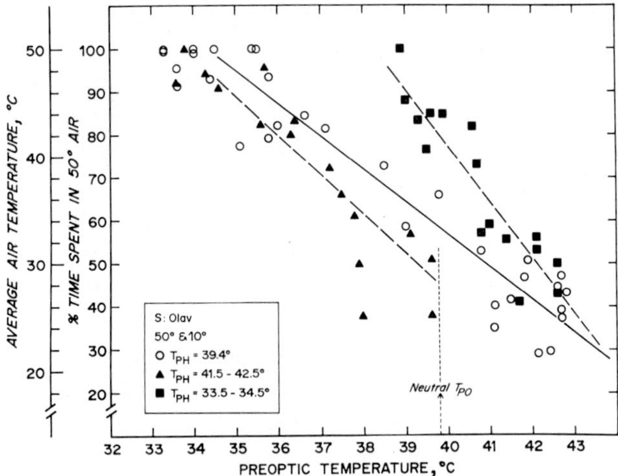

Fig. 6. Behavioral analysis of several experiments in which both preoptic and posterior hypothalamic temperatures were simultaneously displaced. Open circles and solid line show behavioral changes when preoptic temperature alone was varied. Other points show how this function changes when the posterior hypothalamus is simultaneously cooled to 34 °C (squares) or warmed to 42 °C (triangles).

logically. In the squirrel monkey, the posterior hypothalamic area occupies a tissue volume of about 1 mm^3, so it is hardly surprising that thermode placement appears to be critical.[1]

Preoptic-posterior hypothalamic interaction. Additional experiments conducted on one of the positive animals yielded information on the interplay between the two hypothalamic areas in the control of thermoregulatory behavior. The results of one series are summarized in figure 6. As in the previous experiments, local preoptic temperature was displaced to some level other than neutral for 10 min. Simultaneously, the posterior hypothalamic temperature was displaced to 42 °C whenever the preoptic area was cooled, or to 34 °C whenever the preoptic area was warmed. Half the time the posterior hypothalamic temperature was held at the neutral level of 39.4 °C. In figure 6, the percentage of time the monkey

1 X-rays taken postoperatively indicate that the depth of the thermode tip is slightly greater in the two positive than in the two negative animals. Poor thermode perfusion in the positive animals, achieved by shortening the inner thermode tube, eliminates the behavioral responses.

chose 50 °C air is plotted as a function of the average preoptic temperature during that 10-min interval. The open circles describe a function similar to that already seen in figure 5 when preoptic temperature alone is varied. The filled symbols describe functions that result when the preoptic temperature is varied and the posterior hypothalamus is simultaneously heated to 42 °C (triangles) or cooled to 34 °C (squares). The calculated regression lines (dashed lines) have steeper slopes than the original function (solid line). This slope increase is nearly 2-fold when the posterior hypothalamus is cooled simultaneously with preoptic warming. Not so striking is the slope increase when the posterior hypothalamus is warmed simultaneously with preoptic cooling. This is due to two factors, high variability and a ceiling effect that operates for preoptic temperatures below 35.5 °C. Nevertheless, it is clear from these results that during transient thermal stimulation, the posterior hypothalamus can over-ride the preoptic area to control the behavioral response, increasingly so as the preoptic temperature approaches neutrality. The slope changes shown in figure 6 suggest that under the experimental conditions explored here the two brainstem areas interact in a multiplicative fashion.

Body temperature changes in the steady state. From the preceding experiments it was clear that the two hypothalamic areas can interact to control thermoregulatory behavior. It is important to determine whether the nature of that interaction is always multiplicative, as suggested above, or whether a different relation obtains in the steady state. It has been reported previously [1, 2] that when the preoptic area is heated or cooled for 1 h or more, the exaggerated behavioral changes seen in figure 4 do not persist indefinitely. Under these conditions the behavioral response, initially strong to quickly bring the animal into a purposeful thermal balance with his environment, later moderates to regulate the body heat content at the new level.

A series of 20 behavioral experiments was conducted on each of the two positive monkeys to determine the steady-state body temperatures resulting from preoptic and posterior hypothalamic thermal displacements presented both alone and together. The protocol for each experiment included a 1-hour period of preoptic temperature displacement alone, a 1-hour period of posterior hypothalamic temperature displacement alone and a 1-hour period of preoptic and posterior hypothalamic temperature displacements together. Each of these periods was separated by a 30-min return to thermally neutral conditions.

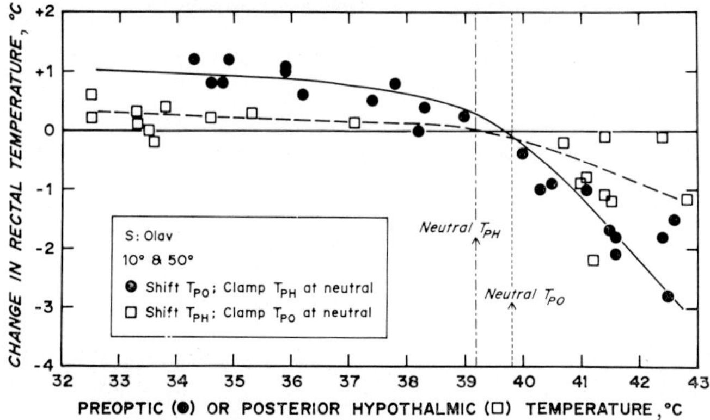

Fig. 7. Steady-state changes in rectal temperature at the end of 1-hour displacements of preoptic (solid circles, solid line) or posterior hypothalamic (open squares, dashed line) temperature.

Figure 7 shows the steady-state change in rectal temperature from neutral when either preoptic (closed circles) or posterior hypothalamic (open squares) temperature alone was displaced. Each plotted point is the mean of the final 30 min of the 1-hour period of local temperature displacement. The range of standard deviations for all points was 0.05–0.3 °C. The figure shows that preoptic thermal stimulation produced greater steady-state changes in internal body temperature (by a ratio of about 3 : 1) than did comparable thermal stimulation of the posterior hypothalamus. For both brainstem areas, heating is a more effective stimulus than cooling to a behavioral alteration of internal body temperature. In fact, posterior hypothalamic cooling produces a barely discernible effect, elevating internal temperature only a few tenths of a degree.

The change in rectal temperature resulting from simultaneous displacements of the two brain areas is shown in figure 8. The solid triangles and solid line represent the change in rectal temperature resulting from changes in preoptic temperature alone. The other symbols on the graph represent the combined effect of preoptic and posterior hypothalamic temperature displacements. The open circles show rectal temperature changes when preoptic temperature was shifted and the posterior hypothalamus was clamped *hot,* i.e. at 41.5 °C; the open squares show comparable changes when the posterior hypothalamus was clamped *cold,* i.e. at 33 °C. It is clear that the original functional relationship has been shifted up or down, depending upon the posterior hypothalamic tempera-

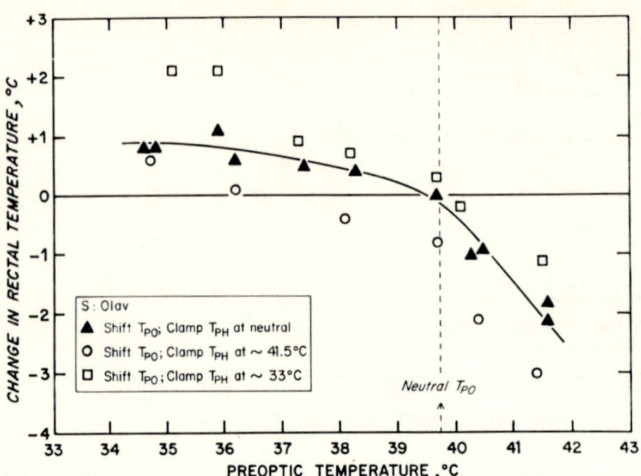

Fig. 8. Steady-state changes in rectal temperature as a function of preoptic temperature when posterior hypothalamic temperature was simultaneously clamped at neutral (solid triangles, solid line), clamped at 41.5 °C (open circles), or clamped at 33 °C (open squares).

ture. It is also apparent that posterior hypothalamic warming modulates the function to a greater extent than does posterior hypothalamic cooling, just as posterior hypothalamic warming by itself produces a greater rectal temperature change than does cooling (fig. 7). Intuitively, these facts suggest that the action of the two brain areas may well be additive in their effect upon behaviorally produced body temperature changes in the steady state.

This notion is further strengthened when one examines the changes in mean skin temperature that occurred during these experiments. Figure 9 shows the change in mean skin temperature from the neutral level as a function of preoptic or posterior hypothalamic thermal displacements alone. Although there is more inherent variability in skin temperature measurements than in rectal temperature measurements, the best-fitting functions bear the same relation to one another as do the two functions in figure 7. Cooling either brainstem area results in a behaviorally produced elevation in skin temperature; warming either area results in an even greater reduction in skin temperature. Once again, thermal stimulation of the preoptic area produces a change of greater magnitude than that produced by comparable stimulation of the posterior hypothalamus. Under the assumption of additivity of action, two theoretical functions

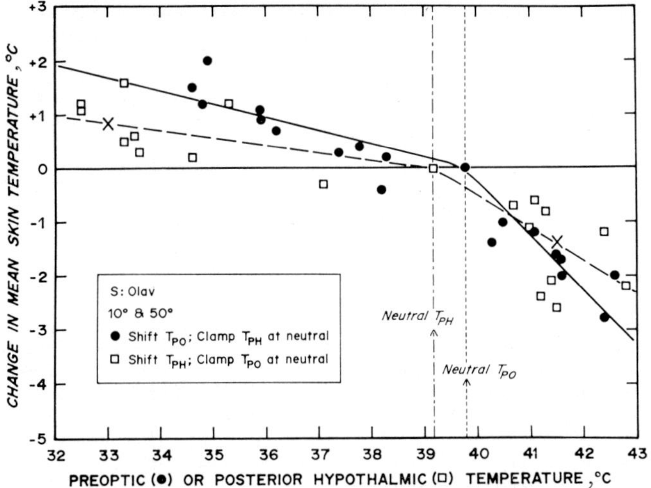

Fig. 9. Steady-state changes in mean skin temperature at the end of 1-hour displacements of preoptic (solid circles, solid line) or posterior hypothalamic (open squares, dashed line) temperature. For explanation of two points marked X, see text.

were determined to predict the experimental results when the temperatures of both areas were simultaneously displaced. The values of +0.85 °C at T_{PH} = 33 °C, and –1.4 °C at T_{PH} = 41.5 °C (indicated by X on the figure) were read from the function in figure 9 that relates the change in mean skin temperature to posterior hypothalamic temperature. These values were algebraically added to the preoptic function at sufficient intervals to determine the two functions plotted in figure 10. The plotted points are the experimental data; the symbols have the same meaning as those in figure 8. Although the data points depart somewhat from the prediction at preoptic temperatures below 36 °C, they are well described by the predicted functions for the rest of the range explored. Additivity of action between the preoptic and posterior hypothalamic areas in the steady state control of behaviorally produced body temperature changes is certainly a tenable assumption.

Implications for a Model of Temperature Regulation

Three important results stand out from these studies and must be considered in any description or model of temperature regulation that

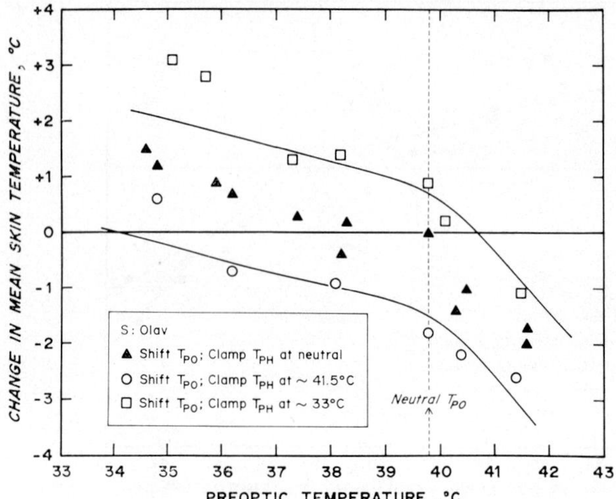

Fig. 10. Steady-state changes in mean skin temperature as a function of preoptic temperature when posterior hypothalamic temperature was simultaneously clamped at neutral (solid triangles), clamped at 41.5 °C (open circles), or clamped at 33 °C (open squares). Solid lines represent predicted relationships based on additivity of action.

incorporates the behavioral loop. Firstly, while no changes in autonomic responses resulting from thermal stimulation of the posterior hypothalamus could be detected, such stimulation can drive behavioral responses nearly as effectively as preoptic stimulation does, particularly when thermal displacements are brief. Secondly, the posterior hypothalamus can over-ride the preoptic area to control thermoregulatory behavior, increasingly so when the preoptic temperature approaches neutrality. Thirdly, posterior hypothalamic thermal stimulation can modify the thermal steady state that results from preoptic heating or cooling, probably in an additive manner.

Figure 11 shows a schematic representation of the neural elements known to be involved in the thermoregulation (both autonomic and behavioral) of endotherms and some of the known and hypothetical links between them. This diagram is similar in many respects to models proposed by HARDY [14] and by CHATONNET and CABANAC [5]. As shown in figure 11, environmental modification, whether a naturally occurring or behaviorally produced physical disturbance, alters the pattern of neural

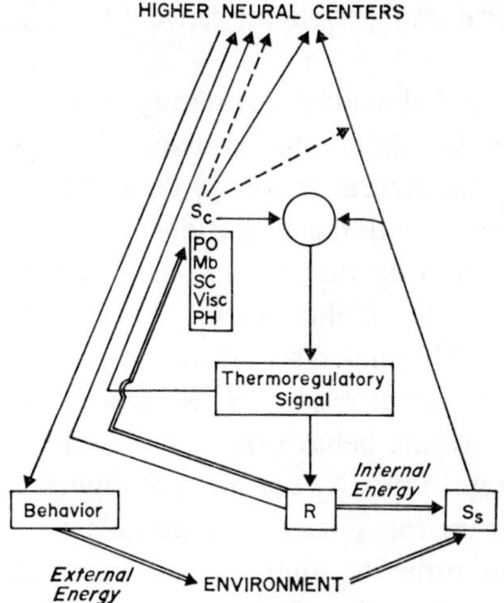

Fig. 11. Schematic representation of the neural elements involved in thermoregulation, both behavioral and autonomic. Arrows denote direction of connecting links characterized as known neural (——), hypothetical neural (– – – –) and energy exchange (═══). S_c = sites located centrally; PO = preoptic area; Mb = midbrain reticular formation; SC = spinal cord; Visc = deep viscera; PH = posterior hypothalamus; S_s = thermosensitive skin structures; R = autonomic responses.

impulses arising in the thermosensitive structures of the skin (S_s). This neural traffic not only goes directly to the higher neural centers (giving rise to thermal sensation in man) but also to a thermoregulatory integrating center, represented by O. The integrating center must also receive neural impulses from many other thermosensitive sites located centrally within the body (S_c). Known central sites that generate afferent signals in response to local temperature change include the preoptic area, the midbrain reticular formation, the spinal cord, the deep viscera, and the posterior hypothalamus. Whether all of these sites (and, indeed, other sites not yet identified) contribute input signals to both the autonomic and behavioral control systems is yet to be determined. The thermoregulatory output signal generated by the integrating center activates the forward controlling elements of the autonomic system, internally energized, in the form of physiological responses, to combat hyperthermia (vasodilation, sweating, decreased metabolic rate, polypnea) or hypothermia (vasoconstriction, piloerection, shivering or nonshivering thermogenesis). The signals related to change in body temperature, the regulated variable, will then be fed back via the central and peripheral sensors to close the loop.

In the freely active animal, a behavioral regulatory system also operates, often to the exclusion of the autonomic system. Many neural elements are common to both systems. Neural pathways providing for the awareness of physiological responses, and the sensations of peripheral stimuli together with their modification by the central signals are represented in figure 11 as converging on the higher neural centers. Current theory holds that, in man at least, 'thermal discomfort' originates here and serves as the drive signal to behavioral responses. No matter at what level the drive originates, it stimulates the behavioral responses by which the organism utilizes physical energy to modify the thermal environment.

Neural impulses from sensors in the posterior hypothalamus must feed into the integrating center in order to alter the thermoregulatory signal derived from other central receptor sites such as the preoptic area. This was clearly demonstrated in both the transient and steady-state experiments reported here. A pathway must also be provided to carry the integrated thermoregulatory signal direct to the higher neural centers, bypassing the autonomic thermoregulatory responses. This is necessary to explain the result that posterior hypothalamic thermal stimulation can alter behavioral, but not physiological, thermoregulatory responses. In fact, the integrated thermoregulatory signal may often travel this route in the organism which possesses full behavioral control over its thermal environment.

Of all the central thermosensitive sites so far studied, the posterior hypothalamus appears to be unique. Only here is there a clear uncoupling of physiological and behavioral responses to local thermal stimulation. Can this be, as HARDY [15] ventures, the locus of integration of the thermoregulatory output signal? If posterior hypothalamic lesions eliminate all thermoregulation, behavioral as well as physiological, then it would seem highly probable that HARDY's view is correct.

Acknowledgements

The research reported in this paper was supported by United States Public Health Service Grant ES-00354. The technical assistance of BARBARA WRIGHT is gratefully acknowledged. A portion of the results was presented at the 25th International Congress of Physiology, Munich 1971 [3].

References

1 ADAIR, E. R.: Displacements of rectal temperature modify behavioral thermoregulation. Physiol. Behav. *7:* 21–26 (1971).

2 ADAIR, E. R.; CASBY, J. U., and STOLWIJK, J. A. J.: Behavioral temperature regulation in the squirrel monkey: Changes induced by shifts in hypothalamic temperature. J. comp. physiol. Psychol. *72:* 17–27 (1970).

3 ADAIR, E. R. and HARDY, J. D.: Posterior hypothalamic thermal stimulation can alter behavioral, but not physiological temperature regulation. Proc. Int. Union physiol. Sci. *9:* 7 (1971).

4 CARLISLE, H. J.: Effect of preoptic and anterior hypothalamic lesions on behavioral thermoregulation in the cold. J. comp. physiol. Psychol. *69:* 391–402 (1969).

5 CHATONNET, J. and CABANAC, M.: The perception of thermal comfort. Int. J. Biometeor. *9:* 183–193 (1965).

6 CRAWSHAW, L. I. and HAMMEL, H. T.: Behavioral regulation of internal temperature in the brown bullhead, *Ictalurus nebulosus*. Comp. Biochem. Physiol. (in press).

7 CUNNINGHAM, D. J.; STOLWIJK, J. A. J.; MURAKAMI, N., and HARDY, J. D.: Responses of neurons in the preoptic area to temperature, serotonin, and epinephrine. Amer. J. Physiol. *213:* 1570–1581 (1967).

8 EDINGER, H. M. and EISENMAN, J. S.: Thermosensitive neurons in tuberal and posterior hypothalamus of cats. Amer. J. Physiol. *219:* 1098–1103 (1970).

9 EMMERS, R. and AKERT, K. A.: A stereotaxic atlas of the brain of the squirrel monkey (*Saimiri sciureus*) (University of Wisconsin Press, Madison 1963).

10 EPSTEIN, A. N. and MILESTONE, R.: Showering as a coolant for rats exposed to heat. Science *160:* 895–896 (1968).

11 GUIEU, J. D. and HARDY, J. D.: Effects of preoptic and spinal cord temperature in control of thermal polypnea. J. appl. Physiol. *28:* 540–542 (1970).

12 HAMMEL, H. T.; HARDY, J. D., and FUSCO, M. M.: Thermoregulatory responses to hypothalamic cooling in unanesthetized dogs. Amer. J. Physiol. *198:* 481–486 (1960).

13 HAN, P. W. and BROBECK, J. R.: Deficits of temperature regulation in rats with hypothalamic lesions. Amer. J. Physiol. *200:* 707–710 (1961).

14 HARDY, J. D.: Thermal comfort and health. ASHRAE J. *77:* 43–51 (1971).

15 HARDY, J. D.: The posterior hypothalamus and the regulation of body temperature. In: Problems in temperature regulation and exercise. Fed. Proc. (in press).

16 HOLMES, R. L.; NEWMAN, P. P., and WOLSTENCROFT, J. H.: A heat sensitive region in the medulla. J. Physiol., Lond. *152:* 93–98 (1960).

17 JESSEN, C. and MAYER, E. T.: Spinal cord and hypothalamus as core sensors of temperature in the conscious dog. I. Equivalence of responses. Pflügers Arch. ges. Physiol. *324:* 189–204 (1971).

18 KELLER, A. D. and HARE, W. K.: Heat regulation in medullary and mid-brain preparations. Proc. Soc. exp. Biol. Med. *29:* 1067–1068 (1932).

19 LIPTON, J. M.: Effects of preoptic lesions on heat-escape responding and colonic temperature in the rat. Physiol. Behav. *3:* 165–169 (1968).
20 LIPTON, J. M.: Behavioral temperature regulation in the rat. Effects of thermal stimulation of the medulla. J. Physiol., Paris *63:* 325–328 (1971).
21 MEYER, H. H.: Theorie des Fiebers und seiner Behandlung. Verh. dtsch. Ges. inn. Med. *30:* 15 (1913).
22 MURGATROYD, D.; KELLER, A. D., and HARDY, J. D.: Warmth discrimination in the dog after a hypothalamic ablation. Amer. J. Physiol. *195:* 276–284 (1958).
23 NAKAYAMA, T. and HARDY, J. D.: Unit responses in the rabbit's brain stem to changes in brain and cutaneous temperature. J. appl. Physiol. *27:* 848–857 (1969).
24 NUTIK, S. L.: Convergence of cutaneous and preoptic region thermal afferents on posterior hypothalamic neurons. J. Neurophysiol. *36:* 250–257 (1973).
25 NUTIK, S. L.: Posterior hypothalamic neurons responsive to preoptic region thermal stimulation. J. Neurophysiol. *36:* 238–249 (1973).
26 SATINOFF, E. and RUTSTEIN, J.: Behavioral thermoregulation in rats with anterior hypothalamic lesions. J. comp. physiol. Psychol. *71:* 77–82 (1970).
27 STITT, J. T. and HARDY, J. D.: Thermoregulation in the squirrel monkey (*Saimiri sciureus*). J. appl. Physiol. *31:* 48–54 (1971).
28 WUNNENBERG, W. and HARDY, J. D.: Response of single units of the posterior hypothalamus to thermal stimulation. J. appl. Physiol. *33:* 547–552 (1972).

Author's address: Dr. ELEANOR R. ADAIR, John B. Pierce Foundation Laboratory, 290 Congress Avenue, *New Haven, CT 06519* (USA)

Recent Studies of Hypothalamic Function
Int. Symp. Calgary 1973, pp. 359–370 (Karger, Basel 1974)

Evidence for the Involvement of Prostaglandins in Fever

W. L. VEALE and K. E. COOPER

Division of Medical Physiology, Faculty of Medicine, University of Calgary, Calgary, Alberta

Monoamines and Body Temperature

Since the mid-1950s, it has been well established that the brainstem contains relatively high concentrations of norepinephrine (NE) and 5-hydroxytryptamine (5-HT) [1, 37]. Further work revealed that the hypothalamus itself is made up of large numbers of cells which contain NE and 5-HT [2, 3, 10]. In 1963, FELDBERG and MYERS [11] proposed the theory that body temperature is regulated by a balance in the release of NE and 5-HT in the region of the anterior hypothalamus [11, 12]. This theory was based on work in the unanesthetized cat in which 5-HT and NE injected either into the cerebral ventricular system or into the tissue of the anterior hypothalamus produced hyper- and hypothermia, respectively. Since that time, support has been obtained for this theory of thermoregulation from many laboratories using several different species of experimental animal. A complicating factor, however, has been the striking differences in the temperature response that is obtained in some species. For example, the rabbit responds to 5-HT with a fall in body temperature but with an increase to NE [7]. For a complete review of the topic of species differences the reader is referred to a recent review by VEALE and COOPER [35]. We would, however, like this opportunity to review one impressive demonstration of the difference in temperature response produced by NE in the rabbit and cat. When imipramine or desipramine is given directly into the lateral cerebral ventricle, hyperthermia is produced in the rabbit and hypothermia produced in the cat [9]. The action of these substances is to inhibit the process of re-uptake of NE into the presynaptic endings, thus allowing the accumulation of endogenous NE at the synapses and producing the changes in temperature. The experiments reviewed above provide strong evidence that NE and 5-HT are involved in the regulation of body temperature and that there appear to be species differences with respect to the direction of the temperature change in response to these substance.

Leucocyte Pyrogen and Body Temperature

When leucocyte pyrogen is injected into the carotid artery, fever is produced in the rabbit after a shorter latency than when a similar injection is made intravenously. This observation led KING and WOOD [24] to suggest that leucocyte pyrogen may be acting directly upon the regions of the central nervous system involved in thermoregulation. Later work in the cat and rabbit revealed that the anterior hypothalamic preoptic area (AH-POA) is the region of the brain from which increases in body temperature can be elicited best [8, 23, 28]. The growing body of evidence suggests that leucocyte pyrogen is the substance which produces fever and that a bacterial pyrogen may stimulate the leucocytes to produce the pyrogenic factor. In contrast to the species differences observed with respect to injections of NE and 5-HT, all species tested to this point with leucocyte pyrogen respond with an increase in body temperature. This difference is consistent with the generally agreed concept that the net effect of pyrogen is not to disorganize thermoregulation itself but to cause an upward shift in the level at which temperature is regulated [6, 19, 38]. Recently, it has been proposed that prostaglandins may be mediators in the fever response to pyrogen [13, 14, 25] and examination of this theory will be the main topic of this paper.

Central Effects of Prostaglandins

The prostaglandins are cyclic, oxygenated, C-20 fatty acids which have been shown to be present [20–22] in hypothalamic extracts as well as cerebrospinal fluid (CSF). MILTON and WENDLANDT [25] introduced prostaglandin E_1 (PGE_1) into the field of thermoregulation when they found that fever was produced in the unanesthetized cat following injection of this substance into the third ventricle. Since that time PGE_1 has been injected into the cerebral ventricles of other species including the rabbit and rat and it has been found that it has the same effect on body temperature [13]. More recently, the anterior hypothalamic area has been identified as the region which is extremely sensitive to local application of PGE_1. In the rabbit, push-pull perfusion of PGE_1 in the AH-POA produces a sharp increase in body temperature with as little as 3 ng perfusion over a 40-min interval [34].

We have 'mapped' the hypothalamus as well as other regions of the rat brain with PGE_1 using the microinjection technique in an attempt to produce hyperthermia [36]. Figure 1 (top) illustrates the mean tempe-

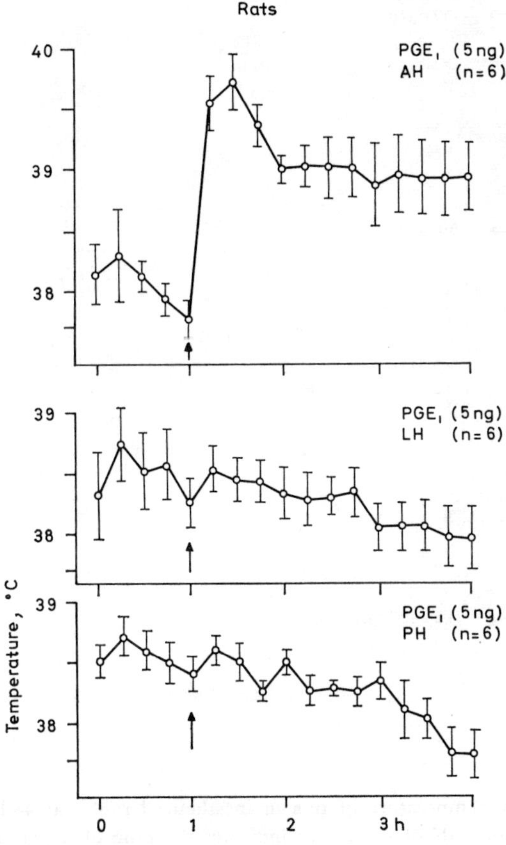

Fig. 1. Records of mean rectal temperature of 3 groups of unanesthetized rats. The arrows indicate injections of 5 ng prostaglandin E_1 (PGE_1) into the anterior hypothalamus (AH, top), lateral hypothalamus (LH, middle) and posterior hypothalamus (PH, bottom). The vertical bars represent twice the standard error.

rature response observed when an injection of 5 ng of PGE_1 is made into the anterior hypothalamic region of the unanesthetized, unrestrained rat. It can be seen that immediately following the injection, body temperature increased sharply from 37.6 to 39.6 °C. Following a slight reduction after the maximum is reached, body temperature remains at approximately 39.0 °C for at least a further 2 h. As is illustrated in figure 1 (middle and bottom), the same amount of prostaglandin given into the

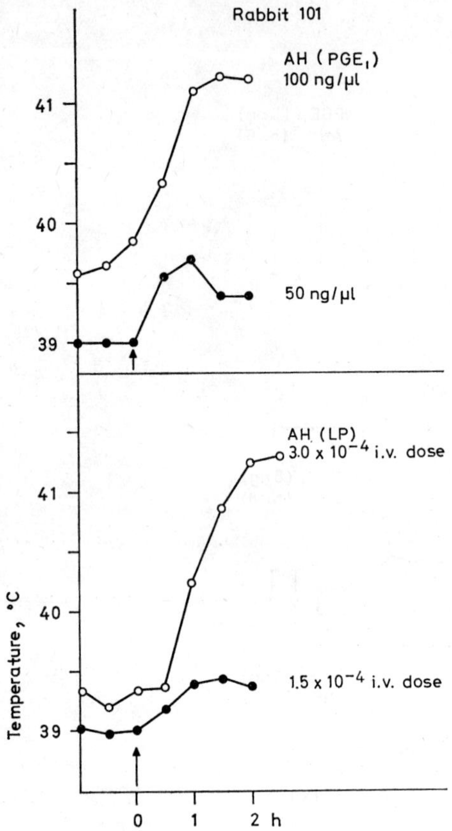

Fig. 2. Records of rectal temperature of one unanesthetized rabbit at 48-hour intervals. The arrow in the top half of the figure indicates the time of injection of 50 ng (●—●) and 100 ng (o—o) into the anterior hypothalamus (AH). In the bottom half of the figure, the arrow indicates the time of injection of 1.5 × 10^{-4} (●—●) and 3.0 × 10^{-4} (o—o) the intravenous dose of leucocyte pyrogen (LP) into the anterior hypothalamus.

lateral or posterior hypothalamic areas produced no change in body temperature. In similar experiments conducted in our laboratory we have found that this site specificity is true for the cat and rabbit as well. In fact, to this point in time, no temperature responses have been produced in regions other than the AH-POA in our experiments in the rat, rabbit or cat. One striking feature of these experiments is the very minute quantity of PGE_1 needed to produce a response. Even though prostaglandin was administered in the quantity of 5 ng in the illustration provided in figure 1, we have observed changes in temperature of 1 °C or more with as little as 100 pg injected bilaterally into the AH-POA of the rat.

In an attempt to determine whether the same regions of the brain are responsive to local injections of leucocyte pyrogen and PGE_1 we have been microinjecting these substances into the brain of the unanesthetized rabbit. The temperature records illustrated in figure 2 are typical of the results of these experiments. It can be seen from figure 2 (top) that 50 ng of PGE_1 injected into the anterior hypothalamic region of the rabbit brain produces an increase in temperature of approximately 0.7 °C. When 100 ng PGE_1 is injected into this same site through the same cannulae 48 h later, body temperature increases more than 1.5 °C. The temperature records shown in figure 2 (bottom) illustrates the response to injections of leucocyte pyrogen made into the same site in the same rabbit with a 48-hour interval between each injection. When 1.5×10^{-4} the amount of leucocyte pyrogen needed to produce a fever when it is injected intravenously is microinjected into the anterior hypothalamic area, a slight increase in body temperature is observed. On the other hand when 3.0×10^{-4} the intravenous dose is injected a sharp increase in temperature of approximately 2.0 °C is produced. At this point in our experiments we have been unable to locate a region of the brain which clearly responds to one of these substances and not the other. As yet, we have not thoroughly examined the region of the midbrain where ROSENDORFF and MOONEY [29] were able to produce a febrile response to the local application of a partially purified leucocyte pyrogen.

Further Support for Prostaglandin-Mediated Fever

In an attempt to find direct evidence in favor of the theory that prostaglandin is a mediator of the fever response to pyrogen, FELDBERG and GUPTA [15] collected CSF from the third ventricle of the unanesthetized cat and assayed it for its prostaglandin-like activity. In these experiments the biological assay, using the rat stomach fundus strip method described by VANE [32], was employed. It is known that the following 4 prostaglandins cause the smooth muscle of the fundus to contract: PGE_1, PGE_2, PGF_1 and PGF_2. The prostaglandin-like activity of the CSF was considered to be due to either PGE_1 or PGE_2, since it appears that only these two prostaglandins produce fever [26]. In their experiments, FELDBERG and GUPTA were able to demonstrate that in samples of CSF collected in the unanesthetized cat from a cannula implanted in the third ventricle with its opening lying close to the anterior hypothalamus, the prostaglandin-like activity could be detected in the quantity of 1.3–10.0 ng/ml. During fever produced by an injection of *Shigella dysenteriae* into the third ventricle, the prostaglandin-like activity rose over 2.5–4.0 times that detected in the cat with a normal temperature. Following intraperitoneal injections of the antipyretic paracetamol, which caused the fever to decline, the prostaglandin-like activity was again similar to that collected in animals with normal temperature, i.e. between 1.5 and 6 ng/ml.

The idea that prostaglandins may be released by pyrogens and thus play the role as the final mediator of fever within the hypothalamus has gained further support from experiments which have demonstrated that anti-inflammatory and antipyretic drugs inhibit the synthetase which synthetizes prostaglandins from the unsaturated fatty acid, arachidonic acid [33]. These important observations, made initially in homogeneates of guinea pig lung and more recently in homogeneates of rabbit and dog brains [18], are compatible with the idea that the rise in temperature in fever is induced by synthesis and release of prostaglandin. Further, the antipyretic substances might reduce temperature in a fever by preventing prostaglandin synthesis and release.

Another line of evidence comes from work in which PGE_1, microinjected into the anterior hypothalamic area of the rabbit, was found to be effective in producing fever, whereas similar injections into the posterior hypothalamus and midbrain reticular formation were not [30, 31]. Consistent with the findings in our laboratory, these investigators have found that PGE_1 is effective in the same region as that found to be sensitive to local application of leucocyte pyrogen. Interestingly, the onset of the febrile response following the central injection of PGE_1 is shorter than that seen following a similar injection of leucocyte pyrogen.

Recently in our laboratory, we have been able to demonstrate in the rat that the magnitude of the febrile response to a microinjection of a few nanograms of PGE_1 into the AH-POA is not altered by ambient temperatures of 3 or 40 °C. Similar evidence has been provided by STITT [30] in the rabbit when he showed that ambient temperature does not alter the magnitude of the response; however, the mechanisms by which PGE_1 fever is produced is indeed altered by the environmental temperature. He points out that in the cold PGE_1 fever was due to increased heat production, while during heat exposure both evaporative and dry heat losses were reduced without significant changes in heat production. At intermediate temperature vasoconstriction was effective in producing fever along with a small increase in heat production. These changes are similar to those observed when fever is produced by an injection of pyrogen. During the fever produced by PGE_1 the AH-POA retained its thermosensitivity as it does with fever produced by pyrogen [30, 31].

Yet another line of evidence has been provided from a series of experiments in which we applied a conjugated antibody to PGE_1 directly to the AH-POA of the unanesthetized rabbit. When slowly infused the antibody produced a slow long-lasting fever. One to ten days after the fever had subsided an intravenous injection of a standard amount of leucocyte pyrogen, which when tested previously, caused a fever of 0.75–1.0 °C, produced either a very much reduced fever or no temperature change et all. While the initial febrile response produced by the injection of the conjugated antibody to PGE_1 cannot be explained at the present time, the reduction or absence of fever following the infusion of the antibody lends support to the hypothesis that leucocyte pyrogen may act be increasing the release of PGE_1 in the AH-POA.

Role of the Ventricular System

Experiments were designed to determine whether or not leucocyte pyrogen, when given intravenously, reaches the tissue of the hypothalamus directly from the bloodstream or through the CSF after being excreted from the choroid plexuses. If leucocyte pyrogen were to be excreted from the plexuses into the CSF and from there into the tissue of the hypothalamus, a rapid washing of the ventricular system from the lateral ventricle to the cisterna magna [13, 16] should interfere with this process. In the rabbit, the fever produced by an intravenous injection of leucocyte pyrogen was compared to the fever produced by the same amount of pyrogen injected similarly during a rapid perfusion of the ventricular system of the unanesthetized rabbit. The onset, magnitude and duration of fever was unaltered by the rapid perfusion of the ventricular system from the lateral ventricle to the cisterna magna [17]. From this work we concluded that leucocyte pyrogen when injected intravenously reaches the tissue of the hypothalamus directly via the bloodstream and that the ventricular system is not an important route of entry for leucocyte pyrogen into the hypothalamus.

The role of the ventricular system in the febrile response produced by intravenous leucocyte pyrogen was further studied by injecting sterile paraffin oil into the lateral cerebral ventricle of the afebrile rabbit in such a way as to fill the entire ventricular system. Such an injection does not alter normal body temperature nor does it alter the time which elapses between an intravenous injection of leucocyte pyrogen and the fever that follows [4, 5]. This would add further support to the finding that the ventricular system appears to be of little importance for the entry of leucocyte pyrogen into the hypothalamus. On the other hand, the height and duration of the fever was greatly increased when leucocyte pyrogen was given after the cerebral ventricles had been filled with sterile paraffin oil. Further, when oil was injected into the cerebral ventricles of rabbits during an intravenous infusion of pyrogen in an amount previously shown to maintain a steady fever level, body temperature commenced to climb again and continued to climb considerably thereafter. We were able to determine that the paraffin oil injected intraventricularly had fully filled the ventricular system but whether it formed an oily barrier on the surface of the ependyma or whether it effectively impeded the flow of naturally secreted CSF is not clear. The exaggerated responses to intravenous leucocyte pyrogen produced by the presence of oil in the ventricular system suggests that its contact with the ventricular wall is of importance in determining the response to intravenous pyrogen. We feel that the most likely explanation is that a normal route of excretion of leucocyte pyrogen or a substance which it releases is through the ependyma into the CSF and that the contact of the oil with the ventricular wall could prevent this route of excretion.

Since FELDBERG and GUPTA [15] found that a prostaglandin-like substance was in greater quantity in CSF during fever than in samples taken during normal resting temperature or following the administration of an antipyretic, we felt it important to determine whether or not PGE_1, when placed in the tissue of the hypothalamus itself, can find its way into the ventricular system. In order to gain this type of information, unanesthetized rabbits were perfused with bilateral push-pull cannulae in the AH-POA with ^{3}H-PGE$_1$. At the same time, the ventricular system was perfused with artificial CSF from the lateral ventricule to the cisterna magna. In this way it was possible to measure the radioactivity of the PGE$_1$ in the CSF at regular intervals during the perfusion experiment. Once control samples had been taken during push-pull perfusion with a modified Krebs solution [27], the AH-POA was perfused with enough ^{3}H-PGE$_1$ to produce an increase in body temperature of 1 °C. Approximately 10 min after the perfusion with ^{3}H-PGE$_1$ was begun, radioactivity could be detected in the effluent from the perfused ventricular system. The concentration of radioactivity appeared to correlate well with the onset of the febrile response [34]. A thin-layer chromographic analysis indicated that the label which was being measured in the CSF taken from the cisterna magna was indeed attached to a prostaglandin. These results clearly indicated that when PGE$_1$ is placed in the tissue of the hypothalamus and a temperature response is produced, this same prostaglandin very rapidly makes it way to the ventricular system in high concentrations. We have interpreted this result as suggesting that PGE$_1$ may be released by leucocyte pyrogen.

Finally, we have recently shown that when reserpine is given intraventricularly in the unanesthetized rabbit, in an amount sufficient to deplete the region of the hypothalamus of its monoamine content, the animal can still produce a fever to both PGE$_1$ given centrally and to leucocyte pyrogen given intravenously even though the animal's ability to thermoregulate is seriously impaired. On the other hand, atropine given into the ventricular system prevents the development of a fever to either PGE$_1$ injected directly into the AH-POA or to leucocyte pyrogen given intravenously.

Summary

The following evidence suggests that prostaglandin may be a mediator in fever produced by leucocyte pyrogen.

1. Prostaglandins are normal constituents of hypothalamic tissue.

2. Injections of very minute amounts of both PGE$_1$ and leucocyte pyrogen into the AH-POA produce fever.

3. In contrast to injections of monoamines, leucocyte pyrogen and prostaglandin produce fever in all species tested so far.

4. Antipyretics interfere with the synthesis of prostaglandin.

5. The latency of the febrile response to PGE_1 is less than that observed when leucocyte pyrogen is given directly into the AH/POA.

6. The conjugated antibody to PGE_1 when placed in the AH-POA interferes with the febrile response to intravenous leucocyte pyrogen.

7. When the ependymal wall is coated with oil, the response to leucocyte pyrogen given intravenously is exaggerated.

8. ^3H-PGE_1, when placed directly in the tissue of the AH-POA, finds its way into the ventricular system.

9. Reserpine does not prevent the febrile response to either central PGE_1 or intravenous leucocyte pyrogen; however, it does interfere with thermoregulation.

10. Atropine given intraventricularly blocks the febrile response to both central PGE_1 and intravenous leucocyte pyrogen.

Acknowledgements

This work was supported by the Medical Research Council of Canada. The authors are grateful to T. MALKINSON and P. REDSTONE for their valuable technical assistance.

References

1 AMIN, A. N.; CRAWFORD, T. B. B., and GADDUM, J. H.: The distribution of substance P and 5-hydroxytryptamine in the central nervous system of the dog. J. Physiol., Lond. *126:* 596–618 (1954).

2 ANDEN, N. E.; DAHLSTRÖM, A.; FUXE, K., and LARSSON, K.: Mapping out of catecholamine and 5-hydroxytryptamine neurons innervating the telencephalon and diencephalon. Life Sci. *14:* 1275–1279 (1965).

3 CARLSSON, A.; FALCK, B., and HILLARP, N. A.: Cellular localization of brain monoamines. Acta physiol. scand. *56:* suppl. 196, pp. 6–28 (1962).

4 COOPER, K. E. and VEALE, W. L.: Potentiation of fever produced by intravenous leucocyte pyrogen, following the injection of paraffin oil into the cerebral ventricles of the unanesthetized rabbit. Experientia *28:* 917 (1972).

5 COOPER, K. E. and VEALE, W. L.: The effect of an inert oil in the cerebral ventricular system upon fever produced by intravenous pyrogen. Canad. J. Physiol. *50:* 1066–1071 (1962).

6 COOPER, K. E.; CRANSTON, W. I., and SNELL, E. S.: Temperature regulation during fever in man. Clin. Sci. *27:* 345–356 (1964).

7 COOPER, K. E.; CRANSTON, W. I, and HONOUR, A. J.: Effects of intraventricular and intrahypothalamic injection of noradrenaline and 5-HT on body temperature in conscious rabbits. J. Physiol., Lond. *181:* 852–864 (1965).

8 COOPER, K. E.; CRANSTON, W. I., and HONOUR, A. J.: Observations on the site and mode of action of pyrogens in the rabbit brain. J. Physiol., Lond. *191:* 325–337 (1967).

9 CRANSTON, W. I.; HELLON, R. F.; LUFF, R. H., and RAWLINS, M. D.: Evidence concerning the effects of endogenous noradrenaline upon body temperature in cats and rabbits. J. Physiol., Lond. *212:* 24–25 (1971).

10 DAHLSTOM, A. and FUXE, K.: Evidence for the existence of monoamine containing neurons in the central nervous system. I. Demonstration of monoamines in the cell bodies of brain stem neurons. Acta physiol scand. *62:* suppl. 232 (1964).

11 FELDBERG, W. and MYERS, R. D.: A new concept of temperature regulation by amines in the hypothalamus. Nature, Lond. *200:* 1325 (1963).

12 FELDBERG, W. and MYERS, R. D.: Effects on temperature of amines injected into the cerebral ventricles. A new concept of temperature regulation. J. Physiol., Lond. *173:* 226–237 (1964).

13 FELDBERG, W. and SAXENA, P. N.: Fever produced by prostaglandin E_1. J. Physiol., Lond. *217:* 547–566 (1971).

14 FELDBERG, W. and MILTON, A. S.: Prostaglandins in fever; in SCHONBAUM and LOMAX The pharmacology of temperature regulation (Karger,Basel 1973).

15 FELDBERG, W. and GUPTA, K. P.: Pyrogen fever and prostaglandin-like activity in cerebrospinal fluid. J. Physiol., Lond. *228:* 41–53 (1973).

16 FELDBERG, W.; MYERS, R. D., and VEALE, W. L.: Perfusion from cerebral ventricle to cisterna magna in the unanesthetized cat. Effect of calcium on body temperature. J. Physiol., Lond. *207:* 403–416 (1970).

17 FELDBERG, W.; VEALE, W. L., and COOPER, K. E.: Does leucocyte pyrogen enter the hypothalamus via the cerebrospinal fluid? Proc. 25th Int. Congr. Physiol., 1971.

18 FLOWER, R. J. and VANE, J. R.: Inhibition of prostaglandin synthetase in brain explains the anti-pyretic activity of paracetamol (4-acetamidophenol). Nature, Lond. *240:* 410–411 (1972).

19 FOX, R. H. and MACPHERSON, R. K.: The regulation of body temperature during fever. J. Physiol., Lond. *125:* 21–22P (1954).

20 HOLMES, S. W.: The spontaneous release of prostaglandins into the cerebral ventricles of the dog and the effect of external factors on this release. Brit. J. Pharmacol. *38:* 653–658 (1970).

21 HOLMES, S. W. and HORTON. E. N.: The identification of four prostaglandins in dog brain and their regional distribution in the central nervous system. J. Physiol., Lond. *195:* 731–741 (1968).

22 HORTON, E. W. and MAIN, I. H. M.: Identification of prostaglandins in central nervous tissues of cat and chicken. Brit. J. Pharmacol. *30:* 582–602 (1967).

23 JACKSON, D. L.: A hypothalamic region responsive to localized injection of pyrogens. J. Neurophysiol. *30:* 586–602 (1967).

24 KING, M. K. and WOOD, W. B.: Studies on the pathogenesis of fever. IV. The site of action of leucocyte and circulating endogenous pyrogen. J. exp. Med. *107:* 291–303 (1958).

25 Milton, A. S. and Wendlandt, S.: A possible role of prostaglandin E_1 as a modulator for temperature regulation in the central nervous system of the cat. J. Physiol., Lond. *207:* 76–77P (1970).

26 Milton, A. S. and Wendlandt, S.: Effects on body temperature of prostaglandins of the A, E and F series on injection into the third ventricle of unanesthetized cats and rabbits. J. Physiol., Lond. *218:* 325–336 (1971).

27 Myers, R. D. and Veale, W. L.: The role of sodium and calcium ions in the hypothalamus in the control of body temperature of the unanesthetized cat. J. Physiol., Lond. *212:* 411–430 (1971).

28 Repin, I. S. and Kratskin, I. L.: An analysis of hypothalamic mechanisms of fever. Fiziol. Zh., SSSR *53:* 1206–1211 (1967).

29 Rosendorff, C. and Mooney, J. J.: Central nervous system sites of action of a purified leucocyte pyrogen. Amer. J. Physiol. *220:* 597–603 (1971).

30 Stitt, J. T.: Prostaglandin E_1 fever induced in rabbits. J. Physiol. *232:* 163–179 (1973).

31 Stitt, J. T. and Hardy, J. D.: Evidence that prostaglandin E_1 may be a mediator in pyrogenic fever in rabbits. Biometeorology *5:* 112 (1972).

32 Vane, J. R.: A sensitive method for the assay of 5-hydroxytryptamine. Brit. J. Pharmacol. *12:* 344–349 (1957).

33 Vane, J. R.: Inhibition of prostaglandin synthesis as a mechanism of action for aspirin-like drugs. Nature New Biol. *231:* 232–235 (1971).

34 Veale, W. L. and Cooper, K. E.: Does the cerebrospinal fluid of the ventricular system serve as a route of egress for prostaglandin from hypothalamic tissue? Proc. Canad. Fed. Biol. Soc. *15:* 330 (1972).

35 Veale, W. L. and Cooper, K. E.: Species differences in the pharmacology of temperature regulation; in Schonbaum and Lomax The pharmacology of temperature regulation (Karger, Basel 1973).

36 Veale, W. L. and Whishaw, I.: Temperature responses produced in the rat by PGE_1 and NE microinjected into the brain (unpublished observations).

37 Vogt, M.: The concentration of sympathin in different parts of the central nervous system under normal conditions and after the administration of drugs. J. Physiol., Lond. *123:* 451–481 (1954).

38 Wit, A. and Wang, S. C.: Temperature-sensitive neurons in preoptic/anterior hypothalamic region. Actions of pyrogen and acetylsalicylate. Amer. J. Physiol. *215:* 1160–1169 (1968).

Authors' address: Dr. W. L. Veale and Dr. K. E. Cooper, Division of Medical Physiology, Faculty of Medicine, University of Calgary, *Calgary, Alberta T2N 1N4* (Canada)

Recent Studies of Hypothalamic Function
Int. Symp. Calgary 1973, pp. 371–390 (Karger, Basel 1974)

Ionic Concepts of the Set-Point for Body Temperature

R. D. MYERS

Laboratory of Neuropsychology, Purdue University, Lafayette, Ind.

I. Introduction

Not surprisingly, there are almost as many concepts of the concept of set-point as there are researchers investigating thermoregulation. At one end of the spectrum is the attractive engineering model, based on physical principles, that defines any sort of set-point in terms of a fixed input or value established external to and independent of the automatic control system [e.g., 7]. At the other extreme is the nihilistic pronouncement that a temperature set-point is not even physiologically necessary and does not exist at all, because the gain of the regulatory system can account for any shift in body temperature [10]. Should the latter statement be true, this paper would necessarily terminate here!

On biological grounds alone, there is simply too much evidence to deny the existence of a set-point function in the maintenance of body temperature and probably for certain other vital functions as well [16]. Consequently, a rational theoretical position on this subject should lie somewhere between the relative extremes cited above. In my opinion it should also incorporate two important points that have been derived independently: (1) the adjustable or variable set-point hypothesis described so lucidly by HAMMEL [6] in which the normal set-point may shift during fever, sleep and other states, and (2) the belief of BENZINGER [1] that 'the set-point must be defined as a temperature-dependent property of a definable anatomical or histological substrate with peculiar physiological, molecular properties.' In accord with numerous experiments carried out in our laboratory, it seems that both of these distinct views have struck home on target.

II. Neurological Assumptions for a Mammalian Temperature Set-Point

Most physiologists consider the set-point to be established inherently as a built-in reference temperature around which regulatory adjustments are continually made. But, how is a value of approximately 37 °C that is so critical for normal metabolic activity assigned in Nature? Based on the voluminous literature published prior to 1970, what are the biological characteristics of the set-point of body temperature? We have evolved the following set of assumptions about the set-point which form the basis of much of the research to be described in this paper:

1. Although thermoregulation is not always fully developed in the mammalian neonate, the set-point value of 37 °C is present at birth.
2. The set-point is universal across all mammals (and perhaps other species), and therefore the mechanism must be based on some very fundamental and intrinsic property of a group of neurons.
3. The set-point value is ordinarily invariant, with exceptions being special physiological conditions such as hibernation, perhaps sleep and the pathological insult of bacteria.
4. An altered set-point level should be one around which the mammal will regulate by means of its normal vasomotor, metabolic and other processes.
5. The set-point is subject to perturbation from endogenous stimuli impinging upon brain tissue or by experimental manipulation.
6. The locus of the set-point in the central nervous system is anatomically separate from the site(s) involved in the regulatory processes.

III. Ionic Basis of the Set-Point

In 1969, we showed that the sodium-induced rise in temperature known for a very long time [e.g., 8] can be blocked by Ca^{++} ions, when these cations are perfused through the cerebral ventricles of the conscious cat [5]. After this, we began to speculate that some sort of an ionic mechanism could be at the basis of a set-point for temperature. After all, the ratio between several essential cations is exceptionally fundamental in a biological sense. For example, such a ratio is common to the extracellular fluid of all mammals, within a species it is relatively stable, and the relationship between ions is an inborn phenomenon whereas the regulatory capacity thought to involve biogenic amines in the hypothalamus is not developed fully until some time after parturition [15].

In 1969, VEALE and I began to search for the locus of a set-point in the central nervous system (CNS), thinking all along that if BENZINGER [1] were correct it would be very difficult to differentiate anatomically. Our reasoning here was simply that a collection of unique set-point neurons, if they did exist, would probably be intermingled with other elements in the anterior hypothalamic, preoptic area, particularly the thermosensitive neurons. After a series of experiments in which circumscribed regions of the hypothalamus of the anesthetized cat were perfused by means of push-pull cannulae, we were dumbfounded by the histological verification of the anatomical sites of perfusion. As portrayed in the inset in figure 1, the region of maximum sensitivity to an alteration in ion concentration was the posterior hypothalamus, not the anterior region as we had anticipated. Perfused at the sites indicated in figure 1, excess Na^{+} ions evoked a sharp rise, whereas excess Ca^{++} ions caused an intense decline in the cat's body temperature.

But how could the posterior hypothalamus be involved in the diametrically opposed responses to these cations? The following sequence of experimental events that ensued in the next two years was directed toward an answer to this perplexing question.

A. Experimental Evidence

If the ratio of Na^{+} to Ca^{++} determines the set-point mechanism located in the posterior hypothalamus, then doubling or halving their absolute concentration simultaneously, or replacing Na^{+} or Ca^{++} ions with isotonic sucrose should have no effect on body temperature. In experiments in which such solutions were used as the perfusate in the push-pull cannulae positioned in the posterior hypothalamus, this conclusion was found to be true [16]. In addition, the selective removal of one ion (Ca^{++}) from the extracellular milieu of the caudal hypothalamus lets the other ion (NA^{+}) predominate. When ethyleneglycol-bis-(β-amino ethyl ether) N, N′-tetra-acetic acid (EGTA) is added to the push-pull perfusate in order to chelate Ca^{++} from local sites in the posterior hypothalamus of the conscious monkey, a sharp rise in temperature, as shown in figure 2, occurs immediately and persists as long as the perfusion of EGTA continues [19]. The relative excess in sodium ions caused by

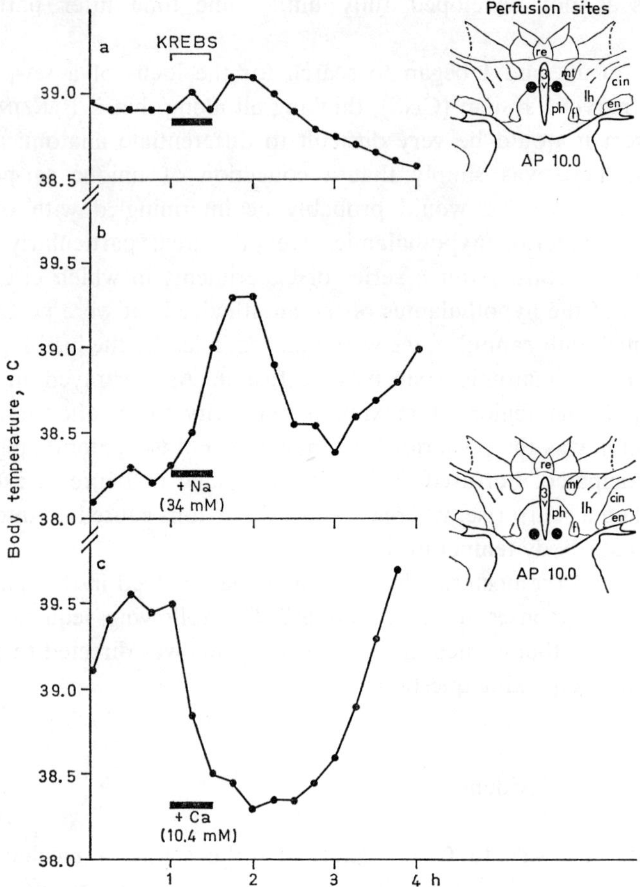

Fig. 1. Temperature records of an unanesthetized cat in response to the local perfusions for 30 min of: *a* Krebs solution alone, *b* Krebs solution plus 34 mM excess sodium, and *c* Krebs solution plus 10.4 mM excess calcium. The sites of the bilateral perfusions are indicated by the filled circles (●) in the inset [16]. cin = internal capsule; en = endopeduncular nucleus; f = fornix; lh = lateral hypothalamic area; mt = mammillothalamic tract; ph = posterior hypothalamic area; re = nucleus reuniens; 3v = third ventricle.

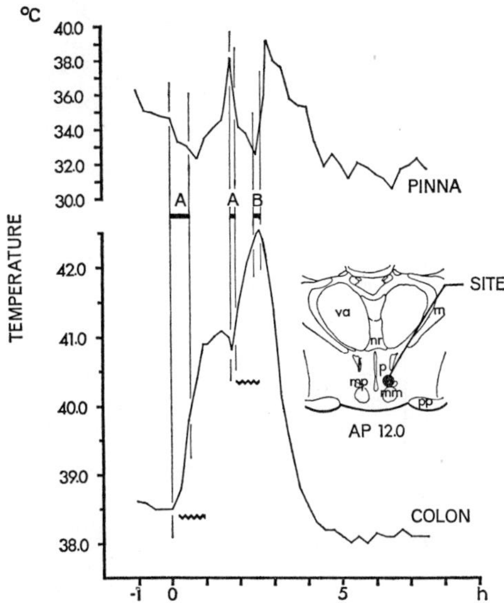

Fig. 2. Temperature records obtained from pinna (top) and colon (bottom) of an unanesthetized monkey in response to perfusions at 50 μl/min (▬) at a site (●) in coronal plane AP 12.0 (inset) of 3.6 mM ethyleneglycol-bis-(β-amino ethyl ether)N,N′-tetra-acetic acid (EGTA) (A) and 23.9 mM excess calcium (B). The zig-zag lines indicate shivering. m = external medullary lamina; mm = mammillary body; mp = mammillary fasiculus princeps; nr = nucleus reuniens; pp = cerebral peduncles; va = ventral anterior nucleus of the thalamus [19].

this depletion of Ca^{++} can drive the monkey's temperature upwards at an astounding rate to a potentially lethal level. This hyperthermia that borders on a runaway hyperpyrexia can be reversed instantaneously as shown in figure 2 by the addition of calcium ions to the perfusate flowing within the posterior hypothalamus. Parenthetically, we have found no other drug, chemical or other central stimulus that leads to hyperthermia with such a pyrexic capability.

If the ratio of Na^{+} to Ca^{++} is altered for any length of time, then the set-point should be able to be reset within metabolic limits for an extended interval. In the conscious monkey, replenishing the excess in Na^{+} or Ca^{++} ions in the posterior hypothalamus at intervals spaced throughout the day results in the resetting of the animal's body temperature for up to half a day [19]. This is shown in figure 3. It should be emphasized that unlike the response to a biogenic amine, there is no

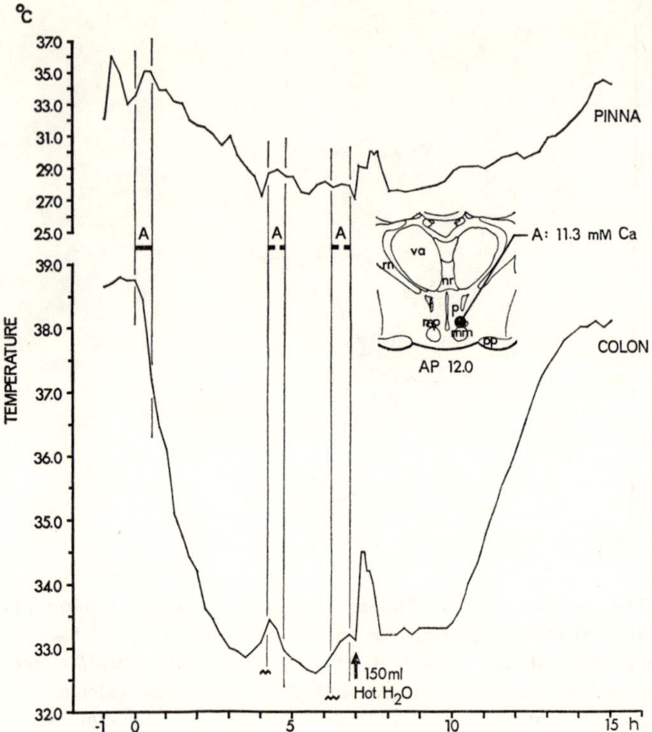

Fig. 3. Temperature records obtained from pinna (top) and colon (bottom) of an unanesthetized monkey in response to repeated perfusions (▬) at 50 μl/min at site A in AP 12.0 (inset) of 11.3 mM excess calcium. At the arrow, 150 ml water heated to 58 °C was given intragastrically. The zig-zag lines indicate shivering. Anatomical abbreviations are the same as in figure 2 [19].

tachyphylaxis to an altered ratio of Na^{+} to Ca^{++} ions in the push-pull perfusion medium.

After the set-point temperature is established at a new level, high or low, the animal is able to thermoregulate normally around that new value. As shown in figure 3, after a solution of Ca^{++} ions was perfused simultaneously at two posterior hypothalamic sites in an unanesthetized monkey and a new set-point level stabilized at about 33 °C, a volume of hot water given intragastrically caused the monkey to vasodilate in an attempt to lose heat, at the same time that its temperature increased over 1 °C for a brief interval. Thereafter, the temperature returned to the new

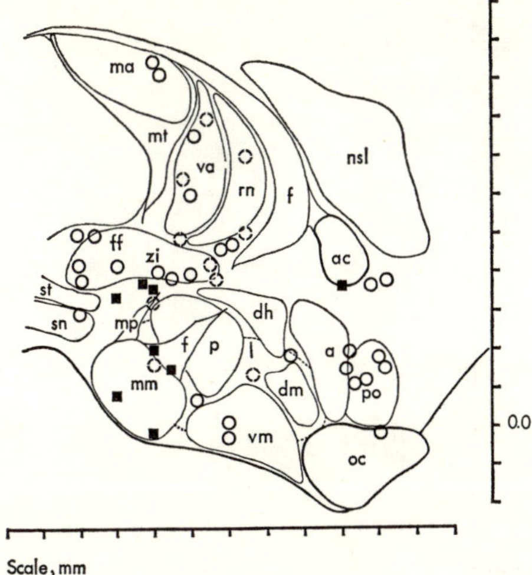

Fig. 4. Sagittal representation of the diencephalic region of the monkey brain portrayed at 1.5–2.4 mm off midline. Length = 1.5–2.5 mm. Each square (■) represents a site where the push-pull perfusion of either excess sodium or calcium ions resulted in an immediate change in body temperature of a least 0.8 °C. Each open circle (o) indicates a site inactive to the perfusion. A dashed circle indicates an inactive site 3.0–5.0 mm off midline. The zero on the vertical scale represents the stereotaxic zero plane 10 mm above the interaural line. a = anterior hypothalamus; ac = anterior commissure; dh = dorsal hypothalamic area; dm = dorsomedial nucleus; f = fornix; ff = fields of Forel; l = lateral hypothalamus; ma = medial anterior nucleus of the thalamus; mm = mammillary body; mp = mammillary fasciculus princeps; mt = mammillothalamic tract; oc = optic chiasm; p = posterior hypothalamus; po = preoptic area; nsl = lateral nucleus of the septum; rn = reticular nucleus of the thalamus; sn = substantia nigra; st = subthalamic nucleus; va = ventral anterior nucleus of the thalamus; vm = ventromedial nucleus; zi = zona incerta [19].

set-point level of 33 °C and remained there for about 2 h until the effects of the excess calcium had dissipated, presumably by means of equilibration. The same regulatory response occurs when cold water is given intragastrically. During a 'super fever' produced by excess Na^+ ions in the hypothalamus, the monkey also regulates similarly in response to peripheral stimuli [19]. In addition to the retention of its thermoregulatory capacity, a monkey, even though deeply hypothermic, nevertheless shows relatively normal EEG activity and can also be induced to eat its monkey chow, albeit somewhat sluggishly [20].

If the hypothalamic regions governing the set-point function and the

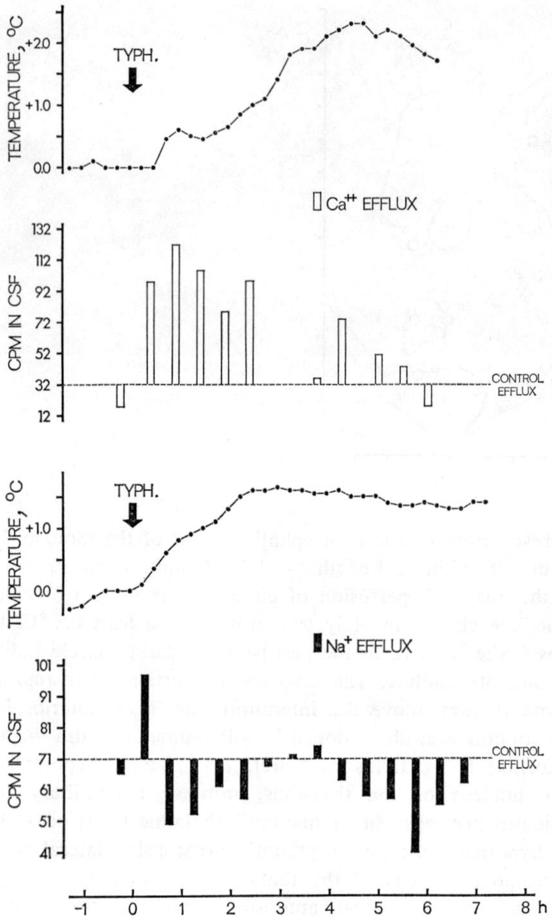

Fig. 5. Fevers produced by 0.2 ml of diluted typhoid vaccine injected into the lateral cerebral ventricles of two cats at zero hour. Each bar represents a 3- to 4-min perfusion of the third cerebral ventricle. The efflux of $^{45}Ca+$ (top) and $^{22}Na+$ (bottom) into cerobrospinal fluid (CSF) is expressed in counts per minute. The value of the sample 1 h before the typhoid (Typh) injection served as the baseline (control) after the rate of efflux in each experiment had become stabilized [14].

regulatory function are indeed anatomically separate, then the morphological distinction should be consistent across species. As shown in figure 4, a sagittal mapping of perfusion sites in the unanesthetized rhesus monkey reveals that the area sensitive to an abberation in Na^+ to Ca^{++} ratio is once again associated with the mammillary complex [19]. This region is homologous to that of the cat [16] and the rabbit [21]. Interestingly, a change in the ratio of Na^+ to Ca^{++} ions in the cerebral ventricles of the rat [12] and golden hamster [13] also evokes a shift in the set-point temperature, the direction of which is identical to the cat, monkey or rabbit. Unlike the response to a biogenic amine, opiate alkaloid and certain other compounds, the hypothermia of Ca^{++} or hyperthermia of Na^+ is thus far remarkably consistent along the phylogenetic scale.

If an exogenous stimulus acts to shift the normal set-point temperature as exemplified by the hyperthermia evoked by pathogenic bacteria [2], then the actual endogenous ratio of Na^+ to Ca^{++} should change within the hypothalamus; further, this shift should correspond to the results obtained when the ratio is altered artificially. This does in fact happen. After stores of the cations have been labeled radioactively with $^{45}Ca^{++}$ or $^{22}Na^+$, typhoid organisms are injected into the ventricles of the cat. As the typically intense fever develops, the normal efflux of $^{22}Na^+$ ions from the hypothalamus into the third ventricle is abolished and the $^{22}Na^+$ ions are retained. At the same time, however, the efflux of $^{45}Ca^+$ ions from the hypothalamus increases into the third ventricle [14]. As illustrated in figure 5, this reciprocal relationship in the extrusion of these 2 cations is one which parallels essentially the slope of the temperature rise. The rates of extrusion change as the asymptote is reached, and again as the fever is sustained (Na^+ efflux) and again as defervescence begins (Ca^{++} efflux).

In the same way, an antipyretic, which lowers at least partially the elevated set-point temperature associated with fever, reverses the ratio of efflux in Na^+ to Ca^{++} ions in the hypothalamus. When 30–36 mg/kg of acetaminophen are given intravenously at a time when a fever has reached the highest peak, the efflux of Na^+ and Ca^{++} ions is opposite to that seen when fever is elicited by a bacterial pyrogen. Figures 6 and 7 show the marked defervescence produced by acetaminophen given systemically to two febrile cats. In the samples collected immediately after the antipyretic was administered, the efflux of $^{45}Ca^{++}$ which had proceeded at a constant rate during the rise in temperature suddenly abated. As soon as the nadir in the temperature curve was reached, the efflux of

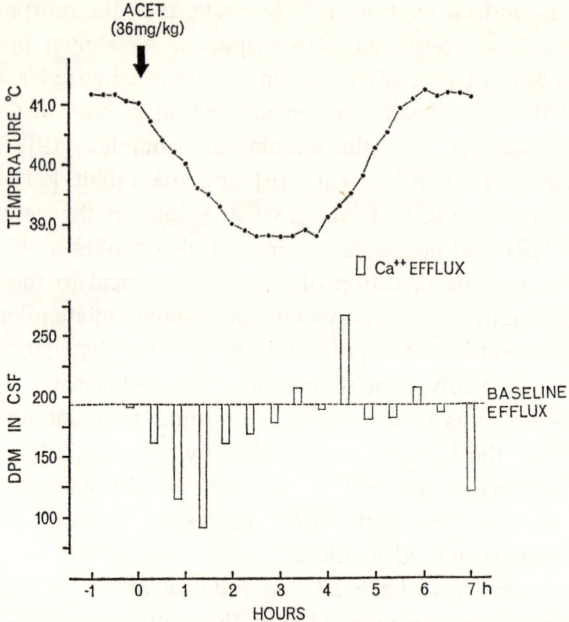

Fig. 6. Antipyretic effect of 36 mg/kg acetaminophen given intravenously (arrow) to an unanesthetized febrile cat. Each bar represents a 4- to 5-min perfusion of the third cerebral ventricle with artificial CSF at 0.1 ml/min. The efflux of $^{45}Ca^{++}$ into the CSF is expressed in disintegrations per min (DPM). The value of the sample collected 1 h before acetaminophen was given served as the baseline control efflux.

$^{45}Ca^{++}$ ions began once again. In a correspondingly opposite fashion, $^{22}Na^{+}$ ions were retained in the hypothalamus as the pyrexia developed but extruded as soon as the antipyretic was given. It should be emphasized that with a lower dose of the antipyretic (e.g., 15 mg/kg) the alterations in efflux are not sharply defined. Nevertheless, the direction of the efflux of the two cations and the temporary reversibility of their extrusion rates by an antipyretic offers compelling evidence of a dynamic mechanism functioning during the resetting of the set-point. Although it is too early to speculate about this mechanism, uncoupling or unbinding from the respective ion complex as well as active or passive transport into the cerobrospinal fluid (CSF) 'sink' could explain these striking changes [3].

If a change in the set-point for body temperature is governed by a

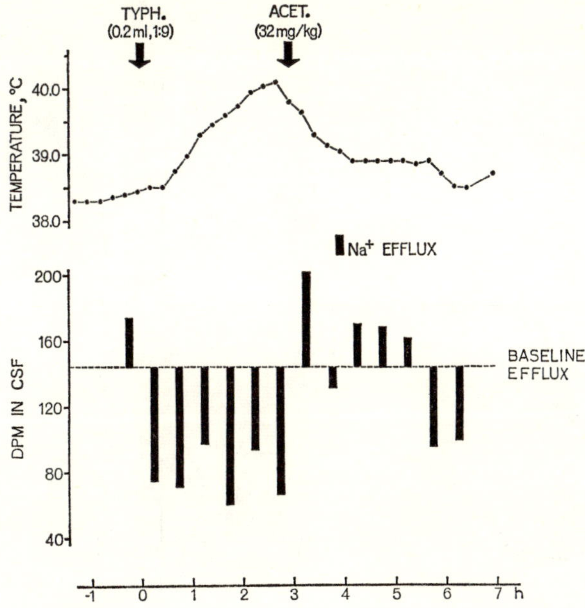

Fig. 7. Antipyretic effect of 32 mg/kg of acetaminophen (Acet) given intravenously (second arrow) to an unanesthetized cat in which 0.2 μl of a 1:9 dilution of typhoid (Typh) vaccine had been injected into the cerebral ventricle (first arrow). Each bar represents a 4- to 5-min perfusion of the third cerebral ventricle with artificial CSF at 0.1 ml/min. The efflux of $^{22}Na^{+}$ in the CSF is expressed in DPM. The value of the sample collected 1 h before acetaminophen was given served as the baseline control efflux.

different central process than a regulatory change, then it may be possible to dissociate the hypothalamic mechanisms. In a series of experiments performed in our laboratory, M. B. WALLER trained the *Macaca nemestrina* to obtain warmth from a heat lamp by pressing a lever. When prostaglandin E_1 is injected in the rostral hypothalamus, not only is the typical prostaglandin fever evoked [4, 9], but the behavioral response for a 4-sec burst of warmth is enhanced dramatically. As shown in figure 8, lever responses for heat vary in the range of between 300 and 700/15-min interval. The maximal rate of response in each period correlates with the period during which the monkey's temperature rises at the fastest rate. In the same monkey, 10.4 mM excess Ca^{++} ions perfused bilaterally at sites in the posterior hypothalamus evoke a typical fall in temperature. In this case, however, lever responses for heat never exceeded 25/15-min

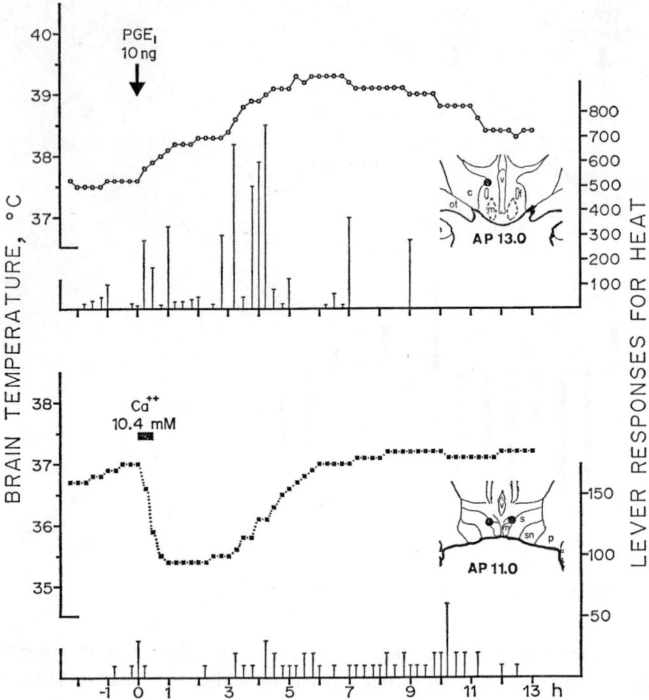

Fig. 8. Lever responses (right axis) for a 4-sec burst of a heat lamp, portrayed by the vertical bars, emitted by a *Macaca nemestrina* monkey seated in a restraining chair. Top: at arrow. 10 ng prostaglandin E_1 was injected in a 1-μl volume at the site indicated by the dot in the inset. Bottom: at bar, bilateral 30-min push-pull perfusion at 50 μl/min of the sites indicated by the dots in the inset of an artificial CSF containing Ca^{++} ions 10.4 mM in excess of the normal value. c = internal capsule; f = fornix; m = mammillary body; ot = optic tract; p = cerebral peduncle; s = subthalamic nucleus; sn = substantia nigra; v = third ventricle.

interval throughout the 6 h during which the *M. nemestrina's* temperature was below normal. Of special significance is the fact that during the phase of recovery from Ca^{++}-induced hypothermia, the monkey apparently did not sense or experience cold, even though the slope of the rising temperature was identical to the rate of increase following the prostaglandin injection. Therefore, it would appear that WALLER's experiments have demonstrated a clear behavioral dissociation between a regulatory hyperthermia and what is believed to be an upward shift in the set-point.

B. Interaction with Ambient Temperature

The set-point of 37 °C is maintained often in the face of a severe challenge from cold or hot environmental temperatures. Therefore, the ratio of Na^{+} to Ca^{++} in the posterior hypothalamus would be expected to remain stable in response to heat or cold stress. On the other hand, the regulatory mechanism within the anterior hypothalamus would be highly active so as to offset the peripheral thermal challenge.

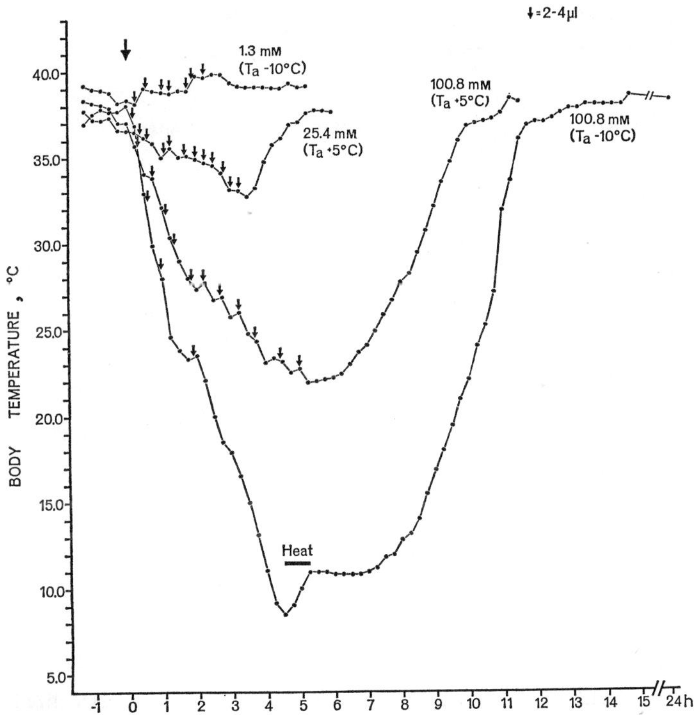

Fig. 9. Colonic temperature of 4 unanesthetized hamsters placed in a chamber at an ambient temperature (T_a) of +5 or –10 °C. An initial infusion was given at zero hour (large arrow) into lateral cerebral ventricles of a 4-μl volume of Ca^{++} ions in excess of their normal level (indicated by millimolar value). Infusions were repeated at intervals indicated on each record by the small arrow. Black bar denotes period during which a heat lamp was turned on above one animal [13].

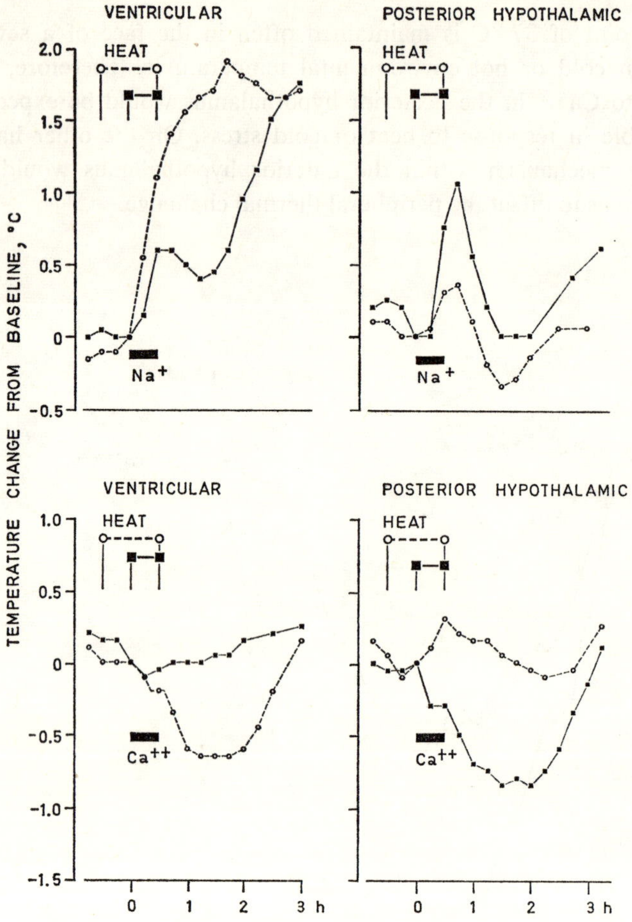

Fig. 10. Deviations from baseline temperature in two unanesthetized cats warmed by hot air just during a perfusion (■–■) or for 30 min before the perfusion plus during the perfusion (○--○). Left panels: perfusions of the third cerebral ventricle with 0.1 ml/min of artificial CSF containing 13 mM excess Na^{+} (top) or 23.4 mM excess Ca^{++} (bottom). Right panels: bilateral push-pull perfusions at 50 μl/min of the posterior hypothalamus with an artificial CSF containing 13 mM excess Na^{+} (top) or 10.4 mM excess Ca^{++} (bottom). In both cases, the horizontal bar denotes the 30-min interval of perfusion.

Last year we began an extensive series of experiments to try to find out whether ambient temperature did influence the artificial resetting of a set-point by the hypothalamic imbalance in the ion ratio. A long-standing interest in the fascinating phenomenon of hibernation led to experiments in which we attempted to induce a hibernator into a state of torpor by altering the Na^{+} and Ca^{++} levels in the CNS. Although this attempt failed, it is possible to lower the body temperature of a hamster to 5 °C, as shown in figure 9, by repeated infusions of a solution containing excess Ca^{++} ions [13]. However, since this low level was attained only when the animal was kept at a very cold temperature – in this case +5 or –10 °C – we began to ponder over the possible interaction between the set-point and the environmental temperature.

Further experiments carried out in our laboratory with a nonhibernator are beginning just now to unravel a puzzle of unimaginable complexity. P. Redgrave and I have found that heating or cooling the unanesthetized cat may influence the typical effects on temperature of excess Na^{+} or Ca^{++} ions, but we were astonished to find that the temperature response depends on two main factors: (1) the specific interval during which the cat is exposed to warm or cold temperatures, and (2) the sites in the CNS at which the ratio of the cations is altered. These two critical factors are illustrated in figure 10 in which all perfusions were started either at the moment that the temperature of the cage was elevated (■–■) or after a 30-min preheating interval (o–o). If warm air is blown into the cat's cage before either excess Na^{+} or Ca^{++} ions are perfused through the third ventricle, the respective hyper- or hypothermia is potentiated. Blowing hot air in the cat's chamber at the same time as the perfusion does not interfere greatly with the usual Na^{+}-induced rise in temperature, although this does attenuate slightly the Ca^{++} fall.

Opposite results are also seen in figure 10 if the same experiments are repeated, but with the exception that the posterior hypothalamus is perfused directly by push-pull cannulae. Quite extraordinarily, preheating the cat with warm air blown into its chamber inhibits or prevents entirely the typical rise or fall in temperature ordinarily induced by an excess of Na^{+} or Ca^{++} ions in the perfusate. Conversely, when the warm air is blown into the animal's chamber just as the posterior hypothalamus is perfused, the stimulus is not of a sufficient duration to disrupt or interfere with the usual hyper- or hypothermic response to excess Na^{+} or Ca^{++} ions in the hypothalamic perfusate. What do these surprising results mean?

In the first instance, it would appear that an excess in either cation affects a variety of physiological systems in the brain. In fact, some sort of generalized excitation and overall depression of neuronal systems may partially occur. In stark contrast, the results of the anatomically specific, hypothalamic perfusions reveal that thermal input could be conveyed to the posterior hypothalamus. In the case of a warm ambient temperature, this input signals the monkey's set-point cells the critical information that, because the environment is already hot, an upward shift in the set-point is not called for. In essence, the hypothalamic set-point mechanism does participate in the defense against an external thermal challenge. It does not operate in an entirely passive manner as we had deduced earlier and quite incorrectly from our findings with ventricular injections of Ca^{++} in the hamster or intraventricular perfusions of Ca^{++} or Na^{+} in the cat. Clearly, it is dangerous to formulate a theoretical viewpoint based solely on experiments involving the perfusion or injection of ions into the cerebral ventricles. Preliminary experiments using cold stimuli tend to bear this out, but far more research is required before the complicated riddle about the origin and nature of the feedback input to the set-point is solved.

C. Cholinergic Link in the Set-Point System

Based on the results of the microinjection of acetylcholine (ACh), the major cholinergic pathway that traverses the hypothalamus from the anterior to posterior regions is apparently delegated to heat production; however, a heat-loss pathway also originates in the posterior hypothalamus [18]. Studies on the evoked release of ACh from isolated sites in the hypothalamus perfused with push-pull cannulae support the microinjection studies. Cooling a monkey enhances the release of ACh at two-thirds of the active sites of the posterior hypothalamus, whereas warming evokes the release at one-third of these sites [17]. If, by virtue of the appropriate balance between Na^{+} and Ca^{++} ions, the constant discharge of a unique group of neurons in the posterior region serves to maintain the set-point level, then ACh may be the transmitter substance that carries the efferent signals to maintain the set-point. This would account for the emergence of ACh releasing sites in the posterior hypothalamus that are homologous to those at which Na^{+} or Ca^{++} exert such profound effects.

IV. Concluding Considerations

A simple explanation of the present findings is that they represent a series of nonspecific effects within the nervous system or are based on neural artifacts. Although, naturally, this 'easy way out' is tantamount to no explanation at all, we nevertheless cannot be certain at this time which of the phenomena are on the side of generalized effects and which are truly specific to the set-point for body temperature. In the last analysis, one can only view all of the evidence collectively (table I), knowing full well that technical mistakes or inadequate experimental controls will surely confound if not confuse the picture. For example, a very high concentration of extracellular Ca^{++} will suppress the function of any cell or it can enhance or reduce transmitter release. In this connection, the method of push-pull perfusion permits the recovery of 90–95 % of the cation and, hence, the millimolar concentrations of Na^{+} or Ca^{++} which actually remain within the hypothalamic tissue are only 5–10 % of that contained in the perfusion medium.

Among the factors illustrated in table I, which mitigate the notion of nonspecificity, are: (1) the anatomical specificity of response to a pertur-

Table I. Evidence for independent ionic set-point mechanism

1. Inborn nature of ionic relationship
2. Stability of Na^{+} to Ca^{++} ratio in ECF
3. Uniformity of Na^{+}–Ca^{++} effect across mammalian species
4. Anatomical isolation in the posterior hypothalamus
5. Relative inactivity of K^{+}, Mg^{++} or inert ions in the posterior hypothalamus
6. Resetting of set-point for indefinite interval
7. Absence of tachyphylaxis to elevated Na^{+} or Ca^{++} levels
8. Normal regulation of temperature after new set-point level is established
9. Hypothalamic retention of Na^{+} and efflux of Ca^{++} during fever
10. Reversal of hypothalamic ion efflux (Na^{+} extrusion, Ca^{++} retention) due to antipyretic
11. Behavioral dissociation between regulatory response to prostaglandin and ion-induced shift in temperature

ECF = extracellular fluid.

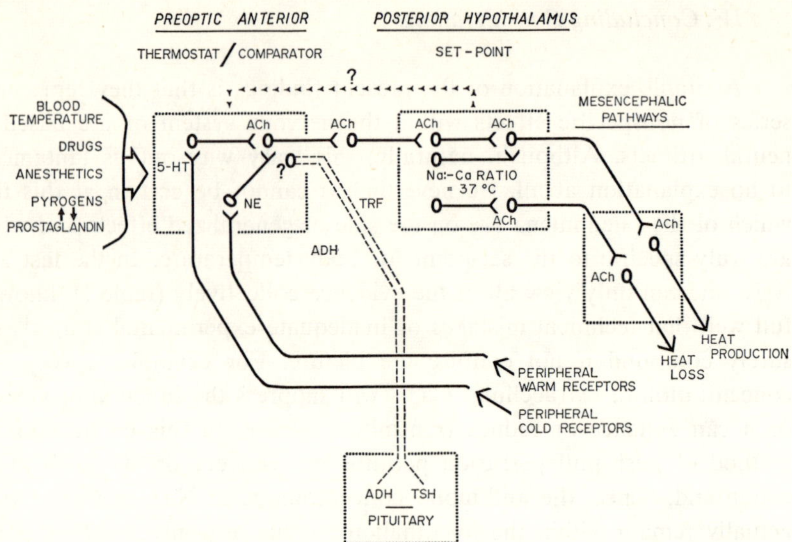

Fig. 11. A schematic diagram of a hypothalamic model of thermoregulation based mainly on experimental findings in the monkey and cat. The anterior hypothalamic preoptic area contains neurons which are thermosensitive as well as a comparator mechanism which contrasts the set-point with the local temperature. The region is also sensitive to a number of substances, as indicated, including pyrogens and prostaglandins which may interact functionally. The release of 5-hydroxytryptamine (5-HT) acts through a cholinergic pathway to signal heat production, whereas the release of norepinephrine (NE) blocks the cholinergic heat production pathway which traverses the posterior hypothalamus. The set-point of 37 °C depends upon the intrinsic ratio of Na^+ to Ca^{++} ions, an aberration of which will evoke a shift in the set-point presumably via independent cholinergic pathways. The regulation of body water and thyroid activity may also be partially mediated by synapses arising in the anterior hypothalamus ACh = acetylcholine; TRF = thyroid releasing factor; ADH = antidiuretic hormone; TSH = thyroid stimulating hormone [modified after MYERS, 11].

bation in the ratio of Na^+ and Ca^{++}; (2) absence of effect of K^+ and Mg^{++} in the same physiological range; (3) the endogenous changes in Na^+ and Ca^{++} efflux from the hypothalamus during a fever produced by a bacterial pyrogen; (4) the reversibility by an antipyretic of cation efflux from the hypothalamus; (5) the capacity of the animal to thermoregulate normally after its temperature set-point is established at a new level; (6) the blockade by a peripheral thermal challenge of a change in set-point induced artificially by a shift in the hypothalamic Na^+ to Ca^{++} ratio;

(7) the normalcy of the EEG and the retention of the capacity to eat following an ion-induced change in an animal's temperature, and (8) the functional specificity in that behavioral changes such as arousal, feeding or rage are evoked by altering the Na^{+} to Ca^{++} ratio in other parts of the diencephalon.

Finally, it should be emphasized that we do not believe that a fluctuating imbalance in ion concentration persists interminably in the hypothalamus. On the contrary, a steady-state constancy in the ratio between Na^{+} and Ca^{++} is envisaged which stabilizes the set-point for most of one's life. During a pathological shift in the set-point, the endogenous mechanism which allows the ion flux to become more labile is undoubtedly a self-limiting one, a common feature in the organization of neural function. This vital self-limiting characteristic ordinarily affords the necessary protection against a lethal deviation in the ratio between the cations. The model shown in figure 11 summarizes pictorially the relationships between the factors that stimulate the anterior hypothalamic, preoptic area and the aminergic pathways that convey the signals for heat production by way of a cholinergic link. The hypothesized set-point neurons in the posterior hypothalamus may share the same cholinergic synapses that transmit the efferent signals for heat production or heat loss. Additional research will clarify the possible feedback loop from the posterior to anterior regions as well as an outflow from structures in the rostral hypothalamus to the pituitary that subserve water balance and long-term metabolic heat production via the thyroid axis.

References

1 BENZINGER, T. H.: Heat regulation. Homeostasis of central temperature in man. Physiol. Rev. *49:* 671–759 (1969).

2 COOPER, K. E.: Central mechanisms for the control of body temperature in health and febrile states; in DOWNMAN Modern trends in physiology, pp. 33–54 (Butterworths, London 1972).

3 COOPER, K. E. and VEALE, W. L.: Potentiation of fever, produced by intravenous leucocyte pyrogen, following the injection of paraffin oil in the cerebral ventricles of the unanesthetized rabbit. Experientia *28:* 917–918 (1972).

4 FELDBERG, W. and SAXENA, P. N.: Further studies on prostaglandin E_1 fever in cats. J. Physiol., Lond. *219:* 739–745 (1971).

5 FELDBERG, W.; MYERS, R. D., and VEALE, W. L.: Perfusion from cerebral ventricle to cisterna magna in the unanesthetized cat. Effect of calcium on body temperature. J. Physiol., Lond. *207:* 403–416 (1970).

6 HAMMEL, H. T.: Neurons and temperature regulation; in YAMAMOTO and

BROBECK Physiological controls and regulation, pp. 71–97 (Saunders, Philadelphia 1965).
7 HARDY, J. D.: The 'set-point' concept in physiological temperature regulation; in YAMAMOTO and BROBECK Physiological controls and regulation, pp. 98–116 (Saunders, Philadelphia 1965).
8 HASAMA, B.: Pharmakologische und physiologische Studien über die Schweisszentren. IV. Mitteilung. Über den Einfluss der anorganischen Kationen auf Wärme- sowie Schweisszentrum im Zwischenhirn. Arch. exp. Path. Pharmakol. *153:* 291–308 (1930).
9 MILTON, A. S. and WENDLANDT, S.: Effects on body temperature of prostaglandins of the A, E and F series on injection into the third ventricle of unanaesthetized cats and rabbits. J. Physiol., Lond. *218:* 325–336 (1971).
10 MITCHELL, D.; SNELLEN, J. W., and ATKINS, A. R.: Thermoregulation during fever. Change of set-point or change of gain. Pflügers Arch. ges. Physiol. *321:* 293–302 (1970).
11 MYERS, R. D.: Hypothalamic mechanisms of pyrogen action in the cat and monkey; in WOLSTENHOLME and BIRCH Ciba Foundation Symposium on Pyrogens and Fever, pp. 131–153 (Churchill, London 1971).
12 MYERS, R. D. and BROPHY, P. D.: Temperature changes in the rat produced by altering the sodium-calcium ratio in the cerebral ventricles. Neuropharmacology *11:* 351–361 (1972).
13 MYERS, R. D. and BUCKMAN, J. E.: Deep hypothermia induced in the golden hamster by altering cerebral calcium levels. Amer. J. Physiol. *223:* 1313–1318 (1972).
14 MYERS, R. D. and TYTELL, M.: Fever. Reciprocal shift in brain sodium to calcium ratio as the set-point temperature rises. Science *178:* 765–767 (1972).
15 MYERS, R. D. and VEALE, W. L.: Body temperature. Possible ionic mechanisms in the hypothalamus controlling the set point. Science *170:* 95–97 (1970).
16 MYERS, R. D. and VEALE, W. L.: The role of sodium and calcium ions in the hypothalamus in the control of body temperature of the unanaesthetized cat. J. Physiol., Lond. *212:* 411–430 (1971).
17 MYERS, R. D. and WALLER, M. B.: Differential release of acetylcholine from the hypothalamus and mesencephalon of the monkey during thermoregulation. J. Physiol., Lond. *230:* 273–294 (1973).
18 MYERS, R. D. and YAKSH, T. L.: Control of body temperature in the unanaesthetized monkey by cholinergic and aminergic systems in the hypothalamus. J. Physiol., Lond. *202:* 483–500 (1969).
19 MYERS, R. D. and YAKSH, T. L.: Thermoregulation around a new 'set-point' established in the monkey by altering the ratio of sodium to calcium ions within the hypothalamus. J. Physiol., Lond. *218:* 609–633 (1971).
20 MYERS, R. D. and YAKSH, T. L.: The role of hypothalamic monoamines in hibernation and hypothermia; in SOUTH Hibernation and hypothermia. Perspectives and challenges, pp. 551–575 (Elsevier, Amsterdam 1972).
21 VEALE, W. L.: Personal commun. (1973).

Author's address: Dr. R. D. MYERS, Laboratory of Neuropsychology, Purdue University, *Lafayette, IN 47097* (USA)

Recent Studies of Hypothalamic Function
Int. Symp. Calgary 1973, pp. 391–398 (Karger, Basel 1974)

Fever, an Abnormal Drive to the Heat-Conserving and -Producing Mechanisms?

K. E. Cooper and W. L. Veale

Division of Medical Physiology, University of Calgary, Calgary, Alberta

The phrase 'fever represents a resetting of the central thermoregulatory thermostat to a higher operating temperature' has been used by many workers in the field of fever research. The mechanism whereby this setpoint is elevated by the products of infecting organisms has received study in the last decade. The concept of altered set-point was clearly an early one [11] and received support from experiments in which sweat rates of mildly febrile subjects at sustained elevated temperatures were compared with those of nonfebrile subjects doing a similar exercise [12], exercise studies [9] and studies of the behavior of the central warm receptor during the plateau phase of fever [7]. In each of these experiments it was possible to demonstrate that at the plateau of the elevated temperature in fever, either produced artificially by the injection or infusion of pyrogens or by naturally occurring disease, a displacement of the body temperature brought about thermoregulatory responses which tended to restore the temperature to the original steady elevated level. These responses seemed quantitatively similar to those observed in afebrile states. The mechanism whereby this temperature 'set-point' is altered in fever has been described by a number of analogs, both electrical and mechanical; mathematical relationships have been derived from these analogs and applied to the thermoregulatory processes in the living organism. More recently the general responses of the thermoregulatory mechanisms have been described in terms of neuronal models [2, 14].

It has been shown [21] that some thermosensitive cells in the hypothalamus change their firing rates in an appropriate manner following the injection of pyrogen into the animal. Thus, warm firing neurons tended to produce lower frequency spikes and cold firing neurons tended to increase the frequency of their firing at any given temperature when the animal was given pyrogen. These observations were made on neurons principally in the anterior hypothalamic-preoptic area (AH-POA) which is generally assumed to be the principle region or site of action of pyrogens [6, 10, 16], though there may also be a contribution to the site of action of pyrogens from the midbrain and possibly from other areas of the central nervous system.

The generally accepted concept of the action of the bacterial pyrogen is that it causes the release from white blood cells, and possibly other tissues, of a polypeptide called *leucocyte or endogenous pyrogen.* Leucocyte pyrogen gets into the brain and there induces the neuronal changes necessary to alter the body temperature. Some evidence has been produced in the cat [20] which suggests that bacterial pyrogens can themselves act within the brain to cause fever. Since these substances are lipopolysaccharides of approximately molecular weight 10^6 it would seem that, though they might on microinjection produce fever when injected into the brain, the more likely substance which would get into the brain in naturally occurring fever would be the circulating leucocyte or endogenous pyrogens. From the delay in the onset of heat-conserving and heat-production mechanisms following microinjection of leucocyte pyrogen into the brain, it has been tempting to suggest that another intermediate substance might be released or synthetized within the brain which would be the ultimate mediator of the neuronal responses. This concept has been supported by the discovery [4, 8, 13, 17] that prostaglandin E_1 can produce fever with very rapid onset when injected into the AH-POA in extremely small amounts. Moreover, the demonstration of the inhibition of synthesis of prostaglandins in tissues by commonly used antipyretics lends further support to such a concept [19]. The origin of this putative mediator of fever has not yet been revealed, and it may come from glial cells, from neuronal tissue or perhaps from leucocytes sequestered in hypothalamic capillaries. There is evidence that prostaglandin E_1 and leucocyte pyrogen act at very similar sites within the brain to produce fever. The action of these pyrogenic substances within the brain may be explained in terms of the direct response of thermosensitive units to these substances, or alternatively the thermosensitive units may themselves be in part driven by other neurons on which the mediators of fever could act.

The site of action of leucocyte pyrogen, as determined by microinjection, could be different from that occurring in naturally occurring fever for two reasons. Firstly, none of the preparations of leucocyte pyrogen which have been used has been completely pure, and thus contaminant substances might have complexed with the leucocyte pyrogen to modify its apparent site of action. Secondly, the rather impure leucocyte pyrogen present in extracts made *in vitro* might not be the same material as that which reaches the brain via the bloodstream, since there could be alterations of the leucocyte pyrogen molecule en route through be endothelial lining of blod vessels. Nevertheless, there has been remarkable coincidence in the results of workers who have microinjected leucocyte pyrogen and those who have microinjected prostaglandin E_1 into the brain in terms of the site of action of the substances. One author who had attempted to label leucocyte pyrogen with a radioactive molecule and then determine its site of uptake into the brain by autoradiography [1] indicated a posterior hypothalamic site for the entry of the labeled material. This work has been criticized on the grounds that the label may been more firmly attached to the contaminants which were present in high concentrations than to leucocyte pyrogen present in minute amounts. It is interesting that the posterior hypothalamus is a site in which modification on the ionic concentrations, particularly of sodium and calcium, can alter the body temperature in such a way that the set-point appears to be altered as judged by the previously mentioned criteria. Nevertheless, injections of either 'leucocyte' pyrogen or prostaglandin into this region do not cause fever.

With this latter series of findings in mind we have performed a series of experiments in which very large lesions have been made in the AH-POA in rabbits. Animals which had these lesions were unable to regulate their temperature in a hot environment and their temperature regulation in a cold environment was impaired. Massive ablation of the AH-POA such as to interfere grossly with body temperature regulation did not prevent the animals from developing fever of the same magnitude following fever due to intravenous leucocyte pyrogen as they had experienced from the same dose of pyrogen before lesioning. However, animals in which the lesions were placed in the posterior hypothalamus did not develop significant fevers following the injection of leucocyte pyrogen intravenously, though they did before the lesions were made. Thus, this is confirming evidence that as essential part of the central fever pathway is anatomically located in the posterior hypothalamus. In the AH-POA lesioned animals, either fever could have been produced by leucocyte pyrogen getting to an extremely small residual undamaged amount of AH-POA tissue or there may have been some other site at which leucocyte pyrogen acted to produce fever. Injection of leucocyte pyrogen into the lesioned AH-POA did not induce fever. Such an idea could be reconciled with the direct microinjection studies if local applications of leucocyte pyrogens in other regions of the brain had started small alterations in body temperature which were rapidly corrected by an intact and active AH-POA. Thus, action of leucocyte pyrogen injected into other sites would be masked by the corrective action of the preoptic area. Which of these two explanations is correct has not yet been established.

That leucocyte pyrogen does occur in naturally occurring fever has received some support from recent work in our laboratory in which heat-labile pyrogen has been demonstrated in the plasma of 4 patients suffering febrile illnesses. The heat-labile pyrogen was detected by microinjecting plasma from febrile patients into the preoptic area of rabbits. We now have reason to believe that this assay could be made more sensitive by filling the ventricular system with inert sterile paraffin oil. Heat-labile pyrogen was not found in the plasma of patients who had been treated and were no longer febrile, or who had no febrile illness.

The ability of the newborn lamb to make endogenous pyrogen has been studied recently [15]. Four hours after birth the lambs did not get fever in response to intravenous bacterial pyrogen, nor did they respond to leucocyte pyrogen made *in vitro*. Thus, at this age, it would appear that a process of maturation of the hypothalamus and possibly other brain areas is necessary before the newborn animal can experience a fever. At 60 h of age, lambs which had not received any bacterial pyrogen previously did not get fever in response to intravenous bacterial pyrogen. Lambs which had been injected with bacterial pyrogen at 4 h *post partum* did get fever when similarly challenged at 60 h *post partum*. Thus, it would appear that the process of formation of leucocyte pyrogen is learned during the first few days of life, or in response to the challenge with endotoxin after birth. The 4-hour-old lambs, and even those *in utero*, when given intravenous bacterial pyrogen experienced a reduction in the number of circulating blood cells. This is characteristic, in the adult animal, of transport of the injected foreign material to the reticulo-endothelial system. It would be reasonable to assume that the white cells circulating in the newborn lambs were capable of recognizing the injected foreign material. Thus, we have evidence that the nervous system of the newborn lamb is not mature enough at birth for the production of fever, and that the lamb white cells, reticulo-endothelial cells or some other system in the body requires sensitizing to injected bacterial pyrogen before fever can occur. The lambs, which at birth failed to respond to intravenous leucocyte pyrogen, did show the ability to maintain their body temperatures, even under quite harsh conditions with atmospheric temperatures around the freezing point, and they were seen to shiver on exposure to cold. So, the maturation necessary in the nervous system to respond to leucocyte pyrogen appears not to be the same development as is necessary for induction of shivering in response to cold.

The concept that monoaminergic pathways may be involved in the induction of fever has been studied [6]. Following apparent total depletion of monoamine content of the AH-POA, by intraventricular reserpine, rabbits still experienced fever following intravenous pyrogen. In more recent work [18], although it was shown that though depletion of the monoamines, noradrenaline or 5-hydroxytryptamine from the AH-POA modified the extent of fever, in no instance was fever totally abolished by very marked depletion of these regions of their monoamine content. Some very recent work in our laboratory has shown that prostaglandin E_1 injceted into the AH-POA in very small amounts produces fever even if the animals have received a dose of reserpine into the lateral cerebral ventricle on the previous day, such as was previously found to deplete the AH-POA of its monoamine content [5]. Such an observation would fit in with the hypothesis that prostaglandin E_1 is released by leucocyte pyrogen entering the brain and that it, too, can act via pathways which are not aminergic. Of further interest, 200 μg atropine injected into the lateral ventricle of rabbits causes a fall in rabbit body temperature and it abolishes the response due to injection of prostaglandin E_1 into the preoptic area. Atropine injection into the same site as that in which the prostaglandin E_1 was injected did not modify the prostaglandin fever. Thus, it would seem that a cholinergic heat-production and heat-conservation pathway may be involved in the fever due to the microinjection of prostaglandin E_1 into the preoptic area, and that the cholinergic synapses are at some distance from the AH-POA. It would seem from previous evidence [18], that the monoaminergic system could modulate the fever produced by the excitation of this cholinergic pathway.

We have investigated the way in which leucocyte pyrogen might get into the brain and how it, or the products which it causes to be released, might get excreted from the brain. Rapid perfusion of the ventricular system with an artificial cerebrospinal fluid CSF from the lateral cerebral ventricle to the cisterna magna does not alter the magnitude, or the time of onset, of fever produced by an intravenous leucocyte pyrogen. This would imply that leucocyte pyrogen reaches the hypothalamus directly via the bloodstream rather than by excretion through the lateral ventricular choroid plexus and absorption through the wall of the third ventricle. However, in experiments designed to test this hypothesis further, the ventricular system was filled with inert pyrogen-free mineral oil and it was found that this resulted in larger and much more prolonged fevers, due to intravenous leucocyte pyrogen, than the same animals experienced without oil in their ventricles [3]. The oil was without effect on body temperature when injected into the lateral cerebral ventricles without any other treatment of the animal. When a steady state of fever was maintained by an intravenous injection followed by an infusion of leucocyte pyrogen, oil introduced into the ventricle caused a secondary rise in body temperature which continued for some hours. Thus, a route of excretion of pyrogenic substances from the AH-POA may be via the CSF.

Our evidence, then, would seem to suggest that leucocyte pyrogen is indeed one of the mediators of fever response and that at least one component of the system may be the ability of the white cell to produce leucocyte pyrogen as a result of exposure to endotoxin shortly after birth in some species. Our work lends support to the hypothesis [8, 13] that prostaglandin E_1 may be the ultimate mediator of fever within the AH-POA. Further, it suggests that prostaglandin E_1 can excite heat production and heat conservation via a cholinergic link which is not at the same site in the hypothalamus as that in which the prostaglandin is released and acts. It also supports the view that both leucocyte pyrogen fever and prostaglandin fever are mediated by a pathway which, while not dependent upon monoaminergic links, can be to some extent modulated by monoaminergic pathways. Again it is suggested that despite the evidence of microinjection studies there may be more than one site of action of leucocyte pyrogen during naturally occurring fever.

The cholinergic heat-conservation and heat-production mechanism which is brought into play by leucocyte pyrogen, may well be driven in a manner proportional to the amount of the prostaglandin E_1 released in the preoptic area. If this drive due to prostaglandin E_1 was maintained at a steady level, and if the sensing units for temperature change and for integrating the peripheral temperature inputs into this change were to utilize monoaminergic pathways, then we could envisage a pathological drive to the thermoregulatory system, in the direction of heat conservation and production, superimposed upon the normal thermoregulatory mechanisms. Such a hypothesis might require that prostaglandin E_1 would not alter the firing rates of the single units thought to be receptors of temperature change. This remains to be determined. In conclusion, we would postulate a pathological chemical drive which plays no part in normal thermoregulation, which would increase body temperature by stimulating heat production and heat conservation. If this stimulus is maintained at a constant level then the behavior of the warm and cold receptors, some of which had been shown to be operative over a very wide range of temperature, would give the appearance of thermoregulation around a new set-point. Thus, a neuronal model with chemical stimuli would obviate the need to think in terms of more complex electrical or mechanical analogs to explain fever. The sites of action of the prostaglandin E_1 *could* be the postulated cholinergic synapses derived from cold-receptive neurons, or the heat production effectors of one neuronal [2] model.

Acknowledgements

Work described in this paper was supported by grants from the Medical Research Council of Canada. The authors wish to acknowledge with thanks the skillful assistance of Mr. P. Redstone and Mr. T. Malkinson in this work.

References

1 Allen, I. V.: The cerebral effects of endogenous serum and granulocyte pyrogen. Brit. J. exp. Path. *46:* 25–34 (1965).

2 Bligh, J. and Maskerey, M.: A possible role of acetylcholine in the central control of body temperature in the sheep. J. Physiol., Lond. *203:* 55 (1969).

3 Cooper, K. E. and Veale, W. L.: The effect of injecting an inert oil into the cerebral ventricular system upon fever produced by intravenous leucocyte pyrogen. Canad. J. Physiol. Pharmacol. *50:* 1066–1071 (1972).

4 Cooper, K. E. and Veale, W. L.: Unpublished observations (1973).

5 Cooper, H. E.; Cranston, W. I., and Honour, A. J.: Effects of intraventricular and intrahypothalamic injection of noradrenaline and 5-HT on body temperature in conscious rabbits. J. Physiol., Lond. *181:* 852–864 (1965).

6 Cooper, K. E.; Cranston, W. I., and Honour, A. J.: Observation on the site and mode of action of pyrogens in the rabbit brain. J. Physiol., Lond. *191:* 325–337 (1967).

7 Cooper, K. E.; Cranston, W. I., and Snell, E. S.: Temperature regulation during fever in man. Clin. Sci. *27:* 345–356 (1964).

8 Feldberg, W. and Saxena, P. N.: Fever produced by prostaglandin E_1. J. Physiol., Lond. *217:* 547–556 (1971).

9 Grimby, G.: Exercise in man during pyrogen-induced fever. Scand. J. clin. Lab. Invest. *14:* suppl. 67, p. 14 (1962).
10 Jackson, D. L.:A hypothalamic region responsive to localized injection of pyrogen. J. Neurophysiol. *30:* 586–602 (1967).
11 Liebermeister, C.: Handbuch der Pathologie und Therapie des Fiebers (Vogel, Leipzig 1875).
12 McPherson, R. K.: The effect of fever on temperature regulation in man. Clin. Sci. *18:* 281–289 (1959).
13 Milton, A. S. and Wendlandt, S.: A possible role of prostaglandin E_1 as a modulator for temperature regulation in the central nervous system of the cat. J. Physiol., Lond. *207:* 76–77 (1970).
14 Myers, R. D.: Hypothalamic mechanisms of pyrogen action in the cat and monkey; in Wolstenholme and Birch Pyrogens and fever. A Ciba Foundation Symposium, pp. 131–146 (Churchill Livingstone 1971).
15 Pittman, Q.; Cooper, K. E.; Veale, W. L., and Petten Van, G. L.: Fever in newborn lambs. Canad. J. Physiol. Pharmacol. (in press).
16 Repin, I. S. and Kratskin, I. L.: On the analysis of hypothalamic mechanism in the pyretic reaction. Fiziol. Zh., SSSR. *53:* 336 (1967).
17 Stitt, J. T.: Prostaglandin E_1 fever induced in rabbits. J. Physiol., Lond. (in press).
18 Teddy, P. J.: Discussion; in Wolstenholme and Birch Pyrogens and fever. A Ciba Foundation Symposium, pp. 124–127 (Churchill Livingstone 1971).
19 Vane, J. R.: Inhibition of prostaglandin syntheses as a mechanism of action for aspirin-like drugs. Nature, Lond. *231:* 232–235 (1971).
20 Villablanca, J. and Myers, R. D.: Fever produced by microinjection of typhoid vaccine into hypothalamus of cats. Amer. J. Physiol. *208:* 703–707 (1965).
21 Witt, A. and Wang, S. C.: Temperature sensitive neurons in preoptic/anterior hypothalamic region action of pyrogen and acetylsalicylate. Amer. J. Physiol. *215:* 1160–1169 (1968).

Authors' address: Dr. K. E. Cooper and Dr. W. L. Veale, Division of Medical Physiology, Faculty of Medicine, University of Calgary, *Calgary, Alberta T2N 1N4* (Canada)

Recent Studies of Hypothalamic Function
Int. Symp. Calgary 1973, pp. 399–407 (Karger, Basel 1974)

Hypothalamic Blood Flow

C. ROSENDORFF

Departments of Physiology and Medicine, University of the Witwatersrand Medical School, Johannesburg

The hypothalamus has for many years featured prominently in modern accounts of the central nervous control of the circulation. There are several mechanisms by which the hypothalamus can affect the cardiovascular system; these include the autonomic control of heart rate, stroke volume and vascular resistance (including that of the skin in thermoregulation). Also, blood volume control depends upon a complex inter-relationship between hypothalamic hormones (such as anti-diuretic hormone), releasing factors (such as corticotropin releasing factor), the adrenal cortex, water and electrolyte handling by the kidney and renal haemodynamics. The importance of the hypothalamus in producing muscle vasodilatation and in the 'alerting' or 'defence' reaction has been emphasized by HILTON [13]. All of these effects can be regarded as the output of a controlling system of which the hypothalamus is at least a part.

If this is so, the hypothalamus must have reference input elements and a feedback signal. Little is known of the controlling input into the hypothalamus in this context. There are hypothalamic neurones which alter their firing rate in response to a temperature rise ('warm-sensitive') or fall ('cold-sensitive'), and these cells have now been extensively studied. There is now convincing evidence that these cells form part of the thermoregulatory system and, as such, will affect cutaneous blood flow. The effect on other vascular beds is less well documented. It has also been shown that there are other thermoregulatory inputs to the hypothalamus, such as those from thermosensors in the brain-stem, spinal cord, abdominal viscera, large vessels and skin. Most theories consider that the central and peripheral inputs in thermoregulation interact at a neural level in the hypothalamus.

In more general terms, where the heart and total peripheral resistance are the controlled system, the controlling input to the hypothalamus is less clear. In the defence reaction there are probably afferent inputs from higher centres, but there are other afferents from the skin [3] via peripheral nerves [2] and in response to cutaneous, auditory and visual stimuli [1]. The possibility of a functional connection of the baroreceptor and chemoreceptor inflows with the hypothalamus has also had some support [10, 14]. HILTON and SPYER [15] have suggested that the whole brain-stem 'depressor' area, from the hypothalamus through the midbrain to the medulla, constitutes a functional unit which integrates the response to baroreceptor afferent stimulation.

Hypothalamic Blood Flow and Temperature Regulation

If the hypothalamus-midbrain-medulla complex integrates the very disparate varieties of input, and produces an appropriate output to the cardiovascular system, and if the major controlled variable is tissue blood flow, then blood flow through this area itself is of considerable interest. It is at least theoretically conceivable that changes in hypothalamic blood flow (HBF) constitute a controlling signal. As an example, the finding of MCCOOK *et al.* [19] may be quoted: these authors showed that carotid occlusion causes an elevation in hypothalamic temperature and it is possible that the hypothalamus is normally cooled by the blood flow through it. Therefore, the hypothalamus might initiate thermoregulatory responses if the HBF alters. Also, it has been suggested that pyrogen might act by altering blood flow; the initial event would be vasodilatation in the hypothalamus, and the increase in blood flow would be associated with an initial fall in hypothalamic temperature followed by a rise in rectal temperature.

All of the above, which sounds plausible, is almost certainly wrong.

1. In the first place, it has been shown [23] that an increase in HBF follows the intravenous injection of bacterial pyrogen (fig. 1), but that there is no detectable fall in hypothalamic temperature at any stage preceding or during the development of the fever [25]. It is probable that the increase in HBF is the result of, rather than the cause of, the fever. It is well-known that bacterial pyrogen causes profound circulatory changes [8] which include not only cutaneous vasoconstriction but also a rise in renal and hepatic blood flow, in the face of a fall of mean arterial

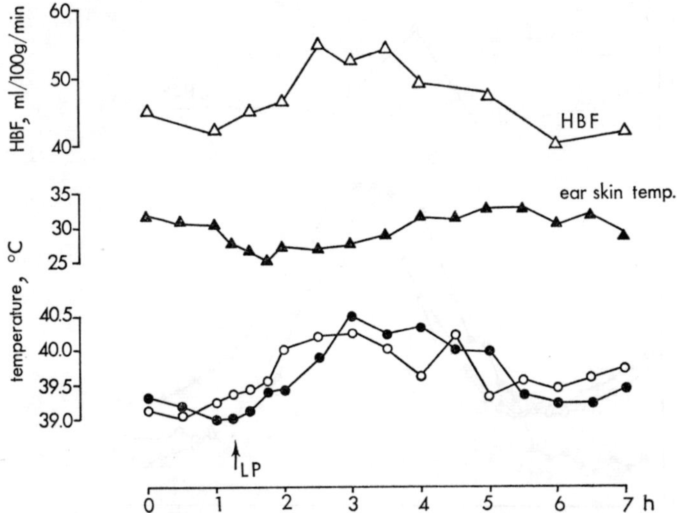

Fig. 1. Effect of local injection of purified leucocyte pyrogen (LP) into the anterior hypothalamus. Abscissa: hours. Ordinate: ear skin, hypothalamic (○) and rectal (●) temperatures (°C), and hypothalamic blood flow (HBF) (ml/100 g/min). At the arrow, 2 μl of LP were injected bilaterally at stereotaxic coordinates aB–16 [20].

blood pressure. However, HEYMAN *et al.* [12] have reported experiments in man in which total cerebral blood flow was unchanged during pyrogen-induced fever.

The mechanism of the hypothalamic vasodilatation during a pyrogen-induced fever is of interest. Heating the animal in another way, namely by raising the ambient temperature (fig. 2), also results in an increase in HBF. Whether the increase in flow seen following intravenous pyrogen is due to the pyrogen, or is a non-specific change due to a rise of body temperature, is not clear from these experiments. In the kidney, there is some evidence to suggest that the rise in body temperature produced by pyrogen is not an important factor in the increased blood flow, since environmental heating may cause a fall in renal blood flow [4, 6, 21]. No such dissociation of the effects of pyrogen-induced fever and temperature rise due to environmental heating has been seen in the present experiments, so that the question remains an open one.

In the rabbit at least, nor-adrenaline (NA), when injected into the hypothalamus, causes a rise in core temperature [7]; this is not associated, at the dosage of NA used, with a fall in HBF but with a rise [24].

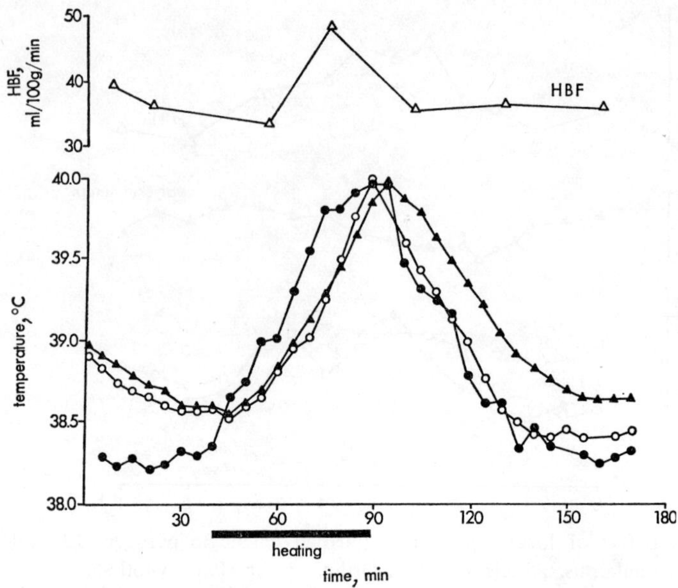

Fig. 2. Effect of increasing the ambient temperature. Abscissa: time in minutes. Ordinate: hypothalamic (○), rectal (▲) and internal carotid artery blood (●) temperature (°C), and HBF (ml/100 g/min) in the rabbit. During the period 'room-heating' the ambient temperature was increased from 21 to about 43 °C (dry bulb), radient heat lamps were turned on and the rabbit wrapped in cotton wool.

HBF and pCO_2

It is possible that an important controlling input in the brain-stem and hypothalamus is the local pCO_2 or pH, and that these values depend upon a balance between local tissue metabolic rate, which in the brain tends to be relatively constant, and tissue perfusion, which is variable. Cortical cells are very sensitive to small changes in pCO_2 [17] and the CO_2 sensitivity of medullary respiratory centres is well-known. There is little or no information on the effects of varying hypothalamic perfusion on local pCO_2.

The reverse of this, namely the CO_2 responsiveness of cerebral vessels, has been extensively studied. It has been shown that a rise in arterial pCO_2 increases total cerebral blood flow (CBF) [11] and HBF [9, 26], but the increase in CBF is limited by sympathetic vasoconstrictor activity [16].

This is consistent with the observation that a rise in arterial pCO_2 causes an increase in activity in the cervical sympathetic nerves [5]. The converse is also true, namely that the effect of stimulating the cervical sympathetic nerves is dependent upon the level of arterial pCO_2 [16]. These results dispose of the idea that the brain takes no part in the integrated vascular response to a rise of pCO_2. Also CO_2 may act not only directly on cerebral vessels but via a reflex arc, one component of which may be the sympathetic efferent nerves.

The Innervation of Hypothalamic Blood Vessels

There is now a large body of evidence from light-microscope, electron-microscope and fluorescent histochemical studies to indicate that cerebral blood vessels have an adrenergic innervation not dissimilar from that in other vascular beds. Adrenergic nerves can be seen on cerebral arterioles down to 10–15 μm, both on the pia and within the brain substance. However, the role of these nerves is controversial; numerous experiments have been performed in which total blood flow has been measured before and after section or stimulation of the cervical sympathetic or vagus nerves, with conflicting and confusing results. A number of reviewers [18, 27] have dismissed the action of cerebral vasomotor nerves as negligible. The negative findings of some authors may be explained on the basis of the limitations of the available techniques; for example, measurement of total CBF only may mask any redistribution of blood flow from one region to another in the brain. We have shown this effect; following pentobarbitone anaesthesia, cortical flow is halved while hypothalamic flow remains unchanged [22]. Also, the few experiments in which adrenergic drugs have been used in an attempt to demonstrate catecholamine sensitivity of cerebral vessels have been vitiated by the systemic effects of these substances on blood pressure. The observed changes in cerebrovascular tone may, therefore, be secondary to the blood pressure changes rather than due to the drugs themselves. Lastly, these drugs have usually been administered into the circulation, and their access to adrenergic receptors sites which may be mainly in the adventitial half of the media is uncertain.

We have approached this problem by using the 133-xenon clearance technique to measure the local HBF response to a variety of aminergic drugs injected into the rabbit hypothalamus [22]. This method allows for

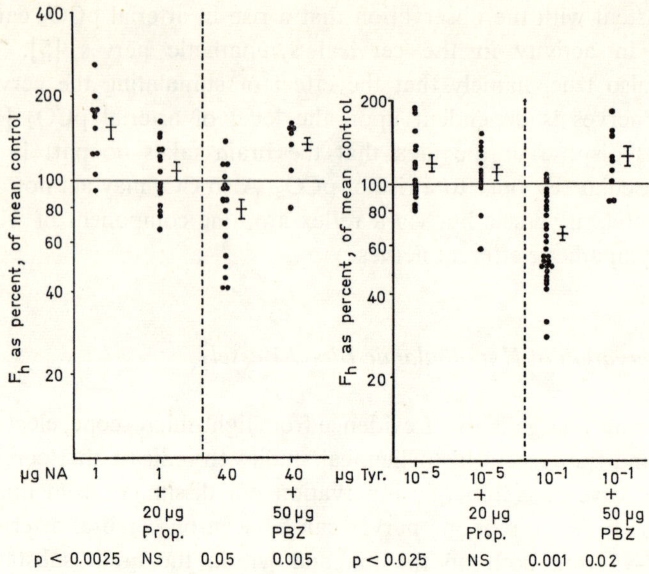

Fig. 3. Left: effect on HBF of noradrenaline (NA) 1 μg per injection and 40 μg per injection, before and during β- and α-adrenergic blockade with propanalol (20 μg Prop.) and phenoxybenzamine (50 μg PBZ), respectively. Ordinate: HBF on the test side as a percentage of the control (untreated) side. Right: effect on HBF of tyramine (Tyr), 10^{-5} μg per injection and 10^{-1} μg per injection, before and during β- and α-adrenergic blockade with propranalol (20 μg Prop.) and phenoxy-benzamine (50 μg PBZ), respectively. F_h = H.B.F.; NS = not significant.

the assessment of the effects of very small amount of these substances on local flow, independent of any systemic effects, in the conscious animal.

Autoregulation of local flow in the hypothalamus could be demonstrated within the mean arterial blood pressure range of 41–140 mm Hg [9].

NA injected into the hypothalamus (1 μg per injection) caused an increase in local flow, while larger doses (10–200 μg per injection) were vasoconstrictive. The vasodilator effect of NA (1 μg) was blocked by propranalol and the vasoconstrictive effect of NA (40 μg) was blocked by phenoxybenzamine (fig. 3). These results are consistent with the presence of α- and β-receptors in the hypothalamic resistance vessels; the β-receptors are activated by smaller doses of NA and the α-receptors by larger doses. The net effect of the larger dose is an α-receptor vasoconstriction [9, 22, 24].

Tyramine, which causes the release of endogenous NA also produces dose-dependent effects on blood flow (fig. 3). Small doses, (10^{-5} μg per injection) increased blood flow, while a larger dose, (10^{-1} μg per injection) reduced hypothalamic perfusion. The vasodilator effect of the smaller dose could be reduced by β-blockade with propranalol, while the vasoconstrictive effects of the larger dose was blocked, even reversed, by α-blockade with phenoxybenzamine (fig. 3).

All of this is consistent with the idea that endogenous NA i.e. NA released from nerve terminals, is vasoactive in 'physiological' concentrations, and is strong support for the importance of these nerves in local cerebrovascular control in the hypothalamus.

Conclusion

The efficiency with which the hypothalamus autoregulates its blood flow in the face of changes in systemic blood pressure would suggest that HBF is not an important controlling system in the autonomic control of cardiac output and total peripheral resistance. It is more likely that the hypothalamus is included in the integrated vascular response to changes in body temperature, pCO_2 and to sympathetic stimulation.

Acknowledgements

Sections of this work were done in collaboration with Dr. G. MITCHELL. We are grateful to the South African Medical Research Council, the Atomic Energy Board and the University of the Witwatersrand Senate Research Committee for financial support.

References

1 ABRAHAMS, V. C.; HILTON, S. M., and MALCOM, J. L.: Sensory connexions to the hypothalamus and midbrain, and their role in the reflex activation of the defense reaction. J. Physiol., Lond. *164:* 1–16 (1962)

2 ABRAHAMS, V. C.; HILTON, S. M., and ZBROZYNA, A.: Active muscle vasodilatation produced by stimulation of the brain stem. Its significance in the defense reaction. J. Physiol., Lond. *154:* 491–513 (1960).

3 BARD, P.: A diencephalic mechanism for the expression of rage with special reference to the sympathetic nervous system. Amer. J. Physiol. *84:* 490–515 (1928).

4 BYFIELD, G. V.; TELSER, S. E., and KEETON, R. W.: Renal blood flow and glomerular filtration as influenced by environmental temperature changes. J. amer. med. Ass. *121:* 118–123 (1943).

5 BISCOE, T. J. and MILLAR, R. A.: Effects of inhalation anaesthetics on carotid body chemoreceptor activity. Brit. J. Anaesth. *40:* 2–12 (1968).

6 COOPER, K. E.; CRANSTON, W. I.; DEMPSTER, W. J., and MOTTRAM, R. F.: Pyrogen-induced vasodilatation in the transplanted kidney. J. Physiol., Lond. *155:* 21–22P (1960).

7 COOPER, K. E.; CRANSTON, W. I., and HONOUR, A. J.: Effects of intraventricular and intrahypothalamic injection of noradrenaline and 5-HT on body temperature in conscious rabbits. J. Physiol., Lond. *181:* 852–864 (1965).

8 CRANSTON, W. I.: Fever, pathogenesis and circulatory changes. Circulation *20:* 1133–1142 (1959).

9 CRANSTON, W. I. and ROSENDORFF, C.: Local blood flow, cerebrovascular autoregulation and CO_2 responsiveness in the rabbit hypothalamus. J. Physiol., Lond. *215:* 577–590 (1971).

10 DJOJOSUGITO, A. M.; FOLKOW, B.; KYLSTRA, P. H.; LISANDER, B., and TUTTLE, R. S.: Differentiated interaction between the hypothalamic defense reaction and baroreceptor reflexes. I. Effects on heart rate and regional flow resistance. Acta physiol. scand. *78:* 376–385 (1970).

11 HARPER, A. M. and GLASS, H. I.: Effect of alterations in the arterial carbon dioxide tension on the blood flow through the cerebral cortex at normal and low arterial blood pressures. J. Neurol. Neurosurg. Psychiat. *28:* 449–52 (1965).

12 HEYMAN, A.; PATTERSON, J. L., and NICHOLS, F. T.: The effects of induced fever on cerebral functions in neurosyphilis. J. clin. Invest. *29:* 1335–1341 (1950).

13 HILTON, S. M.: Hypothalamic regulation of the cardiovascular system. Brit. med. Bull. *22:* 243–248 (1966).

14 HILTON, S. M. and JOELS, N.: Facilitation of chemoreceptor reflexes during the defense reaction. J. Physiol., Lond. *176:* 20–22 (1965).

15 HILTON, S. M. and SPYER, K. M.: Participation of the anterior hypothalamus in the baroreceptor reflex. J. Physiol., Lond. *218:* 271–293 (1971).

16 JAMES, I. M.; MILLAR, R. A., and PURVES, M. J.: Observations on the extrinsic neural control of cerebral blood flow in the baboon. Circulat. Res. *25:* 77–93 (1969).

17 KRNJEVIČ, K.; RANDIČ, M., and SIESJÖ, B. K.: Cortical CO_2 tension and neuronal excitability. J. Physiol., Lond. *176:* 105–122 (1965).

18 LASSEN, N. A.: Neurogenic control of CBF. Scand. J. clin. Lab. Invest. *6:* suppl. 102, *vi:* F (1968).

19 MCCOOK, R. D.; PEISS, C. N., and RANDALL, W. C.: Hypothalamic temperatures and blood flow. Proc. Soc. exp. Biol. Med. *109:* 518–522 (1962).

20 MONNIER, M. and GANGLOFF, H.: Atlas for stereotaxic brain research on the conscious rabbit. Rabbit brain research, vol. 1 (Elsevier, Amsterdam 1961).

21 RADIGAN, L. R. and ROBINSON, S.: Effects of environmental heat stress and exercise on renal blood flow and filtration rate. Amer. J. Physiol. *159:* 585 (1949).

22 ROSENDORFF, C.: The measurement of local cerebral blood flow and the effect of amines; in MEYER and SCHADÉ Progress in brain research, vol. 35. Cerebral blood flow, pp. 115–156 (Elsevier, Amsterdam 1972).

23 ROSENDORFF, C. and CRANDSTON, W. I.: Measurement of hypothalamic blood flow in the conscious rabbit by a radioactive inert gas clearance technique. 5th Europ. Conf. Microcirculation, Gothenburg 1968. Bibl. anat., Vol. 10, pp. 292–297 (Karger, Basel 1968).

24 ROSENDORFF, C. and CRANSTON, W. I.: Effects of intrahypothalamic and intraventricular norepinephrine and 5-hydroxytryptamine on hypothalamic blood flow in the conscious rabbit. Circulat. Res. *28:* 492–502 (1971).

25 ROSENDORFF, C. and MOONEY, J. J.: Central nervous system sites of action of a purified leucocyte pyrogen. Amer. J. Physiol. *220*(3)*:* 597–603 (1971).

26 SCHMIDT, C. F.: The intrinsic regulation of the circulation in the hypothalamus of the cat. Amer. J. Physiol. *110:* 137–152 (1934).

27 SOKOLOFF, L.: The action of drugs on the cerebral circulation. Pharmacol. Rev. *11:* 1–85 (1959).

Author's address: Dr. C. ROSENDORFF, Departments of Physiology and Medicine, University of the Witwatersrand Medical School, Hospital Street, *Johannesburg* (South Africa)

Panel and General Discussion

In the final session of the symposium an attempt was made to summarise recent developments in the field and to provide a forum for clarification of concepts relating to the diverse functions of hypothalamic systems. The session included formal presentations by panel members from endocrine (Dr. J. HAYWARD, Dr. G. GRANT, Dr. A. KASTIN, Dr. J. KRAICER) and non-endocrine (Dr. J. D. HARDY, Dr. P. MACLEAN, Dr. O. SMITH, Dr. C. ROSENDORFF) areas of interests, followed by a general discussion.

Brief summaries of the endocrine and non-endocrine panel presentations and discussions are presented in that order.

Endocrine Panel

The sophisticated studies of Dr. HAYWARD have provided electrophysiological correlates for different dynamic states of the magnocellular neurosecretory cells in the supraoptic nucleus of the unanaesthetised monkey: cells of 3 different spontaneously firing types – 'silent' (S), 'continuously active' (CA) and 'burstor' (B) were suggested to represent, respectively, states of synthesis-transport-secretion (S), 'tonic' secretion (CA) and 'pulsatile' release (B) of, for example, vasopressin. Moreover, links between behavioural input such as slow sleep, nociceptor-induced behaviour or drinking, on the one hand, and endocrine response on the other ('activation' of electrical activity, changes in the release of vasopressin) were demonstrated.

The presentation by Dr. GRANT provided a much-needed up-to-date report on detailed studies of structure-activity interactions of the 3 hypothalamic peptides which have been isolated and synthetised – TRF, LRF and SRIF (somatostatin).

For the TRF and its synthetic analogues, the recent studies at the Salk Institute have already provided chemical criteria enabling to understand parameters of receptor recognition and will enable meaningful studies on, for example, interaction of TRF analogues with receptors, receptor topography.

The studies with the LRF decapeptide and its synthetic analogues have resulted in the production of peptides with enhanced biological potency or other analogues with antagonistic properties. The work with the more recently characterized peptide – SRIF was presented more in the form of a progress report, giving evidence for validation of this peptide as a regulatory agent in the secretion of biologically active growth hormone.

The following presentation by Dr. KASTIN complemented admirably the chemical studies by providing a summary of current use and future prospects for clinical application of the hypothalamic peptides. In addition to diagnostic and therapeutic uses of the TSH- and LH- (or FSH-) releasing peptides, extra-endocrine studies were reported ranging from interaction between brain monoamines and melanocyte stimulating hormone or release-inhibiting hormone (MIF or MRIH) in parkinsonian patients, to the use of the TSH releasing peptide in depressive states.

The last presentation by Dr. J. KRAICER dealt with recent investigations in his laboratory concerned with the elucidation of membrane and ionic mechanisms involved in the release of preformed adenohypophysial hormones after stimulation of the release process by the hypothalamic releasing factors (hormones). The conclusions from these studies confirmed the requirement of Ca^{++} for the release process but indicated that different intracellular pools of Ca^{++} may be involved, depending on the secretagogue used. However, the study of transmembrane potential (TMP) changes in a population of adenohypophysial cells (thyrotrophs) stimulated to secrete increased amounts of hormones did not correspond with the simple release mechanism based on membrane depolarisation predicted by the Douglas hypothesis – a decrease in TMP could not be shown.

The general discussion ensuing from these presentations stressed some of the advances made and the direction which future studies may take.

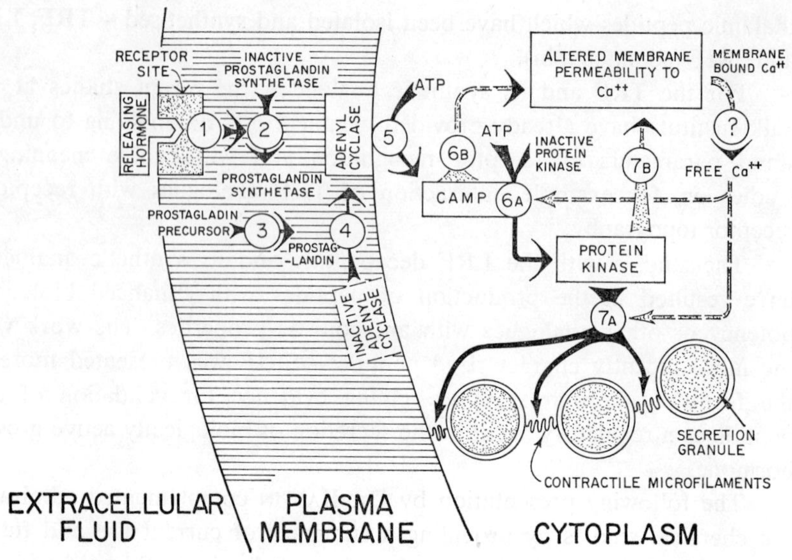

Fig. 1. For abbreviations, see text.

1. What practical benefits in clinical endocrinology and in appropriate non-endocrine areas are likely to accrue from the recent discoveries in the field of the hypothalamic peptides that regulate the secretion of adenohypophysial hormones? Simultaneous multi-hormone diagnostic procedures may be practicable and may lead to considerable economic benefits (Van Loon, Kastin).

2. What is the explanation underlying the differences in the speed of onset and the duration of effect of conventional neurotransmitters and the hypothalamic 'neurohormones' (Scharrer, Hayward, Sachs)? It remains to be resolved which of the hypothalamic hormones require or utilize the neurophysins or neurophysin-like proteins. What, if any, is the physiological function of the 'neurophysins' (and their like), other than their possible role as hormone carriers or precursors?

In response to questions on the current state of knowledge on stimulus-secretion coupling applicable to the anterior pituitary in particular and (possibly) to the release of granule-stored hormones in general, Dr. J. Kraicer proposed the following tentative scheme to describe the series of events initiated by the interaction of the hypothalamic releasing peptides with the adenohypophysial cell plasma membrane, and culminating in the release of pre-formed hormone.

Experimental data which form the basis for this scheme have been reviewed recently [1]. The sequence of events is proposed as follows.

1. The hypothalamic releasing hormone interacts with a receptor site on the plasma membrane of its specific target cell.

2. This interaction results in the activation of prostaglandin synthetase.

3. Prostaglandin synthetase then catalyses the conversion of a prostaglandin precursor to prostaglandin.

4. Prostaglandin 'activates' membrane-bound adenyl cyclase.

5. Adenyl cyclase then catalyses the conversion of adenosine triphosphate (ATP) to 3^1, 5^1-cyclic AMP (cAMP) within the cytoplasm.

6. cAMP: (a) activates a cAMP-dependent protein kinase, and (b) alters membrane permeability to Ca^{++} (membrane-bound Ca^{++} $\rightarrow$ free Ca^{++}?)

7. cAMP-dependent protein kinase: (a) activates, by phosphorylation, a protein moiety involved in the release process. This activated protein moiety may be one element in the contractile cytoskeleton-vesicle complex, which when activated, leads to contraction, with subsequent fusion of membranes and extrusion of hormone-containing granules, and (b) alters membrane permeability to Ca^{++} (membrane bound Ca^{++} $\rightarrow$ free Ca^{++}?)

Two questions may be raised concerning the role of Ca^{++}: (1) at what stage is the process Ca^{++}-dependent, and (2) what is the role of Ca^{++} in the release process? In answer to the first question, Ca^{++} is required at a stage beyond the activation of cAMP since steps 1–5 (fig. 1) can take place in a Ca^{++}-free environment. In answer to the second question, one can only speculate that Ca^{++} is required (a) for the activation of cAMP-dependent protein kinase, (b) and/or for the protein kinase-induced phosphorylation of a protein moiety involved in the release process, and/or (c) beyond this step, in the 'shortening' associated with the contractile cytoskeleton-vesicle complex.

Non-Endocrine Panel

Dr. ROSENDORFF's paper, outlining a technique for assessment of blood flow in small, localised areas of the hypothalamus, and the effects of some putative neuro-transmitter substances thereon was discussed briefly. The first problem lay in the variation from normal blood flow induced by the tissue damage which the cannulae must cause, and the modification to local blood flow measured subsequently in the same site which gliosis, due to the first trauma, would cause. The response to CO_2 and the presence of autoregulation argued for the reasonable normality of the circulation in the region of the Xenon injection probes. Frequent, repeated probing of the same area should be avoided. The question also arose as to whether noradrenaline would stimulate local nervous tissue metabolism, and thereby modify local blood flow as a consequence of a build-up of metabolites. In the discussion, it became clear that there was little evidence to correlate the brain tissue blood flow with local functional activity and, due to the small energy turnover of the cells, doubt was cast on the possible effects of nerve cell metabolism on blood flow. Nevertheless, until techniques had become more refined, such a possibility could not be excluded. It became clear that knowledge of the control of the circulation in relation to all hypothalamic function was scanty, and could be a fruitful subject for investigation.

Following this discussion, Dr. MacLean outlined some major facets of the hypothalamic control of emotional expression. He drew the distinction between the subjective aspect of emotion, or affect, and the expressive aspects such as verbal behaviour, and searching, aggressive, protective, dejective, gratulent and caressive behaviours, which can be associated with desire, anger, fear, sorrow, joy and affection. Dr. MacLean traced historically the evidences leading to the linkage of these emotional expressions with ascending dopamine containing neurones in the walls of the third ventricle, the aqueductal region and the tegmentum, and emphasised the striatal connections involved. His own work, involving destruction of these cells by 6-hydroxydopamine produced cataleptic states in the squirrel monkey. It appears that ascending dopaminergic systems can energise an animal's behaviour. Further study of the literature would suggest a link between the lateral and medial forebrain bundles, connecting with the protoreptilian and paleomammallian parts of the brain and the limbic system with emotional activity. Dr. MacLean's experiments, severing the lateral and medial parts of the forebrain bundles in the squirrel monkey, eliminated their emotional expression and the apparent expression of 'personality'. Thus, such expressions are limited to pathways passing through the hypothalamus, containing fibres passing to and from the limbic system structures. Of clear importance are the connections with the striatal and pallidal systems. Dr. MacLean's experimental lesions in the globus pallidus grossly altered genital display behaviour in the squirrel monkey without causing the animals to have motor impairment. Again electrical stimulation in the region of the pallidohypothalamic tract elicited sexual activity and signs of anger. Further evidence was presented for the participation of an ascending dopaminergic system in the ansa lenticularis, just lateral to the fornix and connected to the globus pallidus, in innate display behaviour and in feeding behaviour.

Thus Dr. MacLean, in linking hypothalamic systems with emotion and behaviour, which responses are known to inter-relate with hormone secretion, body temperature regulation and circulatory control, to name but a few, has drawn attention to the need for more integrative study of hypothalamic function.

To conclude the formal presentations in the discussion, Dr. Hardy reviewed the role of the hypothalamus in thermoregulation. Our knowledge of this connection goes back about 100 years, but despite a large volume of work which had been done in that period there are still large areas of ignorance in our concepts of the hypothalamic control of thermoregulation. Starting with the early work on brain-stem sectioning, and later with electrical stimulation, Dr. Hardy outlined the evidence which linked the hypothalamus with thermosensitivity and control of body temperature. He outlined the classical work which has gone on over many years at the John Pierce Foundation Laboratories in New Haven, in determining the overall thermosensitivity of the hypothalamus, then in the location there of thermosensitive single units and in the role of the hypothalamus in the behavioural control of body temperature. From the hypothalamus-centred study of thermoregulation, further enquiry had located thermosensitive structures in the spinal cord, and in many other areas of the central and peripheral nervous systems. Studies of single units which are specifically thermosensitive and of others which combine thermosensitivity with other sensory modalities provide input and output data which fit models of thermoregulation, whether of the physical or mathematical type, or the neuronal models such as those presented by Dr. Bligh. Further to this, the abnormality of thermoregulation, fever, had come nearer to explanation with the discovery of the possible role of prostaglandins.

Dr. Hardy's group had demonstrated that some prostaglandins were not just stimulators of metabolism with associated thermoregulatory responses, but were indeed capable of manipulating the controls of body temperature. Dr. Hardy's presentation drew attention to the danger of considering a function in the 'isolated' hypothalamus, which now appeared to be not only a very important controlling region, but part of an integrated system for the regulation of body temperature probably involving the entire nervous system.

Reference

1 KRAICER, J.: Mechanisms involved in the release of adenohypophysial hormones; in FARQUHAR and TIXIER-VIDAL Ultrastructure in biological systems. The anterior pituitary (in press).

Author Index

Compiled by LOUISE WORKMAN, Calgary, Alberta

Subject Index

Compiled by Barbara Demeneix, Calgary, Alberta